WHEN BIOSPHERES COLLIDE

A History of NASA's Planetary Protection Programs

WHEN BIOSPHERES COLLIDE

A History of NASA's Planetary Protection Programs

by Michael Meltzer

NASA SP-2011-4234

Library of Congress Cataloging-in-Publication Data

Meltzer, Michael.
  When biospheres collide : a history of NASA's planetary protection
programs / by Michael Meltzer.
      p. cm. -- (NASA SP ; 2011-4234)
  Includes bibliographical references and index.
  1. Space pollution. 2. Space environment. 3. Outer space--
Exploration--Environmental aspects. 4. Environmental protection--
Moral and ethical aspects. I. Title.
  TL1499.M45 2010
  363.739--dc22
                                          2008005759

ISBN 978-0-16-085327-2

9 780160 853272   90000

CONTENTS

LIST OF FIGURES

LIST OF TABLES

●●●

Reference Note

Many footnotes in this book cite Internet-based sources. Inevitably, some pages on the World Wide Web do not last forever. If you are trying to locate a source whose URL no longer works, you may contact the NASA History Program Office at NASA Headquarters for help (e-mail *histinfo@hq.nasa.gov*). Their Historical Reference Collection contains hard copies of many of the online sources cited in this book, and even if they do not have the one you are seeking, they may be able to assist you in locating it.

PREFACE

Col. Ross: "Is there any point to which you would wish to draw my attention?"

S. Holmes: "To the curious incident of the dog in the night-time."

Col. Ross: "The dog did nothing in the night-time."

S. Holmes: "That was the curious incident."

—The Memoirs of Sherlock Holmes
by Sir Arthur Conan Doyle (1893)

Significant challenges abound in the astrobiological study of the solar system, similar to those faced by Sherlock Holmes in another context. In the search for extraterrestrial life, a negative result is nearly impossible to obtain, much less interpret. We are bathed in Earth organisms, which makes finding our own kind of life palpably easy and detecting indigenous life on other worlds much more difficult. We are not exploring the solar system to discover life that we have brought with us from home, and we are aware that Earth organisms (read: *invaders*) could very well erase traces of truly extraterrestrial life.

Likewise, we don't know what would happen if alien organisms were introduced into Earth's biosphere. Would a close relationship (and a benign one) be obvious to all, or will Martian life be so alien as to be unnoticed by both Earth organisms and human defenses? We really have no data to address these questions, and considerate scientists fear conducting those experiments without proper safeguards. After all, this is the only biosphere we currently know—and we *do* love it!

With this volume, Michael Meltzer details the fascinating history of our attempts at planetary protection and those who have worked to protect Earth from otherworldly organisms, while safeguarding other worlds from the all-too-pervasive life on Earth. Such a history is particularly important at this time, since it provides a point of departure for understanding the field as we undertake further explorations. Meltzer's work will help us face what may well be a crucial topic in the future of the science of life and the future of humans in space.

John D. Rummel, Senior Scientist for Astrobiology
NASA Science Mission Directorate, Planetary Sciences Division

FOREWORD

The question of whether there is life on other planetary bodies became directly relevant to astronomy in the Renaissance, when Galileo made his observations confirming that the wandering "planets" might actually be objects similar to our own Earth, traveling around the Sun. (Previously, this question had been the purview of philosophers. In the fourth century BC, Metrodorus of Chios neatly summed up the attitude of his mentor, Epicurus, toward extraterrestrial life: "To consider the Earth as the only populated world in infinite space is as absurd as to assert that in an entire field of millet, only one grain will grow.") In the 19th century, Schiaparelli's description of "canali" (channels) on Mars encouraged Percival Lowell to depict canals, which might have been built by a dying civilization on a nearly dead planet. Such a Martian civilization was represented as hostile by H. G. Wells, who imagined that such a race would covet a warm and wet Earth, and might invade with the intention of conquering our more comfortable planet. In Wells's fantasy, Earth fights back, and the Martians are themselves conquered by the least of Earth's inhabitants: the microbes.

Planetary protection addresses the real consequences that might result from an actual transfer of biological material between different planets if indeed life does exist on other worlds. We know, from many experiences over centuries of transporting organisms from one location to another on Earth, that such transfers are not always harmless. Fleas brought to Europe in the 1300s carried the Black Death plague that caused the most significant reduction of the global human population in documented history. Early interactions with native inhabitants of the "New World" by Columbus and subsequent European explorers inadvertently infected them with European diseases, engendering epidemics that caused the collapse of several civilizations before Europeans were quite aware that they existed much less had any opportunity to study or learn from them. Recently, the transport of "invasive species," such as zebra mussels and purple loosestrife, has been widely recognized to cause significant damage to local ecologies and economies, though not directly causing human diseases. These issues have been in the forefront of public consciousness throughout the last and the current centuries. It was in an effort to minimize the potential for negative consequences from space exploration that planetary protection was first initiated as this volume describes.

Being the NASA Planetary Protection Officer, it is my job to ensure that our space exploration efforts neither unintentionally contaminate other solar system bodies, nor Earth upon spacecraft return. However, my interest in planetary protection is much more personal. As part of my laboratory's research on muscle atrophy, my first spaceflight experiment was flown on the last mission of the Space Shuttle *Columbia* that tragically disintegrated during reentry on 1 February 2003. Surprisingly, when we opened our recovered hardware several months after the accident, many of our experimental animals had survived the crash. Inadvertently, our research had demonstrated that, if properly shielded, even multicellular animals could survive a meteoritic-style entry event, one of the key steps required for the natural transport of living organisms between planetary bodies. This recognition makes it even more critical that we don't carry life from Earth with us on missions to search for life elsewhere—otherwise, if we find some, we might not be able to tell the difference!

When Joshua Lederberg first raised his concerns about planetary protection in the 1950s, we had no evidence that there was another place in the solar system where life could exist. We now know much more about our planetary neighbors, and from the data we have currently, we can say confidently that there are conditions present on or in other solar system bodies that do not exclude the possibility of Earth life surviving there. Although the surface of Mars is cold and dry, the subsurface is likely to be warmer and wetter, and we know that organisms on Earth survive kilometers beneath the surface, gathering energy from the rocks that surround them. Likewise, the surface of Europa, a moon of Jupiter, is frozen and exposed to high levels of hard radiation; however, the subsurface contains a thick (~100 km) layer of liquid water, overlaid by ice and at the bottom in contact with Europa's rocky interior. Tidal stresses appear to keep Europa warm, which raises the possibility that hydrothermal vent systems, known to support abundant life on Earth, may also be present in a Europan ocean. With our increasing knowledge about life on Earth, and about conditions favorable for life on other planets, our responsibility also increases to carry out solar system exploration in a manner that does not contaminate the target locations that we want to explore, nor Earth with samples of planetary materials that we choose to return.

When Biospheres Collide provides an excellent grounding in the history and fundamentals of planetary protection, as well as an overview of recent and ongoing activities within NASA's programs. The public can be proud of the way that NASA has responded to planetary protection concerns, with its efforts to act responsibly, communicate openly, and address effectively the potential for contamination. Those who do not learn from history may be doomed to repeat it, but it is to be hoped that NASA's future activities in planetary protection will demonstrate that those who do learn from history may also repeat at least some of the past's successful efforts. This current volume is a valuable contribution to that capability.

Cassie Conley, NASA Planetary Protection Officer

ACKNOWLEDGMENTS

Many people gave me vital information for this book, and I want to thank them for their time as well as interest in the project. Some of these people include:

John Rummel, who shared many stories and articles from his decades in the planetary protection field.

Perry Stabekis for his insights on the evolution of the field.

Craig Fischer, whose tour of the former Lunar Receiving Laboratory and whose many anecdotes were key in understanding the Apollo mission.

John Stonesifer for his memories of the Apollo mission.

Don Larson for a wonderful tour of Lockheed Martin spacecraft cleaning and assembly facilities.

Ben Clark, for his perspectives on different types of Mars missions.

Margaret Race for her perspectives on the ethical aspects of the field.

Cassie Conley for her views on the importance of planetary protection.

Thanks also to the insights and data provided by Kent Carter, Judy Allton, Martin Favero, John Hirasaki, Richard Green, David Schultz, Wayne Koons, Roger Launius, Israel Taback, Linda Billings, James Young, Dennis Vaughn, Bruce Keller, Bob Polutchko, Tom Young, Brad Perry, and Bob Clark. Also thanks to Bob Fish for his tour of the USS *Hornet*. Several of the people who taught particularly instructive planetary protection-related classes include Jack Barengoltz, Michele Bahr, Amy Baker, Gary Coulter, Margaret Race, John Rummel, Perry Stabekis, and Dave Tisdale.

Johnson Space Center and Langley Research Center archival personnel who helped on the research for this book include Shelly Kelly and Gail Langevin. At NASA Headquarters, thanks to Steve Dick for his guidance and support throughout the project. Archivists Colin Fries and Jane Odom were very helpful in finding key material from the NASA Historical Reference Collection. Steve Garber oversaw the book's production process, while Nadine Andreassen was a valuable point of contact for all things NASA. Many thanks to the fine individuals at Headquarters responsible for the finishing of this book: Stacie Dapoz and Andrew Jarvis carefully copyedited this volume; Chris Yates meticulously handled the layout; and Gail Carter-Kane oversaw this team with unwavering patience and support.

And finally, thank you to my wife, Naisa, for her spiritual advice and affection, and to my daughter, Jordana, for her inspiration. This book is also in remembrance of Robert Meltzer, who brought such joy to the world.

WHY WE MUST PROTECT PLANETARY ENVIRONMENTS

> The 20th Century will be remembered, when all
> else is forgotten, as the century when man burst his
> terrestrial bonds.
>
> —*Arthur Schlesinger, Historian*[1]

From the time that humans first began sending spacecraft out from Earth, the possibility has existed of forever changing the extraterrestrial environments that we visit. If we irrevocably alter the nature of other celestial bodies, we compromise all future scientific experimentation on these bodies and may also damage any extant life there. By inadvertently carrying exotic organisms back to Earth on our spaceships, we risk the release of biohazardous materials into our own ecology. Such concerns were expressed by scientists from our country and from other nations shortly before the launch of Sputnik, and these concerns have continued to be addressed through the present day. Our human race has a great urge to explore the unknown, but this must be done in a responsible manner that considers the potential impacts of our actions on future exploration.

Exploration: A Basic Ingredient of Our Society

Since long before the founding of the United States, the need to explore has permeated our culture. The American colonies were

---

1. Arthur Schlesinger, Jr., quoted in, *Congressional Record*, 94th Cong., 2nd sess., 30 September 1976, p. H11946.

created by adventurers bent on finding and settling new, fertile places, and this spirit of discovery has continued through the years. From 1840 through 1860, the United States spent one quarter of its annual budget on naval expeditions.[2] And in the 1950s, a new frontier was added to those of land and sea exploration: space.

The historian Daniel Boorstin has said that we humans are discoverers with a burning "need to know."[3] It is part of what makes us human. But how we address this need is critical, particularly in our exploration of space, because the choices we make will affect the integrity of the data that scientists in the centuries ahead will use for their investigations. We thus have a responsibility to conduct our explorations of celestial bodies with as much wisdom and consideration of future generations as we can muster.

In the 1800s, land and sea exploration played an important role in maintaining the United States' vitality. In the 20th century after the Space Age began, new imperatives were created. One of these was to investigate other worlds and contrast them with our own.[4]

Jerome B. Wiesner, a Massachusetts Institute of Technology (MIT) professor who advised Presidents Eisenhower and Kennedy on space issues, understood the critical nature of the space challenge. Wiesner chaired a transition team for space that wrote a key report in 1961 for President-elect Kennedy, which recognized the influence that space exploration would have in establishing our country's place in the world. In the report's section on "principal motivations for desiring a vital, effective space program,"[5] the many expected benefits of exploring space were addressed, one of which was the major benefit to our nation expressed by the following statement:

---

2. Steven J. Dick, "Why We Explore," http://www.nasa.gov/missions/solarsystem/Why_We_01pt1.html and http://www.nasa.gov/missions/solarsystem/Why_We_01pt2.html (accessed 2 October 2004).

3. Charles D. Walker, "Why We Explore Space," National Space Society Web site, http://www.nsschapters.org/policy-cmte/files/FinalFrontier_1992.pdf (accessed 2 December 2010).

4. Lawrence B. Hall, James R. Miles, Carl W. Bruch, and Paul Tarver, "The Objectives and Technology of Spacecraft Sterilization," NASA news release, 9 February 1965, folder 006695, "Sterilization/Decontamination," NASA Historical Reference Collection, NASA History Division, NASA Headquarters, Washington, DC.

5. Jerome B. Wiesner, chair, Wiesner Committee, *Report to the President-Elect of the Ad Hoc Committee on Space*, 10 January 1961, NASA Historical Reference Collection and at Air War College (AWC) Web site, http://www.au.af.mil/au/awc/awcgate/report61.htm (accessed 2 December 2010).

First, there is the factor of national prestige. Space exploration and exploits have captured the imagination of the peoples of the world. During the next few years the prestige of the United States will in part be determined by the leadership we demonstrate in space activities.[6]

The author James Michener also understood the importance of space exploration. He believed that we as a culture have to express our need to explore or risk losing a vital momentum. He recognized that at different times in a nation's life, different goals had to be pursued. In an address to Congress, Michener asserted that "there are moments in history when challenges occur of such a compelling nature that to miss them is to miss the whole meaning of an epoch. Space is such a challenge."[7]

As our ability to explore other worlds developed, our "intent to travel in space [increased] with irresistible force."[8] At the heart of this endeavor was, in part, a yearning to establish that we Earthlings are not alone in the universe.[9] In the words of Lawrence B. Hall (when he was NASA's Special Assistant for Planetary Quarantine) and colleagues, "A most urgent, perhaps the most urgent, question to be answered by instrumented flights relates to the existence of life on the planets. An affirmative answer to this question has biological, medical, and even religious implications that far transcend the results to be obtained by mere geographical and physical exploration of the planets."[10]

Randolph et al. expand on this theme in their paper examining the theological and ethical consequences of discovering life on other worlds, when they write about the types of life that might be found. We may one day discover a Martian life-form that is biochemically similar to life on Earth due to the large amounts of material that have been

6. Wiesner, *Report to the President-Elect*.
7. James A. Michener, "Space Exploration: Military and Non-Military Advantages," speech delivered before U.S. Senate Subcommittee on Science, Technology, and Space, Washington, DC, 1 February 1979. Published in *Vital Speeches of the Day*, City News Publishing Company, Southold, NY, 15 July 1979.
8. Hall et al., "Objectives and Technology of Spacecraft Sterilization."
9. J. Roger, P. Angel, and Neville J. Woolf, "Searching for Life on Other Planets," *Scientific American* (April 1996): 60–66.
10. Hall et al., "Objectives and Technology of Spacecraft Sterilization."

exchanged between the planets from meteorite and asteroid impacts. This would indeed be a world-shaking event. But discovering life on Mars that had an *independent genesis* from life on Earth would have considerably deeper scientific, ethical, and religious implications. Such a discovery would suggest that if two independent forms of life could have arisen in just our solar system, and that if Earth-like or Mars-like planets existed elsewhere, then life might well be commonplace throughout our galaxy.[11] A discovery of such ubiquitous life, some of which might have high intelligence, would call into question how special our particular human life-form is. A traditional Judeo-Christian belief is that we humans were created in God's image.[12] This belief might be challenged if space exploration establishes that we are just one of many sentient species in the universe.

Exploring Without Disrupting Zones of Life

Planetary protection is a field concerned with keeping actual or possible "zones of life"[13] pure and unspoiled. A planet's *biosphere* is its complete zone of life, its global ecological system,[14] and includes all its living organisms as well as all organic matter that has not yet decomposed. The term biosphere is frequently used today, but it is not a new concept. It was introduced by Eduard Suess, a professor of geology at the University of Vienna, in his four-volume treatise, *Das Antlitz der Erde* (*The Face of the Earth*), whose English version was published in 1904.[15] The Russian scientist Vladimir Vernadsky then popularized this term in his 1926 book, *The Biosphere.*[16]

---

11. Richard O. Randolph, Margaret S. Race, and Christopher P. McKay, "Reconsidering the Theological and Ethical Implications of Extraterrestrial Life," *Center for Theology and Natural Sciences (CTNS) Bulletin* 17(3) (Berkeley, CA: summer 1997): 1–8.
12. *Genesis*, Chap. I, verse 27, in *The Pentateuch and Haftorahs* (2nd ed.), ed. J. H. Hertz (London: Sonico Press, 1978): 5.
13. University of Florida, Department of Geological Sciences, "The Biosphere," *http://www.geology.ufl.edu/Biosphere.html* (accessed 25 September 2006).
14. Harold J. Morowitz, *Energy Flow in Biology* (Woodbridge, CT: Ox Bow Press, 1968).
15. Eduard Suess, *The Face of the Earth* (*Das Antlitz der Erde*) (Oxford: Clarendon Press, 1904).
16. Vladimir I. Vernadsky, *The Biosphere*, 1926, published in the United States under the name, *The Biosphere: Complete Annotated Edition* (New York: Copernicus, Springer-Verlag, 1998).

◍◍◍

Why Prevent Planetary Contamination?

A Necessary Condition for Good Space Science

Boorstin, Wiesner, and Michener recognized that exploration of new frontiers is a human imperative and that, at times, the health and vibrancy of our civilization may depend on it. But breaking new ground in the quest for human knowledge carries with it certain risks, and we explorers have a responsibility to understand those risks and address them intelligently. Planetary protection concerns itself with the quest to explore space responsibly under conditions of extreme uncertainty. This is no trivial matter, since the possible biospheres that we may encounter, and the impacts on them of our exploration efforts, are extremely difficult to forecast. John Burton Sanderson "J. B. S." Haldane, the British geneticist, biologist, and popularizer of science, expressed the difficulty of envisioning another biosphere's nature when he said, "Now my own suspicion is that the Universe is not only queerer than we suppose, but queerer than we *can* suppose."[17] And yet we still have to act as best we can. John Rummel, NASA's former Planetary Protection Officer (PPO), stated the problem of responsible interplanetary exploration like this: "Ignorance is not bliss. We don't know what we're doing, so let's not do things that we can't account for."[18]

In trying to envision possible forms of life, or environments that could eventually lead to life, we are limited by a lack of perspective. So far, humans have only been able to study Earth life. Although this life can exist in wildly different environments, from volcanic ocean vents

---

17. John D. Rummel, "Mars, Europa, and Beyond," in *Astrobiology: Future Perspectives,* ed. P. Ehrenfreund, W. M. Irvine, T. Owen, Luann Becker, Jen Blank, J. R. Brucato, Luigi Colangeli, Sylvie Derenne, Anne Dutrey, Didier Despois, Antonio Lazcano, and Francois Robert (Dordrecht, Netherlands: Kluwer, 2004), Chap. 7; Helena Sheehan, *Marxism and the Philosophy of Science: A Critical History* (New York: Humanities Press International, 1993), excerpt found at *http://www.comms.dcu.ie/sheehanh/haldane.htm* (accessed 26 February 2005).

18. John D. Rummel, interview by author, Washington, DC, 9 September 2004.

to the insides of ice formations, all of Earth life appears to be related, making use of the same basic biochemistry. In contrast, we have no data whatsoever on alternative chemistries that may also prove feasible for supporting life not as we know it, but of an entirely exotic, unrelated form. The space science community places high value on accurately determining what is currently unknown about other worlds. To accomplish this, we must protect extraterrestrial environments that may harbor undiscovered life-forms, at least until we have thoroughly examined those environments and analyzed any living organisms that exist within them.

Although we don't know what we will find on other worlds, we can posit certain environmental conditions that appear critical to the origin of life-forms. These conditions include the following:

A source of organic compounds.

The potential for energy transfer.

The origin and evolution of functional biomolecules.

The origins of cellularity.[19]

While liquid water is typically mentioned as another essential for life—at least as we know it—some scientists have hypothesized that on very cold planetary bodies, liquid ammonia might have served in place of water to incubate life. Others have suggested that oceans of methane or other hydrocarbons on bodies such as Saturn's moon Titan could have served that purpose.[20]

On Earth, the critical environmental conditions mentioned above might have converged and led to the formation of life in the following way: asteroids, comets, meteorites, and interplanetary dust carried organic compounds to Earth, which were then used in the formation of functional biomolecules, energy transference, and the development of cellularity. Such events should have been possible not only on Earth, but also on other bodies of the solar system. A second theory of life's origins involves endogenous formation of organic molecules (i.e., without dependence on sources external to the planet), driven by either naturally occurring atmospheric energy discharges

19. Rummel, "Mars, Europa, and Beyond," pp. 2, 4.
20. Peter Tyson, "Life's Little Essential," NOVA Science Programming On Air and Online, July 2004, *http://www.pbs.org/wgbh/nova/origins/essential.html* (accessed 14 November 2007).

or through the processing of materials within hydrothermal environments. A version of this theory was espoused by Darwin over a century ago when he envisioned the growth of organic compounds in "some warm little pond."[21]

A combination of these mechanisms could have contributed to the formation and development of life's building blocks on Earth. Recent data such as those from Earth's extremophiles suggest that once life did emerge, it would have used every available environmental niche and energy source, both photosynthetic and chemosynthetic, to support itself.[22]

We cannot expect to find many warm little ponds during our searches for, and attempts to protect, extraterrestrial biospheres or prebiotic environments. We need instead to conceptualize a range of exotic conditions that could lead to organic molecule development followed by the origin of life. Possible energy sources for these processes could range from the freeze-thaw cycles encountered on an ice world to tidal heating to the temperature variations that might be found in deep-sea hydrothermal systems.

Ethical Questions

Today's planetary protection policy is dominated by a very practical consideration—the safeguarding of scientific investigations,[23] and it is debatable to what extent ethical considerations affect the formulation of that policy. According to Pericles "Perry" Stabekis, a long-time NASA contractor who has served on the staff of every one of the Agency's planetary protection officers,

> Planetary protection does not have by policy an ethical component. So as it stands, it is not there to protect the planets for the planets' sake. It is to protect

---

21. Rummel, "Mars, Europa, and Beyond," p. 4.
22. P. Ehrenfreund, W. Irvine, L. Becker, J. Blank, J. R. Brucato, L. Colangeli, S. Derenne, D. Despois, A. Dutrey, H. Fraaije, A. Lazcano, T. Owen, and F. Robert, "Astrophysical and Astrochemical Insights Into the Origin of Life," *Reports on Progress in Physics* 65 (Bristol, U.K.: IOP Publishing, 2002): 1427–1487.
23. Leslie I. Tennen, telephone interview by author, 7 March 2005.

> the planets for preserving them *as a target of biological
> exploration*. [author's italics] . . . So clearly the
> mandate of planetary protection is not one that is based
> on the ethics of the question, but on preserving our
> ability to explore them biologically . . . to keep them . . .
> as pristine as warranted to protect that exploration. . . .
> It's strictly to protect science. Planets for the sake of
> science.[24]

It could be argued, however, that the perceived need to conduct
good scientific investigations is indeed an ethical position, and a
strongly held one, given the amount of money and effort that NASA
has spent on planetary protection.

Our desire to conduct space exploration and search for extraterrestrial life in a scientifically rigorous manner that will produce sup-
portable results dictates that we protect planetary environments until
the period of biological interest is over. But should we also protect
planetary environments beyond that time, and for reasons that are not
as pragmatic as conducting good science? Should we, in fact, preserve
planets in their pristine state because it is the *right* thing to do?

Patricia Sterns and Leslie Tennen, two attorneys writing about
celestial environmental protection, hold a view that resembles a phy-
sician's commitment to do no harm, asserting that "purity of ecosystems throughout the universe must be respected in order to ensure
the natural evolution of the heavens and their diverse planets."[25]
Randolph, Race, and McKay stress the importance of protecting interplanetary diversity, writing that "from an ethical point of view, the
need to preserve a life-form, however lowly, must be more compelling if that life-form represents a unique life-form with an evolutionary history and origin distinct from all other manifestations of life." [26]

---

24. Perry Stabekis, interview by author, Washington, DC, 9 September 2004, and comments
on the manuscript, 21 June 2005.
25. Patricia M. Sterns and Leslie I. Tennen, "Protection of Celestial Environments Through
Planetary Quarantine Requirements," in *Proceedings of the 23rd Colloquium on the Law of
Outer Space* (1981), pp. 107–120.
26. Richard O. Randolph, Margaret S. Race, and Christopher P. McKay, "Reconsidering
the Theological and Ethical Implications of Extraterrestrial Life," *Center for Theology and
Natural Sciences (CTNS) Bulletin* 17(3) (Berkeley, CA: summer 1997): 1–8.

Furthermore, finding a life-form whose origin is distinct from earthly life-forms, write Race and Randolph, has much more significant ethical implications than if the life-form's genetics and biochemistry are closely related to Earth organisms. Exotic life-forms with different evolutionary histories than our own would necessitate that we "begin developing a broader theory of life"[27] and its origins.

President Lyndon Johnson recognized the need to maintain the purity of ecosystems, although it is not clear whether protecting such ecosystems was based solely on scientific and political grounds or on ethical grounds as well. On 7 May 1966, Johnson issued a statement proposing a treaty that would govern space exploration activities in order to, among other objectives, "avoid harmful contamination" of celestial bodies.[28] The United States submitted a draft of this treaty to the United Nations (UN) in June 1966. The language of the draft document was preserved in Article IX of the 1967 UN *Treaty on Principles Governing the Activities of States in the Exploration and Use of Outer Space, Including the Moon and Other Celestial Bodies*, or "*Outer Space Treaty*" (discussed later in the book), which called for avoidance of harmful contamination of the Moon and other celestial bodies, but did not specifically define what would constitute "harmful contamination." Was it that which negatively impacted planetary ecologies, or just scientific endeavors?[29]

This ambiguity continued in more recent treaties. Article 7 of the 1984 UN *Agreement Governing the Activities of States on the Moon and Other Celestial Bodies* (the "*Moon Agreement*") included a directive that specifically addressed the need to preserve nature's equilibrium when it stated that all parties to the agreement "shall take measures to prevent the disruption of the existing balance of [the Moon's] environment, whether by introducing adverse changes in that environment, by its harmful contamination through the introduction

---

27. Margaret S. Race and Richard O. Randolph, "The Need for Operating Guidelines and a Decision Making Framework Applicable to the Discovery of Non-Intelligent Extraterrestrial Life," *Advances in Space Research* 30(6) (2002): 1583–1591.
28. L. I. Tennen, "Evolution of the Planetary Protection Policy: Conflict of Science and Jurisprudence?" *Advances in Space Research* 34 (2004): 2357–2358.
29. United Nations, *Treaty on Principles Governing the Activities of States in the Exploration and Use of Outer Space, Including the Moon and Other Celestial Bodies*, UN doc. no. 6347, January 1967.

of extra-environmental matter or otherwise."[30] But this directive also stopped short of stating why the lunar environment must be protected, whether only for science's sake or also to support an ethical position. The ethical position that the *Moon Agreement* does put forth is the right of all countries to have the same access to the Moon and to all planets, expressed in the statement, ". . . to promote on the basis of equality the further development of co-operation among States in the exploration and use of the moon and other celestial bodies."[31]

The Soviet Union issued statements that also recognized the need to protect celestial ecologies. In its 1971 "Draft Treaty Concerning the Moon"[32] presented to the UN General Assembly, Article VI called for "avoiding the disruption of the existing balance of the lunar environment."[33] Earlier, the USSR had proposed a clearly ethical position regarding space exploration. In March 1962, Chairman Khrushchev wrote to President Kennedy regarding "heavenly matters,"[34] in which he urged that "any experiments in outer space which may hinder the exploration of space by other countries" should be discussed and agreements reached "on a proper international basis."[35] This and other Soviet proposals strongly linked the issue of contamination to the potential for interfering with the rights of states to achieve their own outer space explorations. The USSR focus was not on the negative impacts that might occur to a planet, but rather on the aspirations and objectives of other countries.[36]

These and other ethical concerns regarding the conduct of planetary exploration will be discussed in greater detail in Chapter 8.

---

30. United Nations, *Agreement Governing the Activities of States on the Moon and Other Celestial Bodies*, 1984. Note: The *Moon Agreement* was considered and elaborated by the Legal Subcommittee from 1972 to 1979. The agreement was adopted by the General Assembly in 1979 in resolution 34/68. It was not until June 1984, however, that the fifth country, Austria, ratified the agreement, allowing it to enter into force in July 1984. See *http://www.oosa.unvienna.org/SpaceLaw/moon.htm* for more information.
31. United Nations, *Agreement Governing the Activities of States on the Moon and Other Celestial Bodies*, 1984.
32. Tennen, "Evolution," 2359.
33. Ibid.
34. Chairman Khrushchev to President Kennedy, 21 March 1962, transmitting a letter of 20 March 1962, UN doc. A/AC,105/2 (21 March 1962), as reported in Tennen, "Evolution of the Planetary Protection Policy," 2356.
35. Ibid.
36. Ibid.

IN THE BEGINNING

The Need for Planetary Protection Is Recognized

2

The human species has a vital stake in the orderly, careful, and well-reasoned extension of the cosmic frontier. . . . The introduction of microbial life to a previously barren planet, or to one occupied by a less well-adapted form of life, could result in the explosive growth of the implant. . . . The overgrowth of terrestrial bacteria on Mars would destroy an inestimably valuable opportunity of understanding our own living nature.

—*Joshua Lederberg, Nobel Laureate, Professor of Genetics, and Advocate for Planetary Protection*[1]

From the beginning years of space exploration, space scientists have taken the threat of contaminating celestial bodies very seriously. In 1956, predating the USSR's Sputnik program by a year, the International Astronautical Federation (IAF) at its 7th Congress in Rome voiced its concerns regarding possible lunar and planetary contamination and attempted to coordinate international efforts to prevent this from happening.[2] But the events that most strongly instigated the U.S. space program and, with it, our country's planetary protection

---

1. Joshua Lederberg, "Exobiology: Approaches to Life Beyond the Earth," *Science* 132 (12 August 1960): 398–399.
2. Charles R. Phillips, *The Planetary Quarantine Program: Origins and Achievements 1956–1973* (Washington, DC: NASA SP-4902, 1974), pp. 1–3; "International Institute of Space Law," home page, *http://www.iafastro-iisl.com/*, copyright 2002–2006 (accessed 3 March 2006).

efforts were the 4 October 1957 launch of Sputnik I followed by the 3 November 1957 launch of Sputnik II.

Although the successful Sputnik launches surprised most Americans, proposals and serious studies for such Soviet space missions had begun years before. In 1950, the young Soviet engineer Mikhail Tikhonravov had written a seminal paper on the potential uses of artificial satellites. In 1954, the USSR ordered studies of such projects, including a focus on various engine and propulsion approaches. In 1955 another Soviet engineer, Sergei Korolev, formed a group to coordinate with Tikhonravov in developing the first artificial satellite. The USSR's first official plan for future spaceflight was issued on 30 January 1956, which called for orbiting satellites by 1958, human spaceflight by 1964, and reconnaissance satellites by 1970.[3]

The Sputniks' successful space voyages informed the world that a new era in technological achievement had begun, one whose outcome had serious consequences for the balance of the Cold War. Sputniks' achievements also "shone an unwanted spotlight on American science policy makers."[4] Senate Majority Leader Lyndon B. Johnson thought, after hearing about Sputnik while hosting a barbecue at his Texas ranch, that "somehow, in some new way, the sky seemed almost alien."[5] In response to the Sputnik program, the United States Congress, the military establishment, and much of the American public demanded immediate scientific and technical breakthroughs to match those of the Soviets. Sputnik made the conquest of space a new Cold War imperative for the United States,[6] and with this imperative came concerns about planetary contamination.

3. Asif Siddiqi, "The Man Behind the Curtain," *Air & Space* (1 November 2007), *http://www. airspacemag.com/space-exploration/sputnik_creator.html* (accessed 26 January 2011).
4. Audra J. Wolfe, "Germs in Space: Joshua Lederberg, Exobiology, and the Public Imagination, 1958–1964," *Isis* 93(2) (June 2002): 183.
5. Roger D. Launius, "Sputnik and the Origins of the Space Age," NASA History Office, *http:// www.hq.nasa.gov/office/pao/History/sputnik/sputorig.html* (accessed 21 February 2005).
6. Phillips, *The Planetary Quarantine Program*, pp. 7–10; Perry Stabekis, "Governing Policies and Resources" (presentation given to an unspecified NASA meeting, presentation slides/ notes given by Stabekis to author on 9 September 2004, Washington, DC); Perry Stabekis, "Governing Policies and Resources" (course notes from *Planetary Protection: Policies and Practices*, sponsored by NASA Planetary Protection Office and NASA Astrobiology Institute, Santa Cruz, CA, 19–21 April 2005).

Figure 2.1 Joshua Lederberg in a laboratory at the University of Wisconsin, October 1958.

Figure 2.2 Nobel Laureate Melvin Calvin.

●●●

Planetary Protection Visionaries and Key Organizations

Several scientists raised powerful alarms during the first years of space exploration efforts. As the superpowers of the world raced into space, these scientists feared that the exploration vehicles might well carry not only instruments, but also a mélange of Earth microorganisms. What effect would these bacteria and viruses have on the pristine environments that they invaded?

Scientists such as Stanford geneticist and Nobel Laureate Joshua Lederberg (see figure 2.1), University of California at Berkeley chemist and Nobel Laureate Melvin Calvin (see figure 2.2), and British biologist J. B. S. Haldane identified a very specific danger—that *the potential for scientific discovery could be forever compromised if space exploration was conducted without heed to protecting the environments being explored.* What a tragedy it would be if living organisms were found on a celestial body, but it could not be determined whether they were of alien origin or from Earth, unwanted hitchhikers on the very spacecraft that were there to test for life. Lederberg feared that "the capacity of living organisms [such as from Earth] to grow and spread throughout a new environment" was a matter of the gravest concern in our management of space missions.[7] We did not know if terrestrial organisms would indeed grow on, or within, another body in our solar system. However, if the number of terrestrial organisms on another body did balloon to Earth-like numbers, the effect on the body's environment could be dramatic, changing soil characteristics, the ecologies of any existing life-forms, and even the oxygen, carbon dioxide, and nitrogen ratios in its atmosphere.[8]

After a meeting in November 1957, when another Soviet space spectacular seemed imminent, Lederberg worried about the extent

---

7. Joshua Lederberg, "Exobiology: Approaches to Life beyond the Earth," *Science* 132 (12 August 1960): 393.
8. Lawrence B. Hall, "Sterilizing Space Probes," *International Science and Technology,* April 1966, p. 50, folder 006695, "Sterilization/Decontamination," NASA Historical Reference Collection; Audra J. Wolfe, "Germs in Space: Joshua Lederberg, Exobiology, and the Public Imagination, 1958–1964," *Isis* 93(2) (June 2002).

to which nationalistic motives to dominate the space race sometimes outranked scientific objectives, and he expressed concern that serious lunar and planetary biological contamination would result.[9] He believed that protection of scientific exploration and, in particular, of the ecosystems that might be encountered during voyages into space, was an ethic that should not be compromised. Lederberg articulated his concerns when he commented that "the scientific values of the space program, and the protection of its habitats for life (including our own) would be subordinated to geopolitics."[10]

Lederberg's warnings made an impression on members of the National Academy of Sciences (NAS), including its president, Detlev W. Bronk. NAS expressed apprehension that insufficiently sterilized spacecraft would "compromise and make impossible forever after critical scientific experiments."[11] NAS did not stop there, but adopted resolutions in February 1958 calling for the International Council of Scientific Unions (ICSU) to assist in developing the means of preventing harmful spacecraft contamination that "would impair the unique . . . scientific opportunities."[12] According to Lederberg, Bronk was careful to convey NAS concerns in a manner that left the issue in ICSU's hands, "so it wouldn't become an American proposal that had to be sold to the rest of the world, that it actually could be worked through on an international basis."[13]

CETEX

On 8 February 1958, as a result of NAS's expression of concern, ICSU established an ad hoc Committee on Contamination by Extraterrestrial Exploration (CETEX) to provide preliminary findings on lunar, Martian, and Venusian vulnerabilities to contamination. CETEX

9. Edward Clinton Ezell and Linda Neuman Ezell, *On Mars: Exploration of the Red Planet 1958–1978* (Washington, DC: NASA SP-4212, 1984).

10. Joshua Lederberg, "Apollo is Engineers' Triumph," *Houston Chronicle* (16 August 1969); folder 001275, "Lederberg, Joshua: Biographical Data," NASA Historical Reference Collection.

11. Ezell and Ezell, *On Mars: Exploration of the Red Planet 1958–1978*.

12. Ibid.

13. Steven J. Dick, "Interview with Dr. Joshua Lederberg," Rosslyn, VA, 12 November 1992; "Lederberg Interview," NASA Historical Reference Collection.

held its first meeting at The Hague on 12–13 May 1958, with Marcel Florkin, Belgian biochemist, as its president and Donald J. Hughes of the United States representing the International Union of Pure and Applied Physics. Other organizations at the meeting included the International Astronomical Union, the International Union of Biological Sciences, and the Special Committee for International Geophysical Year. The CETEX meeting produced the following statements and recommendations regarding various aspects of lunar exploration:

The Moon's Atmosphere. The Moon's atmosphere contained such little matter (estimated at only 10 to 100 tons) that the release of tons of volatile material from a flyby spacecraft's operations, such as the setting off of explosives for marking purposes, was likely to alter the atmosphere for very long periods. A landing vehicle would also almost certainly alter the Moon's atmosphere radically, since landing requires deceleration and the release of tons of chemical propellants. CETEX thus recommended that the Moon's atmosphere be studied in the initial phases by orbiting craft and that no flares be lit until extensive atmospheric data had been obtained.

Moon Dust. Space scientists had great interest in the chemical characteristics of particles on the Moon's surface and did not want them contaminated. CETEX thought that rocket impacts and other spacecraft operations would mainly lead to local contamination of lunar dust and thus not be that serious. Nuclear explosions, however, which might be planned in order to yield seismic data on the Moon's interior, did present a danger of contaminating the Moon's dust on a wide scale. Such explosions would release radioactive products that, especially in the extreme vacuum of the Moon, would likely behave as gases and thus be rapidly distributed over large areas through diffusion. Some of the radioactive products would be in a highly reactive form and might well combine with Moon dust to form nonvolatile compounds. The result of these processes could be that the entire lunar surface acquired additional radioactivity that interfered with radiochemical analyses performed to shed light on the Moon's past history.[14]

Cosmic Dust. Scientists examined the possibility that valuable information regarding cosmic (interstellar and interplanetary) dust

---

14. "Development of International Efforts to Avoid Contamination of Extraterrestrial Bodies," News of Science, *Science* 128 (17 October 1958): 887–889.

would be lost if the Moon's surface was disturbed during a mission. Cosmic dust contains a large percentage of low atomic weight elements such as hydrogen, carbon, nitrogen, and oxygen, and studying these elements and their relative abundances can potentially yield information about the evolution of the solar system. However, scientists believed that these low atomic weight molecules in cosmic dust are largely volatized by solar radiation, and thus little would be found on the Moon. Mainly the residues of high atomic weight elements in the dust would remain on the Moon, and this material was not particularly interesting, since similar deposits were found at the bottom of Earth's oceans.[15]

Panspermia Hypothesis. The scientific community considered the possibility that Moon dust might be important in evaluating the panspermia hypothesis—that transport of tiny living spores in cosmic dust was responsible for disseminating life throughout the universe (panspermia is discussed in more detail later in the chapter). But scientists concluded that the same solar radiation that volatized the lighter molecules in cosmic dust would also decompose living spores in the dust. Thus, CETEX did not expect Moon dust to shed light on the panspermia hypothesis.[16]

Contamination of the Lunar Surface by Microorganisms. The scientific community decided that there was little chance that introducing living spores or other microorganisms from Earth would lead to a sustainable form of life on the Moon. No known cells on Earth could, the scientists believed, grow or multiply under lunar conditions of high vacuum and no water.[17]

Complex Molecule Synthesis. The emergence and continuance of life requires the synthesis and replication of complex molecules. To understand the mechanisms by which life might originate, it was necessary to discover processes by which such complex molecules (which on Earth are carbon-based) might be constructed and replicated. Space scientists thought that the analysis of Moon dust might conceivably offer clues regarding these processes. In fact, some processes might have been taking place on the Moon that could eventually, in the right environment, lead to the formation of life. These processes might have been

---

15. Ibid.
16. Ibid.
17. Ibid.

similar to or different from those that took place on Earth. If such lunar processes were going on, then they could have been seriously upset through introduction of Earth chemistries to the lunar surface. While scientists considered the probability of such an occurrence to be remote, the damage done through contamination might conceivably have been very great. To lower the risk of this happening, CETEX recommended limiting the areas on the Moon in which spacecraft were allowed to land, thus localizing any effects of contaminants.[18]

At its May 1958 meeting, CETEX also examined the risks of space expeditions contaminating Mars and Venus. Since scientists believed that water, nitrogen, carbon oxides, and solar energy for photosynthesis were all available on Mars, they perceived a risk that terrestrial microorganisms would be able to grow there. CETEX therefore recommended that precautions be taken against space vehicles landing either accidentally or deliberately on Mars without first sterilizing them of all living organisms. Such precautions were considered for Venus as well. To not follow these safety measures was to jeopardize the search for extraterrestrial life on the planets.

CETEX reasoned that future searches for extraterrestrial life could also be jeopardized by setting off nuclear explosions on Mars or Venus. Introducing a different level of background radiation on these planets might, it was feared, "greatly influence any form of life found there."[19]

In its discussions, CETEX realized that there were inherent conflicts between certain types of experiments such that fulfilling the objectives of one would hinder the goals of another. For instance, an explosion of a thermonuclear device that would, as discussed above, provide valuable seismic data regarding a planet's interior might subsequently ruin the opportunity to conduct meaningful radiochemical analyses.[20]

One outcome of CETEX's contamination concerns was the publishing of two articles that attempted "to set a tone for developing a code

---

18. Ibid.
19. Ibid.
20. Such nuclear explosions on the Moon are now prohibited. The *Outer Space Treaty* of 1967 (discussed later in the book) does not allow the use of nuclear weapons on the Moon. In particular, Article IV includes the following language:

    "States Parties to the Treaty undertake not to place in orbit around the Earth any objects carrying nuclear weapons or any other kinds of weapons of mass destruction, install such weapons on celestial bodies, or station such weapons in outer space in any other manner.

of behavior in space research."[21] While CETEX did not at its inception have "sufficient scientific and technological data to enable it to propose a specific code of conduct,"[22] it recognized that such a code needed to be drafted as soon as possible in order to safeguard future interplanetary research and achieve a suitable compromise between all-out planetary exploration and protection of extraterrestrial environments. CETEX recommended that ICSU gather "interested and expert parties"[23] to address this issue as well as the statements and recommendations that were discussed above. ICSU did eventually gather such parties; what evolved is detailed in the section below on COSPAR.

Creation of the NAS Space Science Board

On 14–17 May 1958, immediately after CETEX's initial meeting, a Satellite-Life Sciences Symposium recommended by NAS's Detlev Bronk was held in Washington, DC. Lederberg wrote about his planetary protection views in a paper presented at this symposium, which was sponsored by NAS as well as the American Institute of Biological Sciences and the National Science Foundation (NSF). Lederberg's views about biological contamination in space carried considerable influence and contributed to NAS's establishment of the Space Science Board (SSB) on 4 June 1958. Lloyd V. Berkner served as SSB's first chairman, and its initial members included Nobel Laureate

---

"The Moon and other celestial bodies shall be used by all States Parties to the Treaty exclusively for peaceful purposes. The establishment of military bases, installations and fortifications, the testing of any type of weapons and the conduct of military maneuvers on celestial bodies shall be forbidden."

U.S. Dept. of State, *Treaty on Principles Governing the Activities of States in the Exploration and Use of Outer Space, Including the Moon and Other Celestial Bodies,* entered into force 10 October 1967.

21. Richard W. Davies and Marcus G. Comuntzis, "The Sterilization of Space Vehicles to Prevent Extraterrestrial Biological Contamination," JPL external publication no. 698, 31 August 1959 (presented at the 10th International Astronautics Congress, London, 31 August–5 September 1959), *http://ntrs.nasa.gov/archive/nasa/casi.ntrs.nasa.gov/19630042956_1963042956.pdf* (accessed 26 January 2011). The two articles mentioned in the text were "Development of International Efforts to Avoid Contamination of Extraterrestrial Bodies," *Science* 128 (1958): 887; and "Contamination by Extraterrestrial Exploration," *Nature* 183 (April 1959): 925–928.

22. "Development of International Efforts to Avoid Contamination of Extraterrestrial Bodies," News of Science, *Science* 128 (17 October 1958): 887–889.

23. Ibid.

chemist Harold Urey (who had conducted research on Earth's primeval atmosphere and prebiotic evolution) as well as Lederberg himself.

NAS tasked SSB with, among other priorities, surveying the scientific aspects of human space exploration and providing advice on all interplanetary contamination issues. Before the year was out, SSB had established the Panel on Extraterrestrial Life to "recommend approaches to the study of extraterrestrial life and, in particular, issues related to the problem of contamination."[24] NAS also directed SSB to "coordinate its work with the appropriate civilian and Government agencies, particularly the National Aeronautics and Space Administration [which was soon to be established], the National Science Foundation, the Advanced Research Projects Agency, and with foreign groups active in this field."[25]

The Establishment of NASA

On 1 October 1958, as a direct result of the political crises arising from Sputnik's 1957 launch, the United States established the National Aeronautics and Space Administration (NASA). President Eisenhower named T. Keith Glennan, who at that time was the president of Case Institute in Cleveland, Ohio, to be NASA's first Administrator.[26, 27]

The Establishment of COSPAR

Almost coincidental with NASA's inception and after vetting the CETEX recommendations described above, ICSU at its Washington,

---

24. David Darling, "Panel on Extraterrestrial Life," in *Encyclopedia of Astrobiology, Astronomy, and Spaceflight*, http://www.daviddarling.info/encyclopedia/P/PanelET.html (accessed 27 November 2006).

25. Mae Mills Link, "The Bioastronautics Mission Emerges," in *Space Medicine In Project Mercury*, NASA History Series (Washington, DC: NASA SP-4003, 1965), Chap. 1, http://www.hq.nasa.gov/office/pao/History/SP-4003/ch1-3.htm (accessed 26 January 2011).

26. NASA History Office, "NASA's Forty-Fifth Anniversary: Pioneering the Future," http://history.nasa.gov/45thann/html/45home.htm (accessed 13 February 2005).

27. John A. Pitts, "NASA's Life Sciences Program," in *The Human Factor: Biomedicine in the Manned Space Program to 1980* (Washington, DC: NASA SP-4213, 1985), Chap. 3, http://history.nasa.gov/SP-4213/ch3.htm (accessed 26 January 2011).

DC, meeting on 2–4 October 1958 established a Committee on Space Research (COSPAR) to encourage international collaboration and information exchange in interplanetary endeavors. This permanent committee worked and continues to work largely through organizing and sponsoring meetings and publishing scientific papers and information.[28] ICSU stressed that COSPAR was to be concerned with fundamental space research rather technological problems (such as propulsion or construction of rockets).[29] COSPAR's purview included the biological aspects of interplanetary exploration, such as spacecraft sterilization and planetary quarantine.[30]

COSPAR held its first meeting on 14–15 November 1958 in Loudon, Switzerland, with the intention of fostering communication between scientists of different countries. It was meant to be an "organization as independent as possible from politics and governments."[31] Several months after the formation of COSPAR, CETEX held its second and last meeting on 9–10 March 1959 in The Hague. CETEX, an interim, ad hoc group, took the position that interplanetary contamination problems were now better addressed by COSPAR, which then took over CETEX's functions.[32] Also in 1959, the United States' SSB endorsed the formation of COSPAR.[33]

From the beginning, a major objective of COSPAR was to open up a dialog between Eastern bloc space scientists and those from the rest of the world, most of whom were from the United States. There were difficulties in doing this. The Soviets objected to "the high proportion

28. UK-COSPAR, "What is COSPAR?" *http://www.cospar.rl.ac.uk/What_is_COSPAR.htm* (last updated 10 February 2004, accessed 22 February 2005).
29. Donald N. Michael, "General Implication for International Affairs and Foreign Policy," in *Proposed Studies on the Implications of Peaceful Space Activities for Human Affairs* (Washington, DC: Brookings Institute, December 1960), footnote 4, *http://www.bibliotecapleyades.net/brooking/brookings_footnotes05.htm* (accessed 24 March 2011).
30. Charles R. Phillips, "Scientific Concern Over Possible Contamination," in *The Planetary Quarantine Program* (Washington, DC: NASA SP-4902, 1974), *http://history.nasa.gov/SP-4902/ch1.htm* (accessed 26 January 2011).
31. Roger Bonnet, interview by *Astrobiology Magazine*, "The United Nations Of Space," Paris, France, *Space Daily* (15 June 2007), *http://www.spacedaily.com/reports/The_United_Nations_Of_Space_999.html* (accessed 26 January 2011).
32. Phillips, *The Planetary Quarantine Program*, pp. 4–5.
33. Stabekis, "Governing Policies and Resources," p. 2.

of Western representation"[34] on COSPAR and insisted that countries within the USSR such as Ukraine and Byelorussia have independent memberships. The United States considered this similar to asking for Texas and New York to have independent memberships. But when the Soviets threatened to pull out of COSPAR, a compromise was reached, allowing any nation interested in and engaged in some way in space activities to be a member. In addition, COSPAR "agreed to accept on its Executive Committee a Soviet vice president and a U.S. vice president, thus assuring both countries of permanent positions on the executive body of COSPAR."[35] As a result of these changes, the USSR remained a member of COSPAR.

Two kinds of memberships in COSPAR emerged—representation from interested scientific professional organizations, such as the Unions of Geodesy and Geophysics, Scientific Radio, Astronomy, and Pure and Applied Physics: and representation from nation states. In the words of Homer Newell, who in 1961 became the director of NASA's Office of Space Sciences,[36] "the ultimate strength of COSPAR lay in the national memberships, for, as with the International Geophysical Year, the individual countries would pay for and conduct research."[37]

One of COSPAR's early actions was to ask the United States and the USSR to examine approaches for avoiding transfer of terrestrial organisms to other planets. As a result, NASA implemented a planetary quarantine program whose main aim was to protect planets of biological interest so that 1) life detection experiments would not be invalidated by contamination and 2) a planet would not get "overgrown by terrestrial life with consequent irreversible changes in its environment."[38]

---

34. Homer E. Newell, "Political Context," in *Beyond the Atmosphere: Early Years of Space Science* (Washington, DC: NASA SP-4211, 1980), Chap. 18, *http://history.nasa.gov/SP-4211/cover.htm* (accessed 26 January 2011).

35. Ibid.

36. David Darling, "Newell, Homer Ex. (1915–1983)," in *Encyclopedia of Astrobiology, Astronomy, and Spaceflight, http://www.daviddarling.info/encyclopedia/N/Newell.html* (accessed 6 May 2005).

37. Homer E. Newell, "Political Context," in *Beyond the Atmosphere: Early Years of Space Science* (Washington, DC: NASA SP-4211, 1980), Chap. 18, *http://www.hq.nasa.gov/office/pao/History/SP-4211/cover.htm* (accessed 26 January 2011).

38. D. G. Fox, L. B. Hall, and E. J. Bacon, "Development of Planetary Quarantine in the United States," in *Life Sciences and Space Research* X (Berlin: Akademie-Verlag, 1972), p. 2.

COSPAR met every year in different venues so as to allow various nations to host the meetings. COSPAR typically divided its meetings into two parts: 1) a technical symposium on recent space science developments and 2) working group discussions on issues of importance to space exploration. Controversial discussions arose at these meetings regarding the undesirable impacts of space missions. In particular, the space science community voiced concerns over space research's possible compromise of other scientific activities—for example, interference of ground-based radio astronomy by radio signals from satellites. To better address such concerns, ICSU passed a resolution in 1961 requesting COSPAR to analyze proposed space activities that adversely impact other scientific experiments and observations and to share the results of its studies with governments and the space science community.[39]

COSPAR responded to ICSU's request by passing Resolution 1 in 1962 that created a Consultative Group on Potentially Harmful Effects of Space Experiments. Vikram Sarabhai, a physicist who would later head India's atomic energy agency, chaired this group. The Consultative Group also included delegates from the two major launching countries, the United States and the USSR, and from more neutral members. COSPAR's president, H. C. van de Hulst, participated as well because of his belief in the importance of the Consultative Group's mandate, which soon included addressing such issues as the effects of rocket exhausts on Earth's atmosphere and of high-altitude nuclear explosions on Earth's radiation belts.[40]

UN Committee on the Peaceful Uses of Outer Space (UNCOPUOS)

Established by the United Nations General Assembly in December 1958, UNCOPUOS reported the following year on contamination dangers to celestial bodies that could result from their exploration as well as on possible contamination of Earth from bringing materials back. UNCOPUOS is discussed further in Chapter 8.

---

39. Newell, "Political Context."
40. Ibid.

SSB and NASA Planetary Protection Actions

In 1959, SSB focused its attentions on the life sciences as its members developed interests in studying "the viability of terrestrial life-forms under extraterrestrial conditions and the implications of contamination."[41] An ad hoc committee, chaired by Joshua Lederberg, concluded that sterilizing spacecraft was technically feasible and effective means of doing so could be developed. SSB then sent communications to NASA and the Advanced Research Projects Agency (ARPA) on 14 September 1959, recommending the development of sterilization procedures.

NASA Administrator Glennan told SSB that the Agency had indeed adopted the policy of space probe sterilization to the extent feasible for all craft intended to pass near or impact upon the Moon or the solar system's planets. Moreover, NASA Director of Space Flight Programs Abe Silverstein requested that the Jet Propulsion Laboratory (JPL), Goddard Space Flight Center (GSFC), and Space Technology Laboratories[42] initiate the coordinated development of sterilization techniques. Also in 1959, after studying CETEX's research findings on the potential impacts of carelessly designed space missions, SSB decided to support the group's recommendations to establish an international code of conduct that would protect against extraterrestrial planetary degradation.[43]

---

41. Edward Clinton Ezell and Linda Neuman Ezell, "The Rise of Exobiology as a Discipline," in *On Mars: Exploration of the Red Planet, 1958–1978* (Washington, DC: NASA SP-4212, 1984).
42. NASA's National Space Technology Laboratories in southwestern Mississippi was the site for testing main propulsion systems, including Space Shuttle main engines. In the early 1960s, the laboratory was run by Marshall Space Flight Center and was known as the Mississippi Test Facility. It served as the engine test site for the Apollo/Saturn lunar landing program. It was designed so that large barges, using the Pearl River and its tributaries, could transport the huge Saturn S-IC and S-II stages and their liquid fuels and oxidizers to the test stands. "National Space Technology Laboratories," *http://science.ksc.nasa.gov/shuttle/technology/ sts-newsref/sts-msfc.html#sts-nstl* (accessed 26 January 2011), in "George C. Marshall Space Flight Center," *NSTS 1988 News Reference Manual, http://science.ksc.nasa.gov/shuttle/ technology/sts-newsref/* (last hypertexted 31 August 2000 and accessed 6 May 2005), on the Kennedy Space Center Web site, *http://science.ksc.nasa.gov/*. Information content for this online manual was taken from the *NSTS Shuttle Reference Manual* (NASA, 1988).
43. Ezell and Ezell; Stabekis, "Governing Policies and Resources," pp. 1–2; Fox et al., p. 1; Charles R. Phillips, "The Planetary Quarantine Program, 1956–1973" (draft), 9 August 1973, p. 4, folder 006695, "Sterilization/Decontamination," NASA Historical Reference Collection; NAS, "Space Science Board, 1958–1974," *http://www.hq.nasa.gov/office/pao/History/*

Figure 2.3 USSR scientist Anatoli A. Blagonravov, who helped to launch Sputnik and worked to establish cooperation in space between the United States and the USSR.

Tensions with the USSR

Although the focus of COSPAR's Consultative Group on Potentially Harmful Effects of Space Experiments started as a scientific endeavor, political squabbles soon arose, notably due to harangues of the United States by the USSR. Such was also the case at UNCOPUOS gatherings. At the May 1963 meetings of the Scientific and Technical Subcommittee of UNCOPUOS, the USSR representative, Anatoli Blagonravov (figure 2.3), a scientist who helped launch Sputnik, complained about the U.S. West Ford experiment, which had injected large numbers of tiny copper needles into orbit in order to gauge their utility for reflecting radio signals from one ground

sputnik/20fe.html (accessed 26 January 2011); David Darling, "Space Science Board," in *Encyclopedia of Astrobiology, Astronomy, and Spaceflight, http://www.daviddarling.info/ encyclopedia/S/SpaSciBoard.html* (accessed 21 March 2005).

location to another. On 21 May 1963, Blagonravov submitted a paper recommending that COSPAR study the potentially harmful impacts of experiments that would contaminate outer space.[44]

COSPAR's Consultative Group spent considerable effort addressing protection of the Moon and planets from biological contamination. Life scientists absolutely did not want contamination from Earth to compromise the search for and possible study of existing extraterrestrial life-forms, evidence of past life, or evidence of how the chemistry of a planet might evolve toward the formation of life.[45]

The Consultative Group's scientists understood that from economic and probably technical points of view, complete sterility of interplanetary spacecraft could not be achieved. Instead, the Consultative Group discussed probabilities of contamination and sought to arrive at acceptably low figures. "Acceptable" meant not setting the contamination probability so low as to make a spacecraft's development costs prohibitively expensive, but setting it low enough to make the chances of compromising scientific research very small. A "low enough probability" is a very subjective concept, and not surprisingly, "there were a great many opinions as to what probabilities were reasonable and as to how to go about the engineering. . . . The interminable discussions of the scientists were a vexation to the engineers who had to translate prescribed standards into engineering criteria."[46]

COSPAR did not agree on an international set of contamination probability objectives for years, although one number that its Subcommittee for Planetary Quarantine recommended in May 1966,[47] which COSPAR reaffirmed at its 12th plenary session in Prague on 11–24 May 1969, was to keep the probability of contaminating the solar system's planets (and in particular Mars) to no more than one in 1,000 during the anticipated period of biological exploration. COSPAR

---

44. Newell, "Political Context."
45. Ibid.
46. Ibid.
47. Charles W. Craven and Robert P. Wolfson, *Planetary Quarantine: Techniques for the Prevention of Contamination of the Planets by Unsterile Spaceflight Hardware*, Technical Report 32-1216, JPL, 15 December 1967, p. 1, folder 006697, "Quarantine/Sterilization," NASA Historical Reference Collection.

estimated this period at 20 years, extending through the late 1980s, and envisioned approximately 100 missions launching during this time.[48]

Questions of how to translate evolving COSPAR planetary protection requirements into space vehicle engineering criteria were hotly debated by different nations. The United States attempted to facilitate this discussion by sharing the details of how its spacecraft were designed and constructed. The United States' openness differed markedly from the USSR's custom of giving few design details of its spacecraft, saying only that the craft were going to be decontaminated. The USSR's close guarding of its spacecraft details resulted in considerable apprehension among other countries' space scientists, because adequate planetary protection could only be achieved with the full cooperation of *all* spacefaring nations. Insufficient cleaning of even one nation's space vehicles could greatly degrade the scientific opportunities available to all nations. And once these opportunities were lost, they would never be fully recoverable.[49]

Joshua Lederberg and WESTEX

Parallel to the efforts of COSPAR in developing the planetary protection field were some important achievements by noted scientists. One of these was Joshua Lederberg, who was instrumental in defining a role for the life sciences in space research. Although NAS leadership had intended its SSB to include life sciences in its oversight of space science, its initial actions focused on the physical problems of space exploration. Of its 10 original committees, only 1—the Committee on Psychological and Biological Research—focused on life science issues. The remaining 10 dealt with non-life-science issues such as geochemistry, astronomy, orbits, vehicle design, and meteorology.[50]

Joshua Lederberg put forth an admirable effort to alter NASA's research priorities and bring exobiology to "the forefront of space

---

48. Newell, "Political Context."
49. Ibid.
50. Audra J. Wolfe, "Germs in Space: Joshua Lederberg, Exobiology, and the Public Imagination, 1958–1964," *Isis* 93(2) (June 2002): 183–205.

policy."[51] Exobiology, a term coined by Lederberg himself, refers to the study of life's origins on Earth and the development of instruments and methods to search for signs of life on other celestial bodies.[52]

Lederberg's work was driven largely by his fears that the space program would result in irreversible contamination of the planets as well as Earth itself. The origin of his interest in outer space, according to Lederberg himself, occurred at a dinner meeting between him and British biologist J. B. S. Haldane in Calcutta, India, at which Haldane mused whether the USSR would set off a thermonuclear explosion on the Moon to commemorate the 40th anniversary of the Bolshevik Revolution. Haldane suggested that they might see a "red star"—a thermonuclear explosion—appear on the Moon. After considering such a possibility, both Lederberg and Haldane bemoaned the potential contamination that the Moon's pristine environment might suffer from such a political demonstration. Lederberg took the situation as "a striking metaphor for the danger that scientific interests would be totally submerged by the international military and propaganda competition,"[53] noting that science objectives had never gained first priority over military considerations and might get totally excluded.

Soon after his meeting with Haldane, Lederberg began his campaigns for international agreements as well as spacecraft sterilization procedures capable of protecting the Moon and other planets from both radiological and biological contamination. Lederberg felt the need to "assure that fundamental biological science was properly represented in the programs of space research that were just emerging."[54] In February 1958, Lederberg helped convince NAS to pass a resolution demanding thorough space mission planning to ensure that "operations do not compromise and make impossible forever after critical scientific experiments."[55]

51. Wolfe, "Germs in Space."
52. Steven J. Dick and James E. Strick, *The Living Universe: NASA and the Development of Astrobiology* (Piscataway, NJ: Rutgers University Press, 2004), p. 29.
53. Joshua Lederberg, "How DENDRAL was Conceived and Born" (ACM Symposium on the History of Medical Informatics, National Library of Medicine, 5 November 1987), pp. 3–4, *http://profiles.nlm.nih.gov/BB/A/L/Y/P/_/bbalyp.pdf* (accessed 26 January 2011).
54. Ibid., pp. 3–4.
55. Dick and Strick, *The Living Universe*, p. 24; Wolfe, "Germs in Space."

Lederberg's passion and deep concern for the well-being of scientific endeavors in space drew other scientists to his cause. From them, he formed the West Coast Committee on Extraterrestrial Life (WESTEX), a part of NAS's Panel on Extraterrestrial Life,[56] to address the protection and preservation of planetary surfaces during space exploration. Lederberg's WESTEX first met at Stanford University on 21 February 1959. Although a group of scientists that called themselves EASTEX (for East Coast Committee on Extraterrestrial Life, which was, like WESTEX, a part of NAS's Panel on Extraterrestrial Life) had met earlier in Cambridge, Massachusetts, WESTEX was the more active of the two, convening at least five times between 1959 and 1960. WESTEX, which focused on issues of planetary contamination, also differed from EASTEX in that the latter group concerned itself almost exclusively with the origin of life.

Lederberg brought high-profile scientists into WESTEX such as Melvin Calvin of the University of California at Berkeley, who would win the Nobel Prize for his work on photosynthesis; Harold Clayton "H. C." Urey of the University of Chicago and University of California at San Diego, who, among many other efforts, conducted important origin-of-life experiments; geneticist Norman Horowitz of the California Institute of Technology (Caltech); molecular biologist Gunther Stent of the University of California at Berkeley; molecular geneticist Matthew S. Meselson of Caltech and Harvard; nuclear scientist turned biophysicist Aaron Novick of the University of Oregon; microbiologist C. B. Van Niel of Stanford; and a young astronomer, Carl Sagan.[57] This group was purposely chosen so as to include a "diversity of interest . . . compactness of size, and convenience of assembly."[58]

In 1958, ICSU's CETEX had outlined potential contamination dangers to the Moon, Venus, and Mars. But Lederberg believed that CETEX's work did not go far enough toward protecting planetary surfaces

---

56. Darling, "Panel on Extraterrestrial Life."
57. Wolfe, "Germs in Space"; James R. Arnold, Jacob Bigeleisen, and Clyde A. Hutchison, Jr., "Harold Clayton Urey," *Biographical Memoirs*, vol. 68 (Washington, DC: National Academies Press, 1995), pp. 363–412, http://www.nap.edu/html/biomems/hurey.html (accessed 26 January 2011); Dick and Strick, *The Living Universe*, p. 25; Norman H. Horowitz, interview by Rachel Prud'homme, Oral History Project, California Institute of Technology Archives, 9–10 July 1984, http://oralhistories.library.caltech.edu/22/ (accessed 26 January 2011).
58. R. C. Peavey to the Space Science Board, memorandum SSB-77, 13 April 1959, correspondence/1959/April (chronological listing), Joshua Lederberg Papers.

(including that of Earth). One of WESTEX's initial tasks was to prepare scientific statements focusing on this area of concern for future CETEX meetings.[59] WESTEX's first meeting (held 21 February 1959 at Stanford University) was devoted to policy regarding "celestial contamination,"[60] and concluded that rigorous decontamination of spacecraft was essential as well as feasible. This position was reaffirmed at WESTEX's second meeting, which was held a month later at JPL in Pasadena.

Theories on the origins of planetary life. WESTEX's desire to protect planetary surfaces from contamination made it necessary to consider the possible pathways for life to secure a foothold on those planets. One early 20th century theory, that of panspermia, held that life-bearing seeds were scattered throughout space and could germinate wherever conditions were favorable. In order to identify these seeds during a space mission's quest to find life, scientists would need to distinguish them from substances of terrestrial origin as well as materials unrelated to life. Keeping planetary surfaces free from terrestrial microorganisms and other contaminants of human origin would aid greatly in isolating the panspermia seeds.

The newly evolving field of molecular biology raised serious misgivings on the likelihood of seeds of life floating endlessly through space and fertilizing previously barren planets. Lederberg did not categorically dismiss the panspermia theory but advised members of WESTEX that it was not a sound basis on which to justify stricter guidelines for space missions. A more current debate on mechanisms of life origination focused on abiotic synthesis as a means of forming organic molecules. Instead of embracing any particular theory of life origins, however, Lederberg chose to plead ignorance regarding a specific mechanism that was responsible. WESTEX proposed that the wisest course, and the one offering the best opportunities for scientific return, was simply to search for the existence of extraterrestrial organic molecules or bacterial organisms. WESTEX's proposal was a politically savvy one, for it sidestepped arguments and disagreements regarding *how* life might have originated in favor of just identifying the existence of life, or potential building blocks of life, on a particular planet.[61]

---

59. Wolfe, "Germs in Space."
60. Peavey, 1959.
61. Wolfe, "Germs in Space."

Figure 2.4 Carl Sagan with a model of the Viking Mars lander in Death Valley, California. Planetary protection for the Viking mission is examined in Chapter 5.

Prioritizing planetary protection. Beginning with WESTEX's initial meetings in 1959–1960, Lederberg as well as Carl Sagan (figure 2.4) advocated that a high priority be placed on preventing planetary probes from carrying terrestrial contamination into space, and nearly as high a priority on the prevention of back contamination from sample return missions. Lederberg and Sagan lobbied to make these priorities NAS SSB's official policy and to present them to COSPAR, which had become an important international forum for exobiology matters. Lederberg and Sagan's lobbying efforts proved very effective at bringing planetary protection issues to the fore and incorporating them into space science policy at all levels. Allan H. Brown, a scientist on NASA's Biosciences subcommittee, also advocated taking steps to prevent back contamination, underlining (as Lederberg and Sagan did) that even if the risk of such contamination was very small, the scale of harm could be huge indeed. In 1963, COSPAR agreed to form an anticontamination panel.[62]

---

62. Dick and Strick, *The Living Universe*, p. 59.

A dissenting voice. Norman Horowitz was a WESTEX member who had quite a different view on back contamination. Although he gave some support to spacecraft sterilization as a means of preventing forward contamination, he considered the attention given to back contamination to be overblown. Belief in the supposed danger of back contamination, according to Horowitz, rested on three assumptions, each of which he thought was improbable:

- Microorganisms will be found on the planets.
- They will prove dangerous to humans.
- We would be unable to cope with them if they were brought back to Earth and escaped from quarantine.

The probability that the microbes would run rampant on Earth could be estimated by taking the product of the probability of each of the bulleted items (in other words, multiplying together the small probabilities of each of the above events should they occur). This would, in Horowitz's mind, produce a *very* small overall probability of contaminating Earth—so small that he stated, regarding sample materials brought back from Mars: "I would be willing to run the risk involved in a premature return trip, if a less bold schedule meant that a sample of Martian soil could not be brought back to earth in my lifetime."[63]

Horowitz argued that the slight risk of introducing a pandemic disease due to back contamination did not outweigh the "potential benefits to mankind of unhampered traffic with the planets."[64] By way of a historical example, he considered the back contamination resulting from Christopher Columbus's voyages, which was severe, but made a case that even this was justified. If Europeans had known that Columbus's crew would bring syphilis back with them and that it would plague Europe for centuries, the voyages might have been canceled. On the other hand, if Europeans had also been aware of the enormous wealth and resources that would be gained from the voyages, they would no doubt have approved Columbus's travels.[65]

Lederberg countered Horowitz's argument with the conviction that "we are in a better position than Columbus was to have our cake

---

63. Norm Horowitz letter to J. Lederberg, 20 January 1960, Correspondence/1960/January (chronological listing), Joshua Lederberg Papers.
64. Ibid.
65. Ibid.

and eat it too."[66] Caution and patience as well as modern analytical techniques, Lederberg believed, could allow planetary exploration to proceed in a manner that was also protective of Earth and its inhabitants. Other members of WESTEX concurred, but they were not nearly as polite as was Lederberg in refuting Horowitz. Aaron Novick of the University of Oregon grew angry, citing as an example the myxoma virus's introduction to Australia's rabbit population. Propagated in the 1950s as a means of controlling Australia's destructive rabbit hordes, it initially killed 99 percent of the population. Novick's point was that the syphilis that Columbus brought back home might well have wiped out just as high a percentage of Europe's population.[67] And by extension, if a space expedition were to bring an extraterrestrial microbe back to Earth, it could have a similar effect.

Building respect for the exobiology field. One of Lederberg's and WESTEX's notable achievements was to garner a certain amount of respect and funding for the new field of exobiology, and this helped give planetary protection considerations more importance as well. This quest for respect was aided in part by Lederberg's stature as a Nobelist as well as WESTEX's strategy of linking exobiological work with the techniques of molecular genetics. According to WESTEX, since exobiology gave high priority to molecular-level analyses, it would eventually generate vital data for ongoing evolution, microbiology, and theoretical biology scientific debates. Exobiology was developing into a transnational field carrying out cutting-edge molecular genetics and biochemistry experiments concerned, among other things, with investigating the uniqueness of living systems based on nucleic acids and proteins, and these endeavors gave the field prestige. They also helped make a strong case for implementing rigorous planetary protection measures on expeditions searching for extraterrestrial life, for these expeditions "demanded an especially cautious approach to the untouched surfaces of other planets"[68] to be successful. Conversely, recognition by the scientific community of the need for planetary protection also helped

---

66. Dick and Strick, *The Living Universe*, p. 60.
67. Dick and Strick, *The Living Universe*, pp. 60–61; Commonwealth Industrial and Scientific Research Organization (CSIRO), "Controlling Wild Rabbits: Need for Integrated Control Strategy," Australia and New Zealand Rabbit Calicivirus Disease Program, CSIRO, 1997, *http://www.csiro.au/communication/rabbits/qa1.htm*.
68. Wolfe, "Germs in Space," 189.

raise support for exobiology programs. Lederberg effectively "used the fear of contamination—of earth and other planets—as a central argument for funding 'nonpolitical,' 'scientifically valid' experiments in extraterrestrial life detection."[69]

Lederberg and his colleagues did well at selling a message of scientific as well as national prestige to NAS SSB and, as a result, reaped some rewards for their own research programs. In 1962, Lloyd Berkner, SSB's first chairman, told the president of NAS that exobiology, rather than geology, meteorology, or radiation studies, was the "most important scientific research program in space."[70] NASA soon became the prime sponsor of exobiological research. Joshua Lederberg and his Stanford laboratory received grants exceeding $500,000 to carry out exobiological studies.

Development of the planetary protection field has clearly involved the work of a mélange of different organizations, agencies, and facilities. The table in Appendix D lists some of these entities that have been important to planetary protection and briefly summarizes the nature of their contributions.

Development of NASA Planetary Protection Policy and Organizational Structure

NASA's First Spacecraft Sterilization Policy Statements

NAS SSB, concerned with the possibility of planetary contamination due to space exploration, requested Joshua Lederberg to set up an ad hoc committee to study issues of spacecraft sterilization. This committee convened at Stanford University on 6–8 June 1959 and included representatives from NASA's Goddard Research Center and Jet Propulsion Laboratory, General Electric Company, and the U.S. Army Biological Laboratories.[71]

---

69. Wolfe, "Germs in Space," 183, 189.
70. Wolfe, "Germs in Space."
71. Phillips, *The Planetary Quarantine Program*, pp. 9–11.

The Stanford meeting led to considerable debate on sterilization matters. One of the basic issues was just how clean a spacecraft must be for planetary protection purposes. In a 30 July 1959 memo, the atmospheric scientist Leslie H. Meredith, who was NASA Goddard Space Flight Center's (GSFC) newly appointed Chief of the Space Sciences Division,[72] asked just how clean a space probe must be, given the enormously larger size of a planet. Meredith also identified a major problem with attaining a useful state of sterilization, writing that "there are obviously some places which could only be 'cleaned' by a very major effort (i.e., the insides of resistors, transistors, relays, etc.). If it is assumed such places are left 'dirty,' how much point is there to making the remainder of the probe completely 'clean?'"[73] In a memo dated 3 August 1959, John W. Townsend, Jr., GSFC's new Assistant Director, Space Science and Satellite Applications,[74] also wondered whether devoting considerable levels of effort to spacecraft sterilization was a worthwhile endeavor. Townsend did not believe, as a scientist, that the state of knowledge was such as to even know if space exploration presented a danger of planetary contamination. Furthermore, he also saw "an almost impossible situation developing from a practical standpoint"[75] if the space science community went ahead with attempts at space probe sterilization. Barriers he identified at GSFC to accomplishing useful sterilization included the following:

- Lack of qualified biologists and chemists to supervise the program.
- Insufficient GSFC personnel to "handle the extra workload"[76] of sterilization activities.
- The required cleanroom facilities were not available.

---

72. William R. Corliss, "Genesis of the Goddard Space Flight Center Sounding Rocket Program," in *NASA Sounding Rockets, 1958–1968: A Historical Summary* (Washington, DC: NASA SP-4401, 1971), Chap. 6, *http://history.nasa.gov/SP-4401/ch6.htm* (accessed 26 January 2011).

73. L. H. Meredith to 9100, "Memo from Posner on Sterilization of Space Probes," 30 July 1959, folder 006696, "Sterilization/Decontamination," NASA Historical Reference Collection, in a series of memos beginning with Abe Silverstein, Memorandum for Director, Goddard Space Flight Center, "Sterilization of Payloads."

74. Corliss, "Genesis of the Goddard Space Flight Center Sounding Rocket Program," 1971.

75. John W. Townsend, Jr., to J. Posner, "Sterilization of Space Probes," 3 August 1959, folder 006696, "Sterilization/Decontamination," NASA Historical Reference Collection, in a series of memos beginning with Abe Silverstein, Memorandum for Director, Goddard Space Flight Center, "Sterilization of Payloads."

76. Townsend, "Sterilization of Space Probes," 1959.

- None of GSFC's construction or assembly people were familiar with hospital sterilization procedures.
- Added costs and personnel time would raise serious issues.

Even if these problems could be surmounted, Townsend was not sure that adequate sterilization could be achieved due to space vehicle procedures that resulted in "repeated opening of payloads and testing up to the final minute of launching."[77]

On 14 September 1959, Hugh Odishaw, Secretary of NAS SSB, wrote to T. Keith Glennan, NASA's first Administrator, declaring that the spacecraft sterilization recommendations of Lederberg's ad hoc committee had SSB's approval and requested that they be followed. Administrator Glennan answered a month later on 13 October 1959, pledging that NASA would indeed do its best to carry out the recommendations.

During the same month, Abe Silverstein, who was NASA's Director of Space Flight Programs at the time, wrote letters to JPL and Goddard Space Flight Center stating NASA's new policy, that space mission payloads "which might impact a celestial body must be sterilized before launching."[78] The letters listed the payloads of concern and recommended particular sterilization approaches. For instance, in a letter sent to the Director of Goddard on 15 October 1959, Silverstein mentioned the need to sterilize the P-14 Lunar Magnetometer Experiment payload and identified gaseous ethylene oxide as the sterilizing agent that NASA considered the most feasible.[79] Some communications from Silverstein also suggested that the U.S. Army BioLabs at Fort Detrick, Maryland, which had experience with sterilization issues, be brought onto the team. These letters from Silverstein were the first official NASA policy directives on spacecraft sterilization. Weeks after they were written, on 12 November 1959, NASA transferred funds to the Army to support its cooperative efforts.[80]

---

77. Ibid.
78. Charles R. Phillips, *The Planetary Quarantine Program: Origins and Achievements 1956–1973* (Washington, DC: NASA SP-4902, 1974), pp. 9–11.
79. Abe Silverstein, Memorandum for Director, Goddard Space Flight Center, "Sterilization of Payloads," 15 October 1959, folder 006696, "Sterilization/Decontamination," NASA Historical Reference Collection.
80. Phillips, *The Planetary Quarantine Program*, pp. 9–11.

Abe Silverstein's Approach to Spacecraft Sterilization and the Difficulties in Implementing NASA Policy

Homer Newell, the former Director of NASA's Office of Space Sciences,[81] considered Abe Silverstein (figure 2.5) to be "a hard-nosed, highly practical, boldly innovative engineer, with a solid conviction . . . that all research had to have a firm justification in practical applications to which it would ultimately contribute."[82] He was also seen as a powerful manager who got results. Former NASA Administrator Daniel S. Goldin called him "a man of vision and conviction" whose "effective leadership, both at Headquarters and at Lewis [NASA's Lewis Research Center (LeRC)], directly contributed to the ultimate success of America's unmanned and human space programs."[83] Silverstein's letters put forth statements of NASA policy that were unqualified in their demand for sterilization of payloads that could contaminate a solar system body, with no room for equivocating. This NASA requirement articulated by Silverstein, however, proved very difficult to implement. Spacecraft were being launched by physical scientists and engineers with little experience in biological methodologies, especially sterilization techniques, and with almost no data on what these techniques might do to the dependable operation of the spacecraft.[84]

Figure 2.5 Abe Silverstein, NASA Director of Space Flight Programs and an advocate of spacecraft sterilization.

The typical model for sterilization was at that time drawn from the procedures used in hospitals to clean small instruments for surgical applications. These procedures employed autoclaving, involving wet steam at pressures 25 psi over atmospheric and carried out in small pressurized chambers. Autoclaving technologies were very

---

81. Darling, "Newell, Homer E. (1915–1983)."
82. Homer E. Newell, "Staffing," in *Beyond the Atmosphere: Early Years of Space Science* (Washington, DC: NASA SP-4211, 1980), Chap. 8, *http://www.hq.nasa.gov/office/pao/History/SP-4211/cover.htm* (accessed 26 January 2011).
83. NASA, "Dr. Abe Silverstein, Early Architect of the Apollo Moon Landing, Dies," NASA news release 01-109, 1 June 2001.
84. Phillips, "Implementation and Policy Directives," in *The Planetary Quarantine Program*.

poorly suited for sterilizing spacecraft.[85] In addition, medical applications required that *surfaces* of instruments such as scalpels be devoid of bacteria, but not necessarily their interiors, which would never touch the patient. But a spacecraft might be subjected to a high velocity impact with a planetary surface, shattering the shell of the craft and its interior components. Unless the exterior *and interiors* of the craft had been adequately sterilized, such a crash might release viable organisms onto, or beneath, the planetary surface. In addition, resistors and capacitors, plastic components, ceramic components, indeed, all parts of the spacecraft, needed to be sterilized not only on their surfaces, but also within their interiors. Achieving this was an awesome task.[86]

The range of techniques employed by the U.S. Army BioLabs were much better suited for sterilizing spacecraft than those employed by hospitals. The Army BioLabs could sterilize objects as large as army trucks and as small as laboratory balances. The Lederberg ad hoc committee on spacecraft sterilization recommended that these capabilities be exploited, and this was repeated in Silverstein's letters.

The Army BioLabs had refined the use of gases for sterilizing equipment, particularly ethylene oxide, which had several important advantages. It was noncorrosive to most equipment and could be stored in simple plastic containers at ambient temperatures and pressures. The gas consisted of small molecules that were soluble in many different materials, including rubber, plastic, and oil. It was quite a penetrating sterilant, able to work its way into the cracks and interstices of many components.[87] One experiment showed that the gas

85. Phillips, "Implementation and Policy Directives," in *The Planetary Quarantine Program.*

86. Lawrence B. Hall, "Sterilizing Space Probes," *International Science and Technology* (April 1966): 51–52, folder 006695, "Sterilization/Decontamination," NASA Historical Reference Collection.

87. Abe Silverstein to Hugh L. Dryden, Deputy Administrator, "Sterilization and Decontamination of Space Probes," 6 August 1959, p. 8, folder 006696, "Sterilization/Decontamination," NASA Historical Reference Collection; Richard W. Davies and Marcus G. Comuntzis, "The Sterilization of Space Vehicles to Prevent Extraterrestrial Biological Contamination," JPL external publication no. 698, 31 August 1959, p. 8 (presented at the 10th International Astronautics Congress, London, 31 August–5 September 1959), *http://ntrs.nasa.gov/archive/nasa/casi.ntrs.nasa.gov/19630042956_1963042956.pdf* (accessed 26 January 2011).

would penetrate to the inside pages of a New York telephone directory, even when an anvil was sitting on the book.[88]

But while gaseous treatment killed microorganisms on and sometimes in the cracks between surfaces, it could not reliably penetrate *through* solids such as the epoxy potting materials, ceramics, or plastics used to protect electronic components. Gaseous sterilants had difficulty penetrating the interiors of resistors and capacitors and even getting to the threads of installed screws and bolts. A hard landing of the spacecraft on a target body might expose these unsterilized regions, leading to contamination of the planet.[89]

But were microorganisms likely to exist within the interiors of spacecraft components? Might their manufacturing processes have accomplished what gas sterilants could not? Unfortunately, this proved not to be the case. In studies evaluating these concerns, the Army BioLab confirmed that some microorganisms survived plastic polymerization processes. In addition, many electronics components such as capacitors, resistors, and transformers were, the BioLab discovered, received from the manufacturer with viable microorganisms inside. When the component was crushed or cracked, these microorganisms were able to multiply, even in components that had received surface sterilization with ethylene oxide.[90]

Two other approaches—radiation and heat—were possibilities for effectively sterilizing buried or covered contamination, but each one had its problems. Penetrating radiation can be effective, although it has far less impact on microorganisms than on higher life-forms. This is because, to kill a unicell microorganism, the radiation has to hit and sufficiently damage a very small target—the structure of the single tiny cell comprising the organism—to the point where the cell cannot repair itself. To kill a higher life-form, however, only requires killing a small percentage of its many billions of cells. For instance, if radiation only damages cells in the intestinal lining of a mammal, but to the point where the intestine cannot take in water or nutrients, the mammal will likely die. Thus the probability is far greater that enough radiation will hit enough vital cells of a higher life-form to kill it. As

---

88. Hall, "Sterilizing Space Probes," p. 52.
89. Phillips, "Implementation and Policy Directives," in *The Planetary Quarantine Program*.
90. Ibid.

a result of this situation, orders-of-magnitude more intense radiation dosages typically have to be used to kill microorganisms than higher life-forms. But the intense radiation levels necessary to kill microorganisms can damage many spacecraft components.[91]

Heat is effective in killing microorganisms, but only if it can reach them. Wet steam autoclaving is useful for sterilizing surgical instruments at temperatures of about 125°C and exposure times of 15 to 20 minutes, but steam is no more able to penetrate into solid materials or screw threads than is ethylene oxide gas. Alternatively, dry heat can be used. Unfortunately, microorganisms are much more resistant to dry heat than to wet steam exposure, so temperatures of 160° to 170°C for 4 or more hours were typically employed. Such high temperatures can, however, severely stress and damage sensitive equipment. In response to this situation, exploratory experiments at the Army BioLabs examined the effectiveness of lower temperatures applied for longer dwell times. The data that were generated suggested that 125°C and dry-heat exposure times of 24 hours might be sufficient for NASA's needs.[92]

●●●

Sterilization Technology Development

At a 1962 conference on spacecraft sterilization, Charles Phillips of the Army BioLabs at Fort Detrick discussed the issues involved with low heat/long exposure approaches, noting that this method used to be considered highly impractical and was, until recently, poorly examined. Phillips explained the dependence of microorganism death on both time of exposure and temperature. He also discussed the abilities of spacecraft materials to withstand the sterilizing environment, noting that at certain temperatures, materials begin to decompose, and this defined an upper limit on the temperatures that could be used.

---

91. John L. Feinendegen, M. Hinz, and R. Neumann, "Whole-body Responses to Low-Level Radiation Exposure: New Concepts in Mammalian Radiobiology," *Experimental Hematology* 35(4) (April 2007): 37–46; Walla Walla [Washington] County Emergency Management Department, "Hazard Identification and Vulnerability Analysis," October 2003, *http://www.co.walla-walla.wa.us/Departments/EMD/Docs/Hazards/Radiological.pdf* (accessed 10 July 2008).

92. Phillips, "Implementation and Policy Directives," in *The Planetary Quarantine Program*.

Phillips did believe, however, that many materials were available that could withstand 125°C almost indefinitely, whereas at this temperature, he thought that bacteria would be slowly killed off. Phillips was of the opinion that given sufficient exposure time, a spacecraft could be sterilized at 125°C.[93]

The Question of Viruses

Some space scientists believed that the sterilization techniques under investigation did not adequately address how to deal with viruses. The techniques, in fact, tended to ignore viruses as a class. Typical sterility tests did not determine whether viruses survived the sterilization procedures or not. Instead, scientists made assumptions that viruses would not present planetary contamination problems. Viruses, biologists believed, need a host cell to propagate. Viruses are also highly specific to particular types of cells. Thus, if the spacecraft was not carrying those types of host cells, the planet being explored would not get infected. Furthermore, to infect a planet, not only would the virus have to get to it alive, the virus would also have to find suitable host cells on that planet in order to propagate. Phillips believed that since the probability was very low that an appropriate host cell for an Earth virus existed on another planet, then it was also very unlikely that it could propagate on that planet. "The only organisms we are worrying about," Phillips said, "are the husky, tough soil organisms which live on the very simplest kind of nutrient and media."[94]

---

93. Freeman H. Quimby, ed., *Proceedings of Conference on Spacecraft Sterilization* (Washington, DC: NASA Technical Note D-1357, 1962), p. 39; Charles R. Phillips, *The Planetary Quarantine Program: Origins and Achievements, 1956–1973* (Washington, DC: NASA SP-4902, 1974); folder 006697, "Quarantine/Sterilization," NASA Historical Reference Collection.

94. Jack Posner, ed., *Proceedings of Meeting on Problems and Techniques Associated with the Decontamination and Sterilization of Spacecraft—June 29, 1960* (Washington, DC: NASA Technical Note D-771, 1961), p. 18.

Clark Randt, who was appointed director of NASA's Office of Life Science Programs in March 1960,[95] did not totally dismiss viruses as a contamination risk. He commented to Phillips that "the question has arisen as to whether or not bacterial and mycotic [fungal] spores could not in themselves provide the cell on which a virus could survive." Another attendee, Stanley M. Levenson of the U.S. Army Medical Research and Development Command, added that if a space mission introduced a virus on another planet, and if the virus found some way to survive, then any terrestrial animal [including humans] introduced to that planet's ecosystem might get infected, and we wouldn't know whether that contamination originated on Earth or on that planet.[96] Randt's and Levenson's observations were important, for they identified potential pathways by which planets could get contaminated and the search for life could be severely compromised.

Sterilization Approaches for the Ranger Program

In developing the Ranger Lunar exploration program, JPL encountered significant difficulties meeting the terms of Silverstein's directives to sterilize all of the spacecraft's thousands of individual parts. Terms such as "sterilize to the extent feasible" began appearing in JPL correspondence.[97] NASA remained firm, however, regarding the importance of adequate sterilization. A 23 December 1960 memorandum to program and field station directors that was signed by Hugh L. Dryden, NASA Deputy Administrator, for T. Keith Glennan, Administrator, affirmed that "effective decontamination and sterilization procedures for lunar and planetary space vehicles are essential"[98] and called for a thorough study to attain this objective. No mission would be flown until adequate sterilization plans and procedures were developed. JPL initiated the requested studies, focusing in particular

---

95. Mae Mills Link, "Office Of Life Sciences Established: 1960," a section in "NASA Long-Range Life Science Program," in *Space Medicine In Project Mercury* (Washington, DC: NASA SP-4003, 1965), Chap. 4, *http://history.nasa.gov/SP-4003/ch4-5.htm* (accessed 26 January 2011).

96. Posner, *Proceedings*, p. 18.

97. Phillips, "Implementation and Policy Directives," in *The Planetary Quarantine Program*, pp. 25–33.

98. Ibid., p. 27.

on dry-heat approaches. NASA also supported a dry-heat research program being carried out by various contractors, beginning with Wilmot Castle Company in 1961.

Sterilization issues were not the only serious problems besetting the Ranger program, but they were particularly thorny ones. NASA's directives to sterilize its spacecraft came too late to build the requirements necessary for reliable sterilization into the Ranger spacecraft's initial designs. Ranger engineers wondered how they would kill all microorganisms on the 326-kilogram (725-pound) craft and keep it sterile through assembly, testing, and launch operations until it ultimately reached the Moon. Killing all organisms on the craft without damaging sensitive equipment appeared especially difficult after Charles Phillips's group at Fort Detrick showed that viable microorganisms existed that were embedded within electronic components. On the basis of Phillips's work, mission staff needed to assume that all materials and components of the spacecraft might contain internal contamination.[99]

George L. Hobby, a research biologist in JPL's Space Sciences Division, commented at a 1960 meeting on decontamination techniques that several sterilization methods could be effective for spacecraft: dry heating, gamma ray irradiation, the sterilization of materials and components during their manufacturing processes, and the sterile assembly of individual parts. He viewed sterile manufacturing, however, as a difficult venture, since it would require that hundreds of different vendors go to great expense to initiate entirely new manufacturing processes for producing spacecraft parts. It would be far easier and cheaper if internal spacecraft sterilization could be achieved using one of the two standard technologies that existed: heat or high energy radiation. Heat treatment, in fact, appeared to be a promising technology for sterilizing most of the spacecraft, since Hobby estimated that approximately 85 percent of the electronic components and 90 percent of the nonelectric components presently on the craft would survive a 24-hour heat sterilization treatment at 125°C.[100] Presumably, new designs could be developed for the components not able to survive this sterilization regime.

---

99. Posner, *Proceedings*, p. 10.
100. Ibid., pp. 10–12.

The effort to sterilize Ranger had two objectives: to prevent lunar contamination and to collect data for developing more effective and reliable methods for sterilizing spacecraft. Attaining this latter objective would be necessary for future planetary exploration.[101] Under the direction of George Hobby, Ranger staff developed a three-step operation that addressed equipment fabrication, assembly, test, and transportation. All equipment components were to be assembled, then subjected to the dry-heating regime mentioned above: 125°C (257°F) for 24 hours. Choosing this environment was an attempt to walk a tightrope between two unacceptable results. At temperatures and times above these levels, electrical equipment often failed, while below these levels some organisms generally survived.

After heating, the sterilized components were assembled and the piece of equipment tested. Typically, assembly operations began on the Ranger spacecraft eight to nine months before launch. After initial assembly of the entire spacecraft, mission staff subjected it to multiple tests in various environments. During these tests, the spacecraft was "probed, changed in configuration, actuated, torn down repeatedly for repair, modification or recalibration and transported around from one test facility to another."[102] All of these operations made effective sterilization extremely difficult, even though mission staff rigorously cleaned the craft with alcohol and shipped it to Cape Canaveral in a carefully controlled environment. Procedures using other, possibly more effective, liquid and grease sterilants were under development but were not available in time for use in the Ranger project. Some of these other sterilants were also severely toxic.[103]

After completing prelaunch tests and shortly before the launch, mission staff inserted the Ranger spacecraft into the launch rocket's shroud, or nose cone, which was a metal chamber with an aerodynamic configuration mounted on the upper end of the launch rocket and meant to protect the payload—in this case the Ranger

---

101. F. H. Quimby, ed., *Proceedings of Conference on Spacecraft Sterilization* (Washington, DC: NASA Technical Note D-1357, 1962), p. 3; Craven, "Part I: Planetary Quarantine Analysis," p. 20.

102. Quimby, *Proceedings of Conference on Spacecraft Sterilization*, pp. 5–6.

103. R. Cargill Hall, "Ranger's Lunar Objectives in Doubt," in *Lunar Impact: A History of Project Ranger; Part I—The Original Ranger* (Washington, DC: NASA SP-4210, 1977), Chap. 4, http://history.nasa.gov/SP-4210/pages/Ch_4.htm#Ch4_Top (accessed 26 January 2011).

spacecraft—during ascent from Earth. The shroud's base was biologically sealed using a plastic bulkhead to prevent contamination from
outside reaching the spacecraft. A sterilant gas mixture containing
450 milligrams per liter of ethylene oxide was piped into the nose cone
cavity through an inlet port near its base and held there for 11 hours.
An outlet port near the top of the shroud allowed expulsion of gases
during the filling operation. This operation had to be carried out in an
explosive-safe structure near the launch pad. Mission staff hoped that
they had been able to adequately sterilize the craft during their previous cleaning operations, and since any subsequent contamination that
the craft had acquired would have only occurred on its surface, this
would be killed by the toxic gas.[104]

While ethylene oxide was effective at killing bacteria that it contacted, the gas had to be in residence in the nose cone for 11 hours,
and this had to be followed by 2 hours of purging using filtration-
sterilized nitrogen gas. Mission engineers worried that any repairs
needed during countdown that required breaching the biological barrier isolating the spacecraft would necessitate repeating the gas sterilization process. If enough repairs and delays occurred, the launch
window would be missed.

JPL submitted its proposed sterilization plan to NASA on 25 May
1961, and NASA Associate Administrator Robert Seamans approved
it in June 1961. Ranger Project Manager Jim Burke, however, did not
give sterilization particularly high priority compared to other responsibilities of the Ranger staff. In listing 16 competing requirements for
Ranger, Burke put reliability 1st in priority, while sterilization was
ranked 14th.[105]

Protests Against Ranger Sterilization Procedures

During the early 1960s, performance problems plagued the Ranger
spacecraft, and many in the space science community laid the blame
on the sterilization procedures used. Ranger 1, launched in August

---

104. Hall, "Ranger's Lunar Objectives in Doubt"; Quimby, *Proceedings of Conference on Spacecraft
     Sterilization*, pp. 6–7.
105. Hall, *Lunar Impact*.

1961, assumed a lower Earth orbit than planned. Ranger 2 failed at its November 1961 launch attempt. Ranger 3, which launched in January 1962 and was to attempt a lunar landing, missed the Moon by 23,000 miles and did not send back usable video images. Ranger 4 went out of control and crashed onto the far side of the Moon in April 1962, failing to send back video pictures. Ranger 5, which launched in October 1962, missed the Moon due to a power loss and also did not send back video images. Three of these spacecraft, Rangers 3, 4, and 5, had been subjected to dry-heat and terminal-gas sterilization procedures.[106] Vehement accusations followed these failures. Government entities as well as the media pointed fingers at the sterilization procedures as the root causes of at least some of the failed missions. Some members of the space science community adamantly insisted that lunar, unlike planetary, exploration did not require spacecraft sterilization. This position was somewhat supported by the Working Group on Biology at the July 1962 NAS Iowa Summer Study, with Allan H. Brown as chairman. The group held that "contamination of the Moon does not constitute as serious a problem as is the case of the planets." The group did, however, add that "lunar contamination should be kept at a feasible minimum."[107]

This new, more lenient attitude toward contamination reflected to some extent the changing priorities of the U.S. space program. President Kennedy had committed the United States to landing a man on the Moon during the 1960s. Many space scientists had accepted the fact that wherever humans went, their microorganisms would have to go as well. The goal of attaining complete sterilization during lunar missions was being replaced, at least in part of the space science community, by the hope that contamination could be kept low enough such that it would not hinder the search for biological material within lunar samples returned to Earth.[108]

Those speaking out against spacecraft sterilization for lunar exploration attained some successes for their efforts. By November 1962, JPL was directed to cease applying dry-heat sterilization to Ranger space vehicle components. The requirement for terminal gaseous

---

106. Phillips, "Implementation and Policy Directives," in *The Planetary Quarantine Program*, pp. 25–33.
107. Ibid., p. 30.
108. Ibid., p. 30.

surface sterilization of the assembled spacecraft was eventually dropped as well. These policy changes became official on 9 September 1963 when NASA's Management Manual NMI-4-4-1, *NASA Unmanned Spacecraft Decontamination Policy*, was issued. The objective was to protect the Moon against "widespread or excessive contamination"[109] until, in the eyes of NASA, a sufficient amount of data concerning the Moon had been obtained. In other words, the policy's focus was on protecting the integrity of scientific exploration. Furthermore, NASA based its policy on prevailing scientific opinion that the Moon's harsh surface environment would make reproduction of Earth microorganisms extremely unlikely, and that if viable Earth organisms were able to penetrate to the lunar subsurface and survive, any propagation would remain exceedingly localized.[110]

NASA's Management Manual NMI-4-4-1 did not lift sterilization requirements for planetary missions as it had done for lunar exploration. The planets were considered potentially contaminatable. The manual stated that NASA policy was to prevent planetary biological contamination "until sufficient information has been obtained . . . to ensure that biological studies will not be jeopardized."[111]

Establishing Key NASA Life Science and Planetary Protection Capabilities

One of the first people to whom NASA gave responsibility for spacecraft sterilization was Gerhard F. Schilling, a German rocket scientist who was brought to the United States after World War II and served as NASA Project Manager for the Atlas-Able Pioneer series of space probe launches and launch attempts that began in 1958.[112] These probes

109. Ibid., pp. 30–31.
110. Ibid., pp. 30–31.
111. Ibid, p. 32.
112. Charles R. Phillips, "NASA and Its Planetary Quarantine Responsibilities," in *The Planetary Quarantine Program* (Washington, DC: NASA SP-4902, 1974), *http://history.nasa.gov/SP-4902/ch3.htm* (accessed 26 January 2011); Sven Grahn, "The Atlas-Able Pioneers," in "Jodrell Bank's Role in Early Space Tracking Activities—Part 1," *http://www.svengrahn.pp.se/trackind/jodrell/Able5.htm* (accessed 31 January 2006).

were key in helping to change the field of astronomy "from an observational to an experimental science,"[113] and the field of life sciences to one concerned with, among other topics, "the quest for extraterrestrial life-forms."[114]

T. Keith Glennan, NASA's Eisenhower-appointed Administrator, realized early on that the new Agency needed to develop its own life sciences capabilities. This was necessary in order to reduce its reliance on military personnel and facilities to provide biomedical research for future human missions. It would also be necessary for carrying out the space agency's planetary protection program, although even before NASA's influence and Abe Silverstein's first official NASA policy directives on spacecraft sterilization, the military did include, in response to science community recommendations, decontamination planning for its space probes.[115]

NASA had to expand its capabilities without alarming the military community or Congress, for the Agency needed their political support. The Agency also needed to address the expectations of the space science community, which was counting on it to carry out extraterrestrial biological investigations as well as biomedicine research. To help him with the delicate tasks of developing the new Agency while working effectively with the defense establishment, Congress, and the academic community, Administrator Glennan asked Clark Randt, a personal acquaintance and respected clinician, biomedical scientist, and medical administrator, to join NASA and serve as a special life sciences adviser.

In addition, Administrator Glennan formed an ad hoc Biosciences Advisory Committee, naming Dr. Seymour S. Kety, a prominent neurologist and researcher with the National Institutes of Health, as its chair. Glennan instructed Kety to invite scientists with stature in the biosciences to serve on the committee. Glennan wanted these professionals to come from diversified backgrounds and to value

---

113. R. Cargill Hall, "The Origins of Ranger," in *Lunar Impact: A History of Project Ranger* (Washington, DC: NASA SP-4210, 1977), Chap. 1, *http://history.nasa.gov/SP-4210/pages/Ch_1.htm* (accessed 26 January 2011).
114. Hall, "The Origins of Ranger," in *Lunar Impact*.
115. Charles R. Phillips, "NASA and Its Planetary Quarantine Responsibilities," in *The Planetary Quarantine Program* (Washington, DC: NASA SP-4902, 1974), *http://history.nasa.gov/SP-4902/ch3.htm* (accessed 26 January 2011).

fundamental research as much as applied research and technical development. The Administrator needed a committee that 1) would provide specific guidelines for building a NASA life sciences program and 2) was responsive not only to the engineering community and the more applied scientific disciplines, but also to the interests of academia and other groups focused on basic research endeavors.[116]

The Kety Committee included among its advisers the strong planetary protection advocate Joshua Lederberg, as well as the chemist Melvin Calvin, another planetary protection pioneer.[117] The committee carried out its study from June to November 1959, presenting its final report in January 1960. Besides urging increased emphasis on basic science (not a surprising conclusion, given Administrator Glennan's guidelines for selecting committee members), the report cautioned against making NASA life science activities merely an adjunct to human spaceflight applications and operations. Instead, the report strongly recommended that an Office of Life Sciences be established as a major NASA division with two broad objectives:

- Study of the biological nature of extraterrestrial environments, which was to include a serious search for extraterrestrial life.
- Scientific investigations related to human spaceflight and exploration.

The Kety Committee report expressed complete support for a human spaceflight program, but stressed that its ultimate objective was to "expand opportunities for extraterrestrial science."[118]

Planetary Protection and the Office of Life Science Programs

The NASA Office of Life Science Programs, which was formed on 1 March 1960,[119] included spacecraft sterilization and exobiology programs within its mandate. Administrator Glennan named his special

---

116. Pitts, "The Human Factor," Chap. 3; Phillips, *The Planetary Quarantine Program*, pp. 9–12; L. E. DeLisi, "Seymour S. Kety MD: The Man and His Accomplishments," *Psychiatric Genetics* 10(4) (December 2000): 153–158.
117. Phillips, "NASA and Its Planetary Quarantine Responsibilities," 1974.
118. Pitts, "The Human Factor," Chap. 3.
119. Mae Mills Link, "Office Of Life Sciences Established: 1960," in *Space Medicine in Project Mercury* (Washington, DC: NASA SP-4003, 1965), Chap. 4.

adviser Clark Randt to head the new office. Randt was director of the neurology division at the medical school of Cleveland's Case Institute (which became Case Western Reserve University in 1967 when it federated with Western Reserve University).[120] He wanted to expand and strengthen biomedical studies within NASA, with an eye to the clinical applications that would hopefully result. He also enthusiastically supported human spaceflight and believed that a successful program required the cooperation of life scientists, engineers, and mission planners. Randt appeared to be "an excellent choice to bridge the gap between academic life scientists and NASA's engineering—and physical science-oriented management."[121] In February 1960, Randt had recommended to Administrator Glennan that NASA's life sciences program provide research, development, and training related to 1) biomedical requirements for human spaceflight, 2) biological effects of the space environment, and 3) the search for extraterrestrial life, outlining a schedule for accomplishing this work.[122]

Memos and letters from Randt's tenure with the Office of Life Science Programs indicate that he also gave importance to planetary protection matters. Several months after he was appointed, he wrote to Abe Silverstein, Director of NASA Space Flight Programs at the time, affirming the need to identify "adequate safeguards against contamination of celestial bodies" and proposing formation of a group that would "take positive action towards insuring that decontamination and sterilizing procedures are developed and effected."[123] George Hobby of the Jet Propulsion Laboratory offered to organize and lead this group, and JPL made plans, which Randt supported, for constructing a laboratory to investigate fundamental issues in decontaminating and sterilizing space probes.[124] While Randt was pleased with JPL's planetary protection research plans, JPL requested in an 11 October 1960 letter that Randt also make funding available from his office for the labora-

---

120. Case Western Reserve University, "Visiting Case," *http://www.case.edu/visit/about.html* (accessed 10 March 2008).
121. Pitts, "The Human Factor," Chap. 3; Phillips, *The Planetary Quarantine Program*, pp. 9–12.
122. Pitts, "The Human Factor," Chap. 3.
123. Both quotes in the sentence are from Clark T. Randt to Abe Silverstein, "Sterilization of Spacecraft," 7 July 1960, folder 006696, "Sterilization/Decontamination Documentation," NASA Historical Reference Collection.
124. W. H. Pickering to Clark T. Randt, 11 October 1960, folder 006696, "Sterilization/ Decontamination Documentation," NASA Historical Reference Collection.

tory. He was not able at that time to do so, although he thought that the matter should definitely be reexamined when his office developed its FY 1963 budget.[125]

Randt began efforts in 1960 to formulate a NASA policy on spacecraft decontamination.[126] His Office of Life Science Programs also sponsored a study, related to both forward and back contamination, that sought to develop procedures and techniques for "sterilizing earth-launched objects so that they do not contaminate an extraterrestrial body with terrestrial micro-organisms and vice versa."[127]

While Randt made grand plans for life science projects that included several important planetary protection efforts, his overall vision did not harmonize with NASA's priorities at the time. The Agency closed the Office of Life Science Programs in August 1961 due to factors that included inadequate funding and inconsistent management and peer support. Randt's programs met resistance from NASA's engineers and physical scientists as well as from Congress and the military. They objected to such a broad, in-house life sciences effort.

Randt's agenda suffered from bad timing. During 1960 and 1961, NASA's primary mission was to place a human in Earth orbit. The Agency had little interest in bioscience, the search for extraterrestrial life, or planetary protection, although it recognized that these areas would become important in the future, as they indeed did.[128]

The NASA Planetary Quarantine Program

By 1963, the importance that NASA gave to planetary protection matters had changed, largely due to the priority that was being given to lunar landing missions. (This change of attitude is discussed in detail in Chapter 4 on the Apollo program.) On 1 August 1963,

---

125. Clark T. Randt to William H. Pickering, 27 October 1960, folder 006696, "Sterilization/Decontamination Documentation," NASA Historical Reference Collection.

126. Clark T. Randt to Carl E. Sagan, 2 November 1960, folder 006696, "Sterilization/Decontamination Documentation," NASA Historical Reference Collection.

127. Clark T. Randt to Commanding General, Army Medical Service Research and Development Command, 14 June 1960, folder 006696, "Sterilization/Decontamination Documentation," NASA Historical Reference Collection.

128. Pitts, "The Human Factor," Chap. 3.

the NASA Administrator asked the U.S. Surgeon General to detail Captain Lawrence B. Hall, a senior commissioned officer in the U.S. Public Health Service (PHS), to the space agency in order to develop a spacecraft sterilization program. Hall became NASA's first Planetary Quarantine (PQ) Officer. NASA's PQ Program operated from 1963 to 1971 under the Agency's Bioscience Programs, Office of Space Science and Applications.[129]

The PQ Office staff remained small and focused largely on directing research activities. NASA arranged with a variety of contractors to conduct a broad spectrum of necessary PQ research. Some examples of these efforts were as follows:

- The Illinois Institute of Technology researched the survival of terrestrial bacteria under simulated extraterrestrial conditions. Of particular interest was the rate of germination of bacterial spores after exposure to a simulated Martian environment. The data helped in estimating contamination probabilities of Mars as well as other planets.

- The Massachusetts Institute of Technology conducted a related study, evaluating microbial survival when exposed to the effects of ultra-high vacuum, extreme temperature, and ultraviolet and gamma radiation.

- Wilmot Castle Company of New York conducted a study relating to effective spacecraft sterilization approaches, examining the dry-heat resistances of a range of spore-forming microorganisms.

- Grumman Aircraft Engineering Corporation of New York carried out an experiment directly relevant to space travel, examining the growth and viability of terrestrial bacteria inside a closed, artificial environment.

- Northrop Corporation of California researched envisioned sterilization issues on a Mars entry probe—in particular, "the ability of various spacecraft components or parts to withstand dry-heat treatments and ethylene oxide exposure."[130]

---

129. Phillips, "Planetary Quarantine Program," pp. 12–13.
130. Charles R. Phillips, "Appendix: Planetary Quarantine Contractual Research," in *The Planetary Quarantine Program* (Washington, DC: NASA SP-4902, 1974), *http://history.nasa.gov/SP-4902/appendix.htm* (accessed 26 January 2011).

- Florida State University developed probability models of decontamination strategies for individual spacecraft.
- Stanford Research Institute (SRI) in California evaluated the probabilistic models of contamination that were currently being used in the PQ Program. SRI also worked on a more detailed model of expected microbial proliferation on Mars.

The PQ Office's work proved important during the last half of the 1960s in the Apollo program, when it was vital to collect and bring back Moon samples uncontaminated by terrestrial material and also to protect Earth against any Moon microbes on or inside samples taken, the spacecraft, and the astronauts themselves. The particular methodologies employed for accomplishing these objectives are examined in Chapter 4.

The December 1970 Reorganization

By December 1970, several factors made it imperative that NASA reexamine its priorities, including those related to planetary protection. The Apollo program had been hugely successful and had achieved its major objectives. But President Nixon wanted NASA to implement "a reduced, multiple-objective, science oriented space program for the 1970s,"[131] and Congress's interest in the space program was declining. As a result, financial support for space activities was reduced, leaving NASA with fewer space program options.

In response to the situation, NASA adjusted its priorities. It limited human space exploration activities, which were extremely expensive, to Earth-orbital operations and research supporting unspecified and unapproved human missions. The Agency also prioritized science-focused robotic planetary exploration, including the search for extraterrestrial life. Exobiological studies and prevention of forward planetary contamination thus gained markedly in importance.

Many NASA life scientists had been pushing for years for closer ties between the Agency's biological programs, and NASA realized that

---

131. John A. Pitts, "Toward an Integrated Life Sciences Program," in *The Human Factor: Biomedicine in the Manned Space Program to 1980* (Washington, DC: NASA SP-4213, 1985), Chap. 11, *http://history.nasa.gov/SP-4213/ch11.htm* (accessed 26 January 2011).

this might be an appropriate time to make such changes. Associate Administrator Homer Newell requested the NAS Space Science Board to review the Agency's life science efforts and identify their needs. NAS in turn appointed a special committee to do this, which found the space agency's life science programs to be deficient in basic science and strongly requiring better coordination with each other. NAS also underlined the need for a NASA reorganization in the management of its life sciences programs.

In December 1970, NASA responded to the above recommendations by announcing a reorganization, which resulted in certain advantages for the PQ Office. NASA considered the PQ Office's coordination with planetary exploration activities important enough to give the PQ Officer access to top management. NASA reasoned that in order for planetary exploration missions to be carried out smoothly and effectively, it was critical for the PQ Office to conduct its regulatory responsibilities so as to avoid any conflict with the Agency's planetary program operational responsibilities. To ensure this, NASA management gave its Director of Life Sciences an overall coordination role and provided the PQ Officer with a direct line of communication to the Associate Administrator, Office of Space Science.[132]

JPL Planetary Quarantine Efforts

In August 1959, JPL began to develop a small in-house biological group led by George L. Hobby, who had come from the National Institutes of Health. Since that time, many JPL personnel have conducted sterilization and planetary quarantine activities. Their efforts gained impetus after NASA Headquarters established its PQ Office,

---

132. Phillips, *The Planetary Quarantine Program*, pp. 12–13; John A. Pitts, "Toward an Integrated Life Sciences Program," in *The Human Factor: Biomedicine in the Manned Space Program to 1980* (Washington, DC: NASA SP-4213, 1985), Chap. 11, *http://history.nasa.gov/ SP-4213/ch11.htm* (accessed 26 January 2011); Linda Neuman Ezell, "Organizational Charts," in *NASA Historical Data Book: Volume III—Programs and Projects 1969–1978* (Washington, DC: NASA SP-4012, 1988), Appendix A, *http://history.nasa.gov/SP-4012/ vol3/app.a.htm#436* (accessed 26 January 2011); NASA History Division, "Neat Stuff About the Right Stuff," *NACA/NASA Records at NARA, http://history.nasa.gov/nara/rightstuff. html* (last updated 21 January 2005, accessed 7 March 2006).

which provided a centralized capability for program planning and funding.

One large JPL research effort that began early in the space program sought to 1) determine which spacecraft components would survive various sterilization treatments and 2) develop new, more durable components when experiments revealed that heat or other sterilizing techniques damaged them. NASA's PQ Office directly funded many JPL research efforts, although a good deal of JPL's PQ and spacecraft sterilization research got intermingled with, and was difficult to separate from, other JPL activities associated with funded missions not directly related to planetary quarantine. For instance, trajectory computations for various missions served a dual purpose. They were critical for guiding the spacecraft where it was supposed to go and for keeping its components, including the payload, from inadvertently impacting a celestial body. Such computations were also critical in *preventing planetary contamination due to such an impact.*[133]

International Scientific Cooperation vs. Military Priorities, and The Effect on Planetary Protection

The free exchange of technical data across international borders has always benefited scientific endeavors. Cross-border agreements are also key to the success of planetary protection, because keeping celestial bodies free of terrestrial contamination requires the cooperation of all spacefaring nations. During the Cold War, however, the open communication needed to implement broad technical and regulatory cooperation often ran afoul of military priorities. The Defense Department, for instance, regarded "space as a future setting for modern warfare"[134] and thus opposed open exchange of data with the USSR.

The U.S. space program began as an outgrowth of military rocket development that took place during World War II and the following decade. While nonmilitary rocket research had been done earlier,

---

133. Charles R. Phillips, "Planetary Quarantine Research," in *The Planetary Quarantine Program: Origins and Achievements 1956–1973* (Washington, DC: NASA SP-4902, 1974), Chap. 5.

134. Wolfe, "Germs in Space," 203.

and while our country would probably have eventually developed a space program without the military's help, the ready availability of rocket engineering know-how and equipment greatly accelerated the program.[135]

The drivers for the space program included not only the direct strategic benefits and the scientific knowledge that could be attained by actual trips into space, but also the technological and political benefits that would indirectly help our position in the Cold War.[136] As John Casani, the first project manager of NASA's Galileo Mission to Jupiter once said, our country's successes in space exploration conveyed to other states a powerful message regarding "how much technological capability . . . was represented here . . . if we could do that with a civilian enterprise, imagine what this would mean in a military" situation.[137] Our space program, in other words, became a peaceful deterrent to future trouble by letting potentially hostile nations know, in a nonthreatening manner, that we were not a nation to be trifled with.

Exobiologists and planetary protection specialists in the United States as well as in other countries had to walk the tightrope between the open international exchange of scientific ideas and respect for military secrets and restrictions on information vital for national defense. Many U.S. scientists and engineers did not want their hands tied by such restrictions, because they knew that unfettered information exchange was important for creative advancements in their fields. They tried to retain their separation from military and political issues so as to protect their scientific objectivity and their right to uncontrolled trading of ideas and data.

This tension between scientists and the military manifested in the continual struggles between SSB, NASA's Office of Life Sciences, and the National Research Council-Armed Forces Committee on Bioastronautics (AFCOB). This committee, which was established by the NAS National Research Council in 1958 at the formal request of

---

135. Phillips, *The Planetary Quarantine Program*, p. 1.
136. Jerome B. Wiesner, chair, Wiesner Committee, *Report to the President-Elect of the Ad Hoc Committee on Space*, 10 January 1961, NASA Historical Reference Collection and at Air War College (AWC) Web site, *http://www.au.af.mil/au/awc/awcgate/report61.htm* (accessed 7 March 2006).
137. John Casani, recorded telephone interview by author, 29 May 2001.

the Department of Defense,[138] was populated largely by military advisers, but it also included scientists such as Melvin Calvin, an early figure in the planetary protection field who held membership on SSB as well. NAS attempted to combine its SSB with the military-oriented AFCOB, in order that the civilian and military branches of the space program might develop a unified front and also eliminate duplications of effort.

Joshua Lederberg and his colleagues were among those strongly opposed to such an action, arguing that one committee was a group of civilians advising civilians, while the other consisted of military personnel studying military problems. Lederberg wanted the separation of tasks to be preserved. One clear advantage of doing so was to avoid the need for civilian space scientists who became engaged in military studies to obtain security clearances, for this would definitely restrict their ability to exchange scientific ideas and technical know-how across borders.[139] Lederberg recognized that an effective international planetary protection network absolutely needed to be built, in part to protect the integrity of the field. If even one spacefaring nation failed to follow strict planetary protection procedures during its missions, this could result in irreversible contamination of target bodies and disastrous impacts to the exobiological studies of *all* nations.

Exchange of state-of-the-art research data on, in particular, microbial survivability and effective sterilization approaches was key in building an effective international network. Western exobiologists thus carried out energetic efforts to engender international cooperation, hoping it would help prevent forward and back contamination. Their efforts could not, however, open the doors to free information exchange with the Soviet Union, partly because of the Russians' own restrictions on scientific exchange. These were probably more strict than those put on U.S. scientists. According to Joshua Lederberg, official U.S. mechanisms of obtaining information were not able to identify critical details about Soviet sterilization techniques.[140] What the West found out about Soviet actions and practices sometimes

---

138. Mae Mills Link, "The Bioastronautics Mission Emerges," in *Space Medicine In Project Mercury* (Washington, DC: NASA SP-4003, 1965), Chap. 1, *http://www.hq.nasa.gov/office/pao/History/SP-4003/ch1-3.htm* (accessed 26 January 2011).
139. Wolfe, "Germs in Space."
140. Ibid.

had to be obtained unofficially from personal contacts and gossip, as well from reading Eastern Bloc newspaper translations. But in general, Western exobiologists had to deal with a distinct lack of data from the Soviet Union on planetary protection, and this worried them.

The USSR's Luna II Mission

Western scientists grew especially concerned about the possibility of forward contamination when, in September 1959, the USSR announced that its Luna II probe had impacted the Moon's surface. This was followed 30 minutes later by the crash of the third stage of Luna II's launch vehicle on the Moon. Since the Soviets had kept their sterilization procedures largely secret, western scientists feared that science returns on future lunar missions might have been compromised by microbes released from the impacts. In a note to Eugene Kinkead of *The New Yorker* magazine, Lederberg discussed the USSR's claim to have decontaminated Luna II, hoping that was indeed true. The Soviets had also claimed "to have ensured the safe landing of their marker plaques."[141] If they had accomplished this second feat but not thoroughly sterilized their spacecraft, a landing soft enough to preserve the marker plaques would likely not have destroyed the microorganisms on board.

Lederberg considered Luna II's mission to have been "more of a stunt and less scientifically valuable than a graze or an orbit around the Moon would be."[142] This supported his belief, discussed earlier in the chapter, that scientific research interests in space were likely to get deprioritized when there was a key strategic advantage to be had.[143]

Hugh Odishaw, Executive Director of SSB, also expressed concern about the uncertain level of contamination of the Soviet probe. He

---

141. Joshua Lederberg to Eugene Kinkead, *New Yorker*, 16 September 1959, Correspondence/1959/ September (Chronological Listing), The Joshua Lederberg Papers, *http://profiles.nlm.nih.gov/ BB/A/J/V/K/_/bbajvk.pdf* (accessed 26 January 2011).

142. Ibid.

143. Joshua Lederberg, "How DENDRAL was Conceived and Born" (ACM Symposium on the History of Medical Informatics, National Library of Medicine, 5 November 1987), pp. 3–4, *http://profiles.nlm.nih.gov/BB/A/L/Y/P/_/bbalyp.pdf* (accessed 26 January 2011).

noted in a letter to NASA and ARPA that "extra-terrestrial contamination could seriously interfere with proper studies of fundamental importance relating to biological developments in the solar system."[144] Odishaw suggested that because sufficient data were currently unavailable on actual contamination levels of space probe parts, SSB should initiate an immediate study to determine sterilization requirements compatible with present design and assembly processes. He further hoped that close international liaisons could be maintained so that other countries (i.e., the USSR) could have access to the study's results.

Efforts by Individual Scientists To Determine USSR Planetary Protection Approaches and To Develop Alliances with Their Biologists

Lederberg, realizing that his personal connections with Soviet scientists might provide the best means of obtaining critical planetary protection data—in particular, the actual level of contamination control employed on Soviet spacecraft—took matters into his own hands on several occasions. He also strove to develop alliances with Soviet biologists, with the aim of attaining higher priorities for biological research goals on future space missions. Lederberg wrote to the microbiologist V. D. Timakov of Moscow's Academy of Medical Sciences of the USSR in May 1958, stating that "the interest of U.S. biologists . . . will be futile unless we can make a common cause with our fellow-scientists in the biological sciences in the USSR."[145] Lederberg expressed concern that the eagerness to "demonstrate the capabilities of interplanetary vehicles,"[146] as well as the emphasis in space research on collecting data for the physical over the life sciences, might well result in *biological objectives* being overlooked during the planning of interplanetary missions. He gave as an

---

144. Hugh Odishaw (SSB) to T. Keith Glennan (NASA) and Roy Johnson (ARPA), 14 September 1959, Correspondence/1959/September (Chronological Listing), The Joshua Lederberg Papers.
145. Joshua Lederberg to V. D. Timakov, 16 May 1958, Correspondence/1958/May (Chronological Listing), The Joshua Lederberg Papers, *http://profiles.nlm.nih.gov/BB/G/C/W/G/_/bbgcwg.pdf* (accessed 26 January 2011).
146. Ibid.

example the "inadvertent contamination of a planet such as Mars by microorganisms carried from the earth"[147] and added that the same danger applied to the Moon.

Similar sentiments were expressed by Lederberg in a July 1958 letter to K. V. Kossikov of Moscow's Institute of Genetics. Lederberg referred to his own deep interest in origin-of-life studies, and communicated his concerns that "programs of space exploration by various nations should not interfere with ultimate biological exploration of the moon and the planets."[148] In a letter sent at the same time to A. I. Oparin of Moscow's Academy of Sciences, Lederberg articulated his view of the critical nature of space biological research and its vulnerability in the face of strategic considerations:

> Unfortunately, although the experimental study of life on other planets is undoubtedly the most compelling scientific objective in space research, biologists have hardly begun to think about this problem in realistic terms. Since space-flight has become deeply entangled with the tragic political and military rivalry of our times, there is great danger that the fundamental scientific objectives will be overlooked . . .[149]

In 1959, Lederberg requested permission to directly contact a Soviet academician that he knew, N. D. Ierusalimskii, even though this was viewed by some in the government as a possible breach of U.S. security. Lederberg wanted to send Ierusalimskii copies of WESTEX meeting minutes, but the Executive Director of SSB, Hugh Odishaw, advised strongly against this course of action, which he felt would provide "the most advanced thinking of our senior scientists with regard to the many biological questions."[150]

---

147. Ibid.
148. Joshua Lederberg to K. V. Kossikov, 1 July 1958, Correspondence/1958/July (Chronological Listing), The Joshua Lederberg Papers, http://profiles.nlm.nih.gov/BB/A/J/W/Q/_/bbajwq.pdf (accessed 26 January 2011).
149. Joshua Lederberg to A. I. Oparin, 1 July 1958, Correspondence/1958/July (Chronological Listing), The Joshua Lederberg Papers, http://profiles.nlm.nih.gov/BB/A/L/D/C/_/bbaldc.pdf (accessed 26 January 2011).
150. Wolfe, "Germs in Space," 200.

In approximately 1960, Lederberg obtained a security clearance for A. A. Imshenetsky, director of the Institute of Microbiology of the Soviet Academy of Sciences, to visit restricted locales in California. At a Stanford University lunch meeting with Imshenetsky, Lederberg was able to glean possible indications of "the Soviet Union's lack of progress in exobiology,"[151] although this impression might have been due to Imshenetsky's lack of understanding of the English language, or to his intentional evasion of questions.

Other space scientists also used their personal influences to try to obtain information about Soviet spacecraft sterilization processes. In November 1959, Carl Sagan reported a personal attempt to augment official channels of information exchange by conducting a conversation with G. F. Gause of the Soviet Academy of Medical Sciences at the Darwin Centennial Celebration. Sagan's aim was to obtain details on Soviet space probe sterilization procedures. Gause assured Sagan that mission staff had sterilized Luna II's instrument package and its third stage carrier rocket, both of which had impacted the Moon. But when pressed for details, Gause said some curious things, claiming "that his knowledge of the sterilization did not exceed that published in *Izvestia*. He said that the sterilization methods were those known to every graduate student in microbiology at the University of Chicago, and to every manufacturer of canned food, but he would not be more specific than that. He explained his reluctance by the analogy that Abbott Laboratories in Chicago would not divulge trade secrets to competing pharmaceutical firms."[152] Sagan believed that Gause knew far more about the actual sterilization procedures used, but he could get little further information from him.

Eventually, some encouraging data were obtained about Soviet planetary protection activities. Lawrence Hall, NASA's Planetary Quarantine (PQ) Officer, wrote a memo in June 1972 referring to a presentation given at a recent COSPAR meeting[153] by V. I. Vashkov of the USSR, in which he discussed a methyl bromide–ethylene oxide gas mixture used

---

151. Ibid., 201.
152. Carl Sagan, abstract of letter summarizing his conversation with G. F. Gause, November 1959, Abstracts (Chronological Listing), The Joshua Lederberg Papers, *http://profiles.nlm.nih.gov/ BB/G/C/X/P/_/bbgcxp.pdf* (accessed 26 January 2011).
153. This meeting was of the COSPAR Planetary Quarantine Panel, held in Madrid on 19 May 1972.

to sterilize Soviet spacecraft parts. After his presentation, Vashkov answered "with apparent frankness questions on the measures used to sterilize"[154] Soviet Mars-bound spacecraft. Vashkov claimed that individual parts of landers were sterilized by heat (although he did not give the temperature or duration of the procedure) or by a radiation dose of 2.5 millirads,[155] depending on the characteristics of the part. The landers were assembled in cleanrooms, and individual parts were carefully cleaned, probably with hydrogen peroxide. UV light was applied during assembly as an additional sterilant. Finally, before launch, Soviet mission personnel exposed the entire lander to a methyl bromide–ethylene oxide gas mixture for 6 hours at 50°C.

Vashkov also stated that extensive bioassays were carried out on a model of the spacecraft in order to verify the effectiveness of the planetary protection techniques. These assays employed numerous samples taken by a technique called "mopping" that Hall believed was similar to NASA's swabbing procedure (which is discussed in Chapter 7 regarding Mars missions). To prevent recontamination of the Soviet lander, Vashkov mentioned it remained "enclosed in a plastic bag in which a low concentration of the sterilant gas was maintained until the spacecraft had left the earth's atmosphere."[156]

Hall's confidence in the validity of Vashkov's statements was enhanced by their agreement with several published documents. In a study by Vashkov, Rashkova, and Shcheglova,[157] reference was made to experiments with electronic components such as diodes, triodes, capacitors, resistors, relays, sockets, and plugs, showing that they were effectively sterilized using 2.5 millirad of gamma radiation. The study mentioned assembly under sterile conditions, sterilization of some parts using hydrogen peroxide sprays, and final sterilization of

---

154. Lawrence B. Hall, "Soviet Planetary Quarantine Sterilization of Mars 1 and 2," 1 June 1972, NASA Historical Reference Collection.

155. A millirad (mrad) radiation dose is equal to one-thousandth of a rad, the dose causing 0.01 joule of energy to be absorbed per kilogram of matter. Normal background radiation in the United States varies from about 50 to 200 mrad per year.

156. Hall, "Soviet Planetary Quarantine Sterilization," 1 June 1972.

157. V. I. Vashkov, N. V. Rashkova, and G. V. Shcheglova, "Kavantin Planet: Printsipy, Metody, i Problemy," in Osnovy, *Kosmicheskoy Biolgii i Meditsiny*, Chap. 4, vol. 1, part 3 (1970), pp. 3–156, and its English translation, "Planetary Quarantine Principles, Methods, and Problems," NASATT F13, 769 (1971), 113 pages, a summary of which is attached to the above-cited Lawrence B. Hall memo of 1 June 1972.

the spacecraft by filling it to excess pressures with methyl bromide–ethylene oxide gas. Also described was a bioassay procedure, used as a means of verifying the sterility of spacecraft.

Another study by Imshenetsky[158] and Abyzov[159] referred to experiments at the USSR's Institute of Microbiology of the Academy of Sciences that sought to clarify the best procedure for sterilizing "infected" materials, and the result that reducing the temperature and lengthening the exposure period created "effective methods of sterilization for the spacecraft as a whole." This was a result that NASA arrived at as well for terminal sterilization, which is discussed in detail in Chapter 7 of this book. Imshenetsky and Abyzov also mentioned the use of "radiation sterilization," sterile assembly techniques, and final sterilization of the spacecraft using gas. Another study by Vashkov[160] discussed heat sterilization performed at various temperatures and durations and specifically mentioned that this technique was used on Soviet Venus spacecraft. In addition, Vashkov referred to favorable analyses of UV-lamp, hydrogen peroxide, and gas sterilization techniques for spacecraft.

Impacts of U.S. Restrictions on Open Information Transfer

USSR limits on free information transfer regarding its planetary protection protocols have been discussed above. It is also important to mention that *U.S. restrictions* significantly limited data exchanges as well. This was especially so regarding the sharing of certain biohazard data. U.S. restrictions on information transfer definitely reduced the amount of data distributed from projects that involved the Army Biological Warfare Laboratories at Fort Detrick, Maryland.

---

158. The transliteration of his Russian name is alternately spelled "Imshenetsky," "Imshenetski," and "Imshenetskii" in various references. In this book, his name will always be spelled "Imshenetsky."

159. A. A. Imshenetsky and S. Abyzov, "Sterilization of Spacecraft," NASA TT-710 (November 1971): 230–252, a summary of which is attached to the above-cited Lawrence B. Hall memo of 1 June 1972.

160. V. I. Vashkov, "Modern Methods and Means of Sterilization of Spacecraft," NASA TT-710 (November 1971): 207–219, a summary of which is attached to the above-cited Lawrence B. Hall memo of 1 June 1972.

Beginning in 1959, SSB had advocated cooperation between space scientists and Fort Detrick staff, whose experience and equipment for detecting and safely handling minute quantities of microorganisms, some of them extremely dangerous, were excellent foundations on which to develop spacecraft sterilization techniques as well as biological detection instruments. WESTEX also supported partnerships with Fort Detrick, especially for developing sterilization techniques based either on ethylene oxide gas or exposure to radiation. Lederberg was interested in Fort Detrick's capability for testing bacterial survivability in hostile environments such as chemical explosions, for this would have direct applications for identifying effective sterilization techniques.[161]

The cooperative agreement with Fort Detrick produced powerful benefits, but it also raised barriers to the life science community's repeated attempts to establish open, free exchange of data with other countries. Research undertaken at Fort Detrick biological warfare facility generated highly restricted, classified results that could not be openly shared with Soviet or other international colleagues, for to do so would have potentially given the USSR key information on very potent weapon systems.[162]

●●●

To What Extent Did Speculative Literature Influence the Planetary Protection Field?

No technical field exists that is totally disconnected from the society around it and that society's mores and fears. The priorities of the society in which a field of inquisition is embedded cannot help but have some influence on the field's practitioners, whether through budget pressures or due to direct relationships with its members. The planetary protection field was no exception to this.

The exploration of space is an endeavor whose success depends not only on good science, but also on good imagination. NASA scientists and engineers continually strive to do what has never been done before, and this requires originality of thought and leaps of creativity.

---

161. Wolfe, "Germs in Space."
162. Wolfe, "Germs in Space," 192–193.

Science fiction writers, on the other hand, seek to create that which has not yet been envisioned. These writers, like space scientists, search for pathways to places as yet unexplored. Thus, it is not surprising that one field can inform and inspire the other.

Several works of science fiction have resonated so powerfully with the psyches of their audiences, and have been so widely read or listened to, that they have strongly affected the public's beliefs regarding space exploration and, in particular, the need for planetary protection. The earliest of these influential works was Herbert George "H. G." Wells's *The War of the Worlds*, written in 1898. Two decades before this book's publication, the Italian astronomer Giovanni Schiaparelli had thought he saw a series of lines on the surface of the planet Mars that he called *canali*, an Italian word meaning "channels." The astronomer Percival Lowell also observed these lines and, in his 1895 book, *Mars*, mused that "Their very aspect is such as to defy natural explanation."[163] He concluded that they were artificial canals and that "upon the surface of Mars we see the effects of local intelligence."[164] Such a possibility inflamed the public's imagination. Perhaps influenced by Lowell's work as well as by major events of the time—in particular, German unification and the threat of a war that would engulf all of Europe[165]—Wells wrote a book that responded to the questions: What if intelligent creatures did indeed exist on Mars who had far surpassed our technological achievements, and what if those creatures felt hostile toward us?[166]

In *The War of the Worlds*, Martian life definitely was not benign. The creatures from the Red Planet destroy a large part of London and the English countryside, as well as many British citizens. Our planet's

---

163. Percival Lowell, *Mars* (Cambridge, MA: Riverside Press, 1895), *http://www.wanderer.org/references/lowell/Mars/* (accessed 26 January 2011).
164. Ibid.
165. Paul Brians, Department of English, Washington State University, introduction to "Study Guide for H. G. Wells: The War of the Worlds (1898)," *http://www.wsu.edu:8080/~brians/science_fiction/warofworlds.html* (accessed 16 May 2005).
166. NASM, "Percival Lowell and Mars," 12 September 1997, *http://www.nasm.si.edu/ceps/etp/mars/percival.html*; "Exploring the Planets: Mars" Web site, National Air and Space Museum (NASM), *http://www.nasm.si.edu/ceps/etp/mars/* (accessed 16 May 2005); John D. Gosling, "Herbert George Wells: A Biography," on the Web site, "The War of the Worlds—Invasion: The Historical Perspective," *http://www.war-ofthe-worlds.co.uk/index.html* (accessed 16 May 2005).

bacteria, to which we humans have built up a resistance, eventually defeat and kill the Martians. The invaders bring along an insidious red weed that roots in Earth soil, but this too falls prey to Earth's environment, dying from common plant diseases that don't exist on Mars. Nevertheless, the Martian attacks scared many readers. H. G. Wells's book planted fears in people that the ordinary life from one planet will be antagonistic, and even murderous, to those from another planet. The biggest impact of the book was not on its readership, however, but on those who listened to an Orson Welles adaptation of it for a radio play. Orson Welles and his theater company broadcast the play as a Halloween special on 30 October 1938 and aired it over station WABC and the Columbia Broadcasting System's coast-to-coast network.[167] The play was given the format of an emergency news program and listened to by over six million Americans. Almost two million of these thought that the play was the real thing and panicked at the supposed invasion. People who had heard the broadcast "packed the roads, hid in cellars, loaded guns, even wrapped their heads in wet towels as protection from Martian poison gas."[168]

The public's panic reaction to a radio adaptation of H. G. Wells's classic was not limited to U.S. audiences. In February 1949, over 10 years after Orson Welles's broadcast, Radio Quito in Ecuador presented a similar play, but this time set in the Andean country rather than in the United States. A Saturday-night music program was interrupted in mid-song with "a long and frightening description of how Martians had landed twenty miles south of the city, near Latacunga."[169] The announcement told listeners that Latacunga had already been destroyed, and aliens were moving toward Quito in the shape of a cloud. This was followed soon after by another announcement claiming that an air base had been destroyed, leaving many dead and wounded. Then actors talented at imitating well-known public figures spoke, posing as the Minister of the Interior, the mayor of Quito, and others.

167. "Radio Listeners in Panic, Taking War Drama as Fact," *New York Times* (31 October 1938), *http://www.war-of-the-worlds.org/Radio/Newspapers/Oct31/NYT.html* (accessed 10 July 2008).

168. Ken Sanes, "War of the Worlds, Orson Welles, and the Invasion from Mars," Transparency Web site, *http://www.transparencynow.com/welles.htm* (accessed 16 May 2005).

169. Don Moore, "The Day the Martians Landed," *Monitoring Times* (October 1992), *http://members.tripod.com/donmoore/south/ecuador/martians.html* (accessed 10 July 2008).

Thousands ran from their homes, some in pajamas. When Radio Quito admitted its hoax, many in the panicked populace were enraged and converged on the El Comercio building where the radio station was housed. They stoned the building, then set it aflame. When the smoke finally cleared, the building was destroyed and, far worse, 20 people had died in the fire.

Panic reactions were also triggered following a broadcast of the Welles play in Santiago, Chile, in 1944, Providence, Rhode Island, in 1974, and in northern Portugal in 1988.[170] Perhaps more than any other work of science fiction, *The War of the Worlds* left its listeners and readers with images of life's dangerous fragility to exotic species.[171] Such images strongly communicated the potential risks involved when one planet's culture and ecology interfaces with those of another.

Although radio stations, the news media, and the government quickly tried in each instance to convince the populace that there was no reality to an alien invasion, the fear that someday a real war of the worlds might occur did seem to linger. Such fears certainly flared up as the day approached for sending humans to the Moon to collect alien samples and then bring them back to Earth. The extreme measures to which the Apollo mission went to avoid back contamination of Earth (these are discussed later in this book) were certainly driven by scientific considerations, but also were needed to assuage the public fear of an extraterrestrial epidemic taking root on our planet.

What the *War of the Worlds* broadcasts did was trigger a deep terror that may have resided in our collective psyches for many centuries: fear of the outsider whom we don't understand and who wishes us ill. Because such a fear can be so easily activated and is periodically set off by disturbing stories in the media, both nonfictional and fictional, dealing with such terrors has to be an integral part of developing a planetary protection policy. In 1950, Ray Bradbury wrote *The Martian Chronicles* partly to counteract the *War of the Worlds* notion that Mars could be a menace to Earth. In Bradbury's novel, it is humans from Earth who play the role of harmful "invaders from outer

---

170. Robert E. Batholemew, "The Martian Panic Sixty Years Later," *Skeptical Inquirer* (November/December 1998), *http://www.csicop.org/si/9811/martian.html* (accessed 10 July 2008).
171. Audra J. Wolfe, "Germs in Space: Joshua Lederberg, Exobiology, and the Public Imagination, 1958-1964," *Isis* 93(2) (June 2002).

space,"[172] wiping out a wise species of sentient beings from whom we could have learned a great deal. Although *The Martian Chronicles* did not instigate widespread panic or drive a large part of the populace into the streets, it appears to "have deeply affected [Joshua] Lederberg and his WESTEX colleagues."[173] One of the book's similarities to H. G. Wells's novel is the portrayal of exotic life-forms as terribly vulnerable to Earth microbes. In *The Martian Chronicles*, most of the Red Planet's inhabitants get sick and die from a terrestrial disease resembling chicken pox, after which people on Earth destroy themselves through nuclear war. Although the documents produced by Lederberg and WESTEX do not explicitly refer to the above novels, "the themes of unintended consequences and self-annihilation echo throughout their discussions."[174]

The War of the Worlds and *The Martian Chronicles* influenced many journalists reporting on the new field of exobiology, supplying them with the images and vocabulary by which to imagine and articulate the threats that could result from space exploration. Americans learned about exobiology from widely read periodicals such as *Newsweek, Time,* and *Esquire,* as well as from local newspapers. Space scientists gave warnings that planetary exploration might result in back contamination of Earth, and these statements inspired journalists to write speculative pieces which often blended real science narratives with science fiction themes. A 1961 *Time* article entitled "Danger From Space" reported that

> The human species seems about to master the solar system. The contrary may be the truth. Last week Lockheed Aircraft Corp. announced that it has a team of scientists hard at work, hoping to find a way to foil invasions of the earth that may well start from space. The invaders most to be feared will not be little green Venusians riding in flying saucers or any of the other intelligent monsters imagined by science fictioneers.

---

172. Paul Brians, Department of English, Washington State University, introduction to "Study guide for Ray Bradbury's *The Martian Chronicles* (1950)," *http://www.wsu.edu:8080/~brians/ science_fiction/martian_chronicles.html* (accessed 16 May 2005).
173. Wolfe, "Germs in Space."
174. Wolfe, "Germs in Space."

> Less spectacular but more insidious, the invaders
> may be alien microorganisms riding unnoticed on
> homebound, earth-built spacecraft.[175]

In this story, *Time* effectively used science fiction themes to communicate real science issues. Many other popular articles about exobiology also began by evoking visions of alien life, sometimes dangerous, and then followed this with the actual scientific basis of the space exploration projects. Sometimes, though, the repeated implications—that serious space science endeavors resembled fiction novels—damaged space scientists' credibility. Journalists frequently chose titles for their articles that could have been applied to science fiction yarns. For instance, *Esquire* named a December 1963 article on space exploration, "Someone Up There Like Us?"[176] while newspapers chose headlines such as "Invasion From Mars? Microbes!"[177] and "Space Academy Board Warns of Microbe Attack From Space."[178] Lederberg and his colleagues struggled to legitimize the emerging field of exobiology, but their attempts continued to be plagued by news story lead lines such as, "A 'war of the worlds' is now conceivable."[179] This sensationalism was by no means limited to the United States. A cartoon in a Danish magazine, for example, lampooned Lederberg and exobiology in its depiction of three hospital patients afflicted with, respectively, Moon measles, Mars flu, and Venus warts.[180]

Although Lederberg expressed his chagrin with articles that linked the serious science of exobiology with popular culture and, in particular, the tropes of science fiction, many of these same articles communicated legitimate exobiological concepts to the lay public in a highly readable and entertaining manner, and by doing so helped to keep readers intensely interested in NASA's space missions. The engaging narrative styles employed in these articles also helped achieve planetary protection goals by enabling exobiologists to widely publicize their

---

175. "Danger from Space?" *Time* (17 November 1961).
176. Wolfe, "Germs in Space."
177. Ibid.
178. Ibid.
179. Ibid.
180. Ibid.

pleas to implement dependable contamination prevention measures into space exploration.[181]

Michael Crichton wrote *The Andromeda Strain* in 1969 about a plague brought to Earth on a returning space probe. This book about the possible consequences of back contamination became a runaway best seller. Its sales certainly were not hurt by its release's coinciding nicely with the first human Moon landing and the return of lunar samples to Earth. A gauge of the book's effectiveness in communicating the potential dangers of back contamination to readers was that when it was first published, many people thought it was not a made-up story but was actually based on real events.[182] The book elicited "a fear in the public mind that extraterrestrial specimens might introduce a new and unstoppable plague to the Earth."[183] Although events in the book were fiction, its premise has become a metric for worst-case scenarios resulting from back contamination. *The Andromeda Strain* is often referred to in this context by space science professionals and journalists as the outcome that everyone least wants from a sample return mission.[184]

Although *The Andromeda Strain* was hardly the first publication to address back contamination dangers, it put that issue in the public's minds and, by so doing, in front of the scientific community as well. For example, Abigail Salyers, a professor of microbiology at the University of Illinois, wrote about efforts to find life on Mars and other planets. In her articles addressing the safety of bringing Mars samples back to Earth, she used *The Andromeda Strain,* as other space scientists have done, as the starting point of her discussion, then identified occurrences that were quite possible from those that were highly improbable. She asked whether a microbe brought back from a planet without humans could

---

181. Ibid.
182. Michael Crichton, *The Andromeda Strain*, official Web site of Michael Crichton, *http://www. crichton-official.com/strain/* (accessed 20 June 2005).
183. Abigail A. Salyers, Department of Microbiology, University of Illinois at Urbana-Champaign, "Looking for Life on Mars and Beyond," American Institute of Biological Sciences, August 2004, *http://www.actionbioscience.org/newfrontiers/salyers2.html* (accessed 20 May 2005).
184. Margaret S. Race and John D. Rummel, "Bring Em Back Alive—Or At Least Carefully: Planetary Protection Provisions for Sample Return Missions," *Ad Astra* (January/February 1999), *http:// www.astrobiology.com/adastra/bring.em.back.html* (accessed 10 July 2008); "QED," *Telegraph. co.uk, http://www.telegraph.co.uk/digitallife/main.jhtml?xml=/connected/2004/01/21/ecrqed18. xml* (accessed 10 July 2008); Brendan I. Koerner, "Your Planetary Protection Officer—What on Earth Does He Do?" *Slate.com, Washington Post, Newsweek Interactive Co. LLC,* 20 April 2004, *http://www.slate.com/id/2099222/* (accessed 11 July 2008).

cause serious infections in our species. Her answer was definitely yes. But would organisms that evolved in the low-temperature environments of Mars or Europa do well in the relatively high temperatures of our bodies? Probably not. The point is that *The Andromeda Strain* has become a touchstone for planetary protection specialists in discussing matters of back contamination and the dangers that they present.[185]

Red Mars, written by Kim Stanley Robinson in 1993, imagined possible colonization experiences on the planet and the conflicts arising from different visions of what was appropriate. The book articulated important questions about what on a planet should be protected. In the author's words, "the question of whether or not to terraform Mars . . . [was] one of the great engines" of writing *Red Mars* and the two books in the series that followed it. The terraforming question is one that continues to be debated by scientists and ethicists. Do we have the right to someday transform Mars, or some other planet, into an Earth-like planet? Or only if we find no life there? Or should we leave Mars in its natural state even if it proves to be devoid of life?

One of the views expressed in *Red Mars* pertained to "an awfully big and beautiful, sublime landscape"[186] that turned out to be lifeless. According to this point of view, even if we should find that Mars's vast deserts and enormous mountains and canyons contain no lifeforms, we still are not free to tear the planet apart and reconstruct it as we like. We need to keep the planet unspoiled. This is a position that Henry David Thoreau[187] or John Muir, who both revered the wild places, might have taken. Muir once said that wilderness was a sacred area "that should be cherished because of its very existence."[188] This view was also articulated by David Brower, executive director of the Sierra Club from 1952 to 1969 and a founder of Friends of the Earth,

185. Salyers, "Looking for Life on Mars and Beyond."
186. Donna Shirley, "The Great Terraforming Debate-Part 7: Questions: The Martian Future," Science Fiction Museum and Hall of Fame, Seattle, WA, 20 June 2004, *http://www.sfhomeworld.org/make_contact/article.asp?articleID=162*.
187. Henry David Thoreau, "Walking," *Atlantic Monthly* 9 (June 1862): 657–674.
188. Brian Manetta, "John Muir, Gifford Pinchot, and the Battle for Hetch Hetchy," Ithaca College History Department, *Ithaca College History Journal* (spring 2002), *http://www.ithaca.edu/hs/history/journal/papers/sp02muirpinchothetchy.html* (accessed 22 May 2005).

who believed that we must "save wilderness for its own sake, for the mysterious and complex knowledge it has within it."[189]

The book *Red Mars* expanded the forum for discussing these ideas versus the views of those who say that we have a right, perhaps even a duty, to colonize and transform Mars into something good for our species. These various views continue to be discussed in the space science community. Christopher McKay,[190] for instance, of NASA Ames Research Center, argued that "terraforming Mars would be permissible, provided the planet is sterile,"[191] while the aerospace engineer Robert Zubrin took a giant step further, holding that "failure to terraform Mars constitutes failure to live up to our human nature and a betrayal of our responsibility as members of the community of life itself."[192] A more detailed discussion of the ethics of space exploration is included in Chapter 8.

Robinson's Mars trilogy did something else besides bringing up ethical issues that was no less important an accomplishment. It gave Mars a reality in the mind of the reader, transforming the planet from a concept to an actual terrain that could be seen and almost felt—expanses of rust-colored sand, an endlessly rising volcano, enormous gashes in the surface, and a landscape bashed by "meteor strikes so large that pieces from the collisions flew all the way to our Earth and moon."[193] Making a planet more real allows the reader to think about issues such as contamination prevention in a different light, because the planet has become a tangible object, potentially worth protecting. *Red Mars* was a very widely read book. I believe its biggest contributions to the planetary protection field were to give the planet a bona fide existence and a beauty that will affect the decisions made about her in the future.

---

189. Tom Thomson, "David Brower," *http://www.netwalk.com/~vireo/brower.htm*, from the Web site, "Earth talk!"—The quotations of over 200 American naturalists and scientists with brief biographies, *http://www.netwalk.com/~vireo/* (accessed 23 May 2005).
190. C. P. McKay, "Does Mars Have Rights? An Approach to the Environmental Ethics of Planetary Engineering," in *Moral Expertise*, ed. D. MacNiven (New York: Routledge, 1990), pp. 184–197.
191. Martyn J. Fogg, "The Ethical Dimensions of Space Settlement," International Academy of Astronautics, IAA-99-IAA.7.1.07, 1999, *http://www.users.globalnet.co.uk/~mfogg/EthicsDTP. pdf* (accessed 26 January 2011).
192. Robert Zubrin, *The Case for Mars: The Plan to Settle the Red Planet and Why We Must* (New York: Simon & Schuster/Touchstone, 1996), pp. 248–249.
193. S. Troy, "Red Mars: Kim Stanley Robinson-Nebula 1993," Award Winner's Review, Sustainable Village Web site, *http://www.thesustainablevillage.com/awrbooks/html/BooksinHTML/ redMars.html* (accessed 23 May 2005).

DEVELOPING EFFECTIVE PLANETARY PROTECTION APPROACHES

3

> There has been much concern . . . over the fact that sterilization procedures represent a hazard to the success of the total mission. . . . Approach it from the standpoint of not what should we do, but what *can* we do, and *how* should we do it.
>
> —*Freeman Quimby, Editor of 1962* Proceedings of Conference on Spacecraft Sterilization[1]

Abe Silverstein, NASA's Director of Space Flight Programs, issued an edict in October 1959 requiring that "payloads which might impact a celestial body must be sterilized before launching."[2] Sterilization, interpreted strictly, meant that *every last microorganism in the space vehicle had to be either killed or removed.* But NASA quickly realized that meeting this standard would be prohibitively difficult.[3]

In its effort to achieve absolute sterilization—the destruction or removal of all microorganisms—NASA mission staff attempted to walk a tightrope between two extremes: 1) subjecting spacecraft components to severe enough environments to remove all living material and 2) avoiding overly harsh conditions that would damage sensitive equipment and possibly cause a mission to fail. Accomplishing both objectives proved untenable. Even hospital sterilization procedures do not kill *all* living material.

1. Freeman H. Quimby, ed., *Proceedings of Conference on Spacecraft Sterilization* (Washington, DC: NASA Technical Note D-1357, December 1962), p. 2.
2. Charles R. Phillips, *The Planetary Quarantine Program: Origins and Achievements, 1956–1973* (Washington, DC: NASA SP-4902, 1974), pp. 10, 25.
3. Phillips, *The Planetary Quarantine Program*, p. 25.

NASA was developing its contamination control program under enormous time pressure. It had to meet President Kennedy's directive to land a man on the Moon in the decade of the 1960s. The Agency saw that it needed to modify its standard for preparing a spacecraft to visit another world. NASA mission staff thus set a more realistic goal for dealing with spacecraft contamination: attain a sterilization level *sufficient to avoid compromising biological studies* of the Moon or a planet. In doing this, the Agency shifted its metric of successful planetary protection from a standard of absolute sterilization to one *dependent on the limitations of biological sensing technologies.*[4]

Probabilistic Approaches to Spacecraft Contamination Control

As early as 1960, space scientists suggested that a probabilistic approach be used for designing contamination control procedures.[5] During a NASA-sponsored conference held on 29 June 1960 entitled "Problems and Techniques Associated with the Decontamination and Sterilization of Spacecraft," participants recommended that research be conducted to determine probabilities of accidental spacecraft impacts, coupled with statistical studies of viable organism implantations on celestial bodies.[6]

At another conference on spacecraft sterilization held 9 July 1962 by NASA Biosciences Programs, participants considered tolerance levels for the probability that a single microbe would survive after sterilization of a spacecraft or one of its components. Discussion at this conference centered around studies conducted by L. D. Jaffe of JPL (but on temporary assignment to the NASA Lunar and Planetary Office). Jaffe's studies considered various scenarios regarding

---

4. Phillips, *The Planetary Quarantine Program*, pp. 25–33.
5. Charles W. Craven, "Part I: Planetary Quarantine Analysis," *Astronautics & Aeronautics* (August 1968): 20. Also available in folder 006695, "Sterilization/Decontamination," NASA Historical Reference Collection.
6. Jack Posner, ed., *Proceedings of Meeting on Problems and Techniques Associated with the Decontamination and Sterilization of Spacecraft—29 June 1960* (Washington, DC: NASA Technical Note D-771, January 1961).

permissible risks of a robotic expedition contaminating Mars. One of his reports, which he circulated at the conference, estimated the lowest probability level of microbe survival that could currently be achieved with the best sterilization methods available as 10^{-4}. This was equivalent to no more than one in 10,000 landings or planetary impacts containing a microorganism. Conference participants adopted the 10^{-4} probability limit as a desirable goal for planetary flights.[7]

L. D. Jaffe spent considerable effort developing acceptable contamination risk recommendations for spacecraft landing on Mars. He estimated that on the basis of tests with terrestrial organisms under simulated Martian conditions, the probability of microbe growth if released on Mars was approximately unity. He also examined the engineering problems of spacecraft sterilization for robotic missions, as well as aseptic methods for spacecraft assembly and the dangers of recontamination. He concluded that the probability of introducing a viable microbe onto Mars should be held to 10^{-4}, and techniques for accomplishing this were discussed.[8]

During this time, SSB was evaluating permissible probabilities of planetary contamination for flyby trajectories, as alternatives to sterilizing the spacecraft. In its 1962 study, *A Review of Space Research,* SSB recommended a probability of no more than 10^{-4} of an unsterilized flyby impacting (and potentially contaminating) the target planet.[9]

L. D. Jaffe's work focused not only on Mars, but also on the degrees of assurance advisable for missions to the Moon and Venus. He held that sterilizing a spacecraft was probably not necessary for expeditions to the Moon, citing 1960 research by Carl Sagan as well as 1962 work by Soviet microbiologist A. A. Imshenetsky indicating only a remote chance that terrestrial organisms would grow and reproduce on or near the lunar surface. Sagan had discussed the "three major hazards for survival of terrestrial life on the Moon—temperature, corpuscular radiation, and solar

7. Morton Werber, *Objectives and Models of the Planetary Quarantine Program* (Washington, DC: NASA SP-344, 1975), pp. 9–11.

8. Ibid.; L. D. Jaffe, "Sterilizing Unmanned Spacecraft," *Astronautics and Aerospace Engineering* 1 (1963): 22, as reported in summaries of the paper in Joe W. Tyson and Ruby W. Moats, "Exobiology: An Annotated Bibliography—1951–1964" (stamped as "unpublished preliminary data") (Washington, DC: George Washington University, NASA CR-53806, 1 March 1964), http://ntrs.nasa.gov/archive/nasa/casi.ntrs.nasa.gov/19640013479_1964013479.pdf (accessed 26 January 2011), p. 12, and in Werber, *Objectives and Models* (accessed 26 January 2011), p. 11.

9. Werber, *Objectives and Models,* p. 11.

electromagnetic radiation"[10] as probably precluding microbial reproduction on the lunar surface. Nevertheless, Sagan believed that contamination of the Moon with terrestrial microbes and organic matter would constitute "an unparalleled scientific disaster,"[11] interfering with research on the early history of the solar system, chemical compositions of matter in the remote past, and the origin of Earth life. He strongly felt that such contamination should be avoided. Lederberg and D. B. Cowie in 1958[12] had also suggested taking steps to avoid contaminating the Moon, so that any organic chemicals found there would not be confused with those transported from Earth. Jaffe accordingly recommended that the maximum probability of finding a terrestrial microbe on the Moon be limited to no more than 10^{-6} per square centimeter of surface. Rather than sterilizing lunar spacecraft, however, he thought that application of cleanliness procedures would be sufficient for meeting his recommendation.[13]

Regarding the allowable contamination probability assigned to Venus, Jaffe realized that he needed to take into account the likelihood that the planet's surface, which was very hot, would be hostile to terrestrial organisms. The regions most liable to support microbe growth were in Venus's upper atmosphere, but since Earth microbes did not appear to multiply in our own atmosphere, Jaffe thought "there was little chance they would do so in that of Venus."[14] He estimated the chance of a terrestrial microbe finding conditions suitable for growth on Venus to be about 10^{-3}. Jaffe divided the desirable probability limit on a mission contaminating the target planet (10^{-4}) by the probability of finding suitable growth conditions on Venus (10^{-3}) to yield a recommended 10^{-1} probability limit on a mission releasing viable organisms into Venus's upper atmosphere.

Jaffe's recommendations regarding Mars and Venus went through various revisions at NASA's 1962 Conference on Spacecraft Sterilization as well as the Agency's Iowa City Office of Space Science Applications Summer Study. The final guidelines included the following points:

10. Carl Sagan, "Biological Contamination of the Moon," *Proceedings, National Academy of Sciences* 46 (1960): 397.
11. Sagan, "Biological Contamination," 396.
12. J. Lederberg and D. B. Cowie, "Moondust," *Science* 127 (1958): 1473.
13. Werber, *Objectives and Models*, pp. 11–12.
14. Werber, *Objectives and Models*, pp. 11–12.

1. Spacecraft buses[15] and booster last stages must either be sterilized or trajectories must be chosen for them that will ensure a probability of Mars impact no greater than 10^{-4} and a probability of Venus impact no more than 10^{-2}.
2. A Mars entry capsule needs to be sterilized and afterwards handled aseptically, in order to limit the probability to no more than 10^{-4} that a single viable organism will be released on the planet surface.
3. Lunar contamination, while it does not constitute as serious a problem as for the planets, still needs to be kept to a "feasible minimum" in order not to "seriously interfere with future biological and chemical surveys."[16]

Note that these guidelines did not mandate a universal standard of planetary protection but were instead tied to the perceived likelihood that a particular celestial body could support life. Similarly, the *NASA Unmanned Spacecraft Decontamination Policy* (NMI-4-4-1), issued 9 September 1963, stated that the Agency's lunar policy was based on the prevailing scientific opinion that reproduction of terrestrial microorganisms on the Moon's surface would be extremely unlikely.

COSPAR conducted its own deliberations on acceptable contamination levels at a meeting in Florence in May 1964, convening a study group chaired by C. G. Heden of the Karolinska Institute in Stockholm on spacecraft sterilization standards. Representatives from the United States, the USSR, France, Belgium, and the United Kingdom also attended. At this meeting, the participants formulated the first quantitative objectives for planetary protection. Each participant gave a best estimate for parameters used in the analytic framework for approximating acceptable levels of contamination.[17] These discussions led to recommendations that were adopted by COSPAR's executive council, resulting in Resolution 26.5. This resolution established a probabilistic framework for developing planetary protection standards, advocating "a

---

15. The spacecraft bus is the portion of the spacecraft that supplies the support functions (i.e., power, attitude control, etc.) necessary for the payload to meet mission objectives. "Glossary," GSFC, *http://gpm.gsfc.nasa.gov/glossary.html* (accessed 16 November 2007).
16. Both quotes, as well as material for the above guidelines, were taken from Freeman H. Quimby, ed., *Proceedings of Conference on Spacecraft Sterilization* (Washington, DC: NASA Technical Note D-1357, December 1962), pp. 79–81.
17. Carl Sagan, Elliott C. Levinthal, and Joshua Lederberg, "Contamination of Mars," *Science* 159 (15 March 1968): 1191–1196.

sterilization level such that the probability of a single viable organism aboard any spacecraft intended for planetary landing or atmospheric penetration would be less than 1×10^{-4}, and a probability limit for accidental planetary impact by unsterilized flyby or orbiting spacecraft of 3×10^{-5} or less . . . during the interval terminating at the end of the initial period of planetary exploration by landing vehicles."[18]

The term "sterilization" in the COSPAR resolution referred (and refers throughout this book) to reduction of spacecraft microorganism counts to defined levels, as distinguished from absolute sterilization—the complete elimination of all microbes.[19] In May 1966, the COSPAR Subcommittee for Planetary Quarantine recommended that the probability a planet of biological interest will be contaminated within the period of biological exploration be kept down to no more than 10^{-3} and that this standard be adopted by all states engaging in the exploration of space.[20] COSPAR's probabilistic approach was used until 1982, when SSB suggested a nonquantitative policy that is discussed later in this book.[21]

On 6 September 1967, NASA Policy Directive 8020.7 superseded the NMI-4-4-1 standard on robot spacecraft decontamination, specifying that the probability of terrestrial organism transport to planets be kept below certain levels. Furthermore, the directive recognized that viable microorganisms would land on the Moon and stipulated a management plan for them, ordering that they be "identified, quantified and, insofar as possible, located"[22] in order that they could be recognized as terrestrial in origin if found in any return samples.

---

18. Donald L. DeVincenzi, Margaret S. Race, and Harold P. Klein, "Planetary Protection, Sample Return Missions and Mars Exploration: History, Status, and Future Needs," *Journal of Geophysical Research* 103(E12) (25 November 1998): 28578.

19. COSPAR, Resolution 26, Fifth International Space Science Symposium, Florence, Italy, COSPAR Information Bulletin 20: 25–26, Committee on Space Research, Paris, France, 1964, as reported in DeVincenzi et al., "Planetary Protection, Sample Return Missions and Mars Exploration."

20. Charles W. Craven and Robert P. Wolfson, *Planetary Quarantine: Techniques for the Prevention of Contamination of the Planets by Unsterile Spaceflight Hardware*, Technical Report 32-1216, JPL, 15 December 1967, p. 1, folder 006697, "Quarantine/Sterilization," NASA Historical Reference Collection.

21. Perry Stabekis, "Governing Policies and Resources" (presentation given at an unspecified NASA meeting, presentation slides/notes given by Stabekis to author on 9 September 2004, Washington, DC), p. 3.

22. NASA, *Outbound Spacecraft: Basic Policy Relating to Lunar and Planetary Contamination Control*, NPD 8020.7, 6 September 1967, folder 009901, "Lunar Quarantine and Back Contamination," NASA Historical Reference Collection; Phillips, *The Planetary Quarantine Program*, p. 33.

NASA issued another directive based on probabilistic calculations on the same day as the above one, 6 September 1967. This second directive, NPD 8020.10, put forth a guiding criterion for exploration of planets and their satellites deemed important in the search for extraterrestrial life, precursors of life, or remnants of life. According to NPD 8020.10, the probability that a planet or satellite of interest will be contaminated by such exploration during the period of biological interest must not, as in the COSPAR recommendation of 1966, exceed one in 1,000 (1×10^{-3}).[23]

Assessing Contamination Probabilities and Allocating Mission Risk Constraints

NASA developed mathematical models that aided in evaluating factors that could potentially contaminate a target planet. The Agency used data generated from these models to help establish appropriate mission risk constraints, $P(N)$s. These were limits on the probabilities that individual missions would contaminate the target planet, with N specifying the particular mission.[24]

Prior to 1966, COSPAR set the $P(N)$ values for planetary missions. After this, however, COSPAR transferred some of the decision-making to the countries responsible for missions. COSPAR suballocated the 1×10^{-3} overall limit on P_c, the probability that target planet "C" would be contaminated during the period of biological interest, to the different nations launching spacecraft. These suballocations are listed in table 3.1. Note that the United States and USSR each had a suballocation of 4.4×10^{-4} while the total suballocation for all other spacefaring nations was 1.2×10^{-4}, adding up to a total allocation of 1×10^{-3}. The NASA Planetary Quarantine (PQ) Officer, using data supplied by various consultants,

---

23. Basic quarantine policy for planetary missions appeared in NPD 8020.10 (6 September 1967) and was updated by NPD 8020.10A, *Outbound Planetary Biological and Organic Contamination Control Policy and Responsibility* (1 August 1972). Both documents contained the following provision regarding biological contamination: "The basic probability of one in one thousand (1×10^{-3}) that a planet of biological interest will be contaminated shall be used as the guiding criterion during the period of biological exploration of Mars, Venus, Mercury, Jupiter, other planets and their satellites that are deemed important for the exploration of life, life precursors or remnants thereof." Charles R. Phillips, *The Planetary Quarantine Program* (Washington, DC: NASA SP-4902, 1974), Chap. 6.

24. D. G. Fox, L. B. Hall, and E. J. Bacon, "Development of Planetary Quarantine in the United States," *Life Sciences and Space Research* X (Berlin: Akademie-Verlag, 1972): 1–3.

Table 3.1 Suballocations by nation—probability limits on contaminating a target planet during the period of biological interest.

Nation	Probability Limit
United States	4.4×10^{-4}
USSR	4.4×10^{-4}
All Other Nations (total probability of contamination)	1.2×10^{-4}
Total	**1×10^{-3}**

suballocated the U.S. probability limit on P_c, 4.4×10^{-4}, to various individual missions, according to a schedule that considered the type of mission (planetary flyby, orbiter, lander, or lander-orbiter combination) and the anticipated number of similar missions that would be carried out.[25]

One of the PQ Officer's consultants, Exotech Systems, Inc., performed the compilation of information necessary to maintain a database providing current summaries on the probabilities that missions already flown had contaminated the target planet, as well as calculations of projected suballocations to missions in the planning stages. The PQ Officer sometimes modified suballocation values for projected missions as new data became available. For instance, the two Viking missions each received initial suballocations of 7.2×10^{-5}, but because of previous successful Mariner missions to Mars, this was augmented to a combined suballocation for the two Viking flights of 2×10^{-4}.[26]

A Planetary Protection Example: Analyzing Potential Contamination Sources in a Robot Expedition to Mars

During the 1960s, NASA developed plans for future Mars expeditions that included considerable thought on how to prevent planetary

---

25. Task Group on Planetary Protection, "Assessment of the 1978 Report," in *Biological Contamination of Mars: Issues and Recommendations*, SSB, National Research Council, National Academies, Washington, DC, 1992, *http://www7.nationalacademies.org/ssb/bcmarssummary.html*, and *http://www7.nationalacademies.org/ssb/bcmarsch1.html* through *http://www7.nationalacademies.org/ssb/bcmarsch6.html*. Also available from Steven Dick Unprocessed Collection, NASA Historical Reference Collection; Fox et al., "Development of Planetary Quarantine," 1–3.
26. Fox et al., "Development of Planetary Quarantine," 1–3; Pericles D. Stabekis, e-mail comments on author's draft manuscript, 21 June 2005; Phillips, "Planetary Missions," in "Program Accomplishments," in *The Planetary Quarantine Program*, Chap. 7.

contamination. E. M. Cortright, who served as the director of Langley Research Center from 1968 to 1975,[27] envisioned a robotic Mars expedition that involved two lander capsules containing surface analysis equipment and, in particular, equipment for biology experiments.[28] Charles Craven and Robert Wolfson examined such a project through the lens of planetary protection, using systems analysis techniques and experiments to identify and evaluate potentially significant sources of biocontamination. Key to Craven and Wolfson's approach, which was applicable to other target planets as well, was to divide up the difficult challenge of analyzing an extremely complex mission into a series of simpler tasks, including the following:[29]

1. Isolate every possible source of planetary contamination by thoroughly examining all aspects of the flight program.
2. Investigate each contamination source in order to build an in-depth understanding of its mechanism.
3. Construct, where possible, mathematical models useful for calculating contamination probabilities.
4. In situations where meaningful mathematical models cannot be developed, estimate suitable bounds on the probabilities of contamination.[30]

Isolating and Examining Mechanisms of Contamination

In carrying out the above tasks, Craven and Wolfson identified and analyzed these potential mechanisms of planetary contamination:

- Inadequate sterilization of a lander.
- Recontamination of a sterile lander.

27. Lane E. Wallace, "Addressing the New Challenges of Air Transportation: The TCV/ATOPS Program," in *Airborne Trailblazer: Two Decades with NASA Langley's 737 Flying Laboratory* (Washington, DC: NASA SP-4216, 1994), Chap. 2, *http://oea.larc.nasa.gov/trailblazer/SP-4216/chapter2/ch2.html* (accessed 7 March 2006).

28. E. M. Cortright, "The Voyage to the Planets: The Voyager Program" (Fifth Annual Goddard Symposium, Washington, DC, March 1967).

29. Charles W. Craven and Robert P. Wolfson, *Planetary Quarantine: Techniques for the Prevention of Contamination of the Planets by Unsterile Spaceflight Hardware*, Technical Report 32-1216, JPL, 15 December 1967, folder 006697, "Quarantine/Sterilization," NASA Historical Reference Collection.

30. Ibid., p. 2.

- Accidental impacts of a launch vehicle or its components.
- Accidental impacts of unsterilized spacecraft components.
- Accidental impacts of microorganism-bearing spacecraft ejecta.
- Inadequately sterilized propulsion systems.

Characteristics of each of these mechanisms are described below:

Inadequate sterilization of a lander. Planetary contamination could occur if viable organisms are present on the outside of the lander capsule or within the lander vehicle itself. In particular, organisms could be released due to 1) disintegration of the lander during impact or 2) erosion of the lander over time.

Recontamination of a sterile lander. This could occur 1) prior to launch due to improper handling; 2) during the launch or during the interplanetary voyage, by such occurrences as damage to the biological barrier isolating the sterile lander within the rocket's nose cone shroud; 3) during release of the barrier; and 4) during lander release. Other spacecraft components such as the bus or orbiter craft could be the sources of recontamination.

Accidental impacts of a launch vehicle or its components. Through improper maneuvering of the launch vehicle, failure of its propulsion system, or its detonation, the vehicle or components and pieces of it (such as clamps, rings, and bolts) could be accidentally propelled onto an impact trajectory with the target planet.

Accidental impacts of unsterilized spacecraft components. If mission staff lose control of the spacecraft after injection onto an impact trajectory with the planet, then all components of the craft will impact the body, including typically unsterilized components such as the bus that were not intended to do so. This scenario can also occur through a miscalculation in determining the craft's trajectory. In addition, planetary environmental factors such as a thicker atmosphere than predicted may cause the spacecraft's orbit around a planet to decay sooner than expected, resulting in an impact.[31]

Trajectory biasing is an approach for preventing accidental spacecraft impacts on a target body. In this technique, mission control staff deliberately offset (bias) the initial spacecraft trajectory in a direction that will ensure the craft misses its target by hundreds to thousands

---

31. Craven and Wolfson, *Planetary Quarantine*, pp. 2–3.

of kilometers, depending on the size of the target body. As the mission proceeds, mission controller knowledge of the spacecraft's actual trajectory gets more precise due to continued radio tracking and optical navigation. The controllers then can employ the spacecraft's propulsion system to reduce trajectory bias in progressive steps until the desired trajectory is attained. The advantage of trajectory biasing is that if mission staff lose the ability to control the spacecraft for any reason before the bias is removed, the spacecraft will not impact and contaminate the target body.[32]

Accidental impacts of microorganism-bearing spacecraft ejecta. A spacecraft typically releases certain types of ejecta during a mission that may carry microorganisms. These ejecta can include propulsion system gases, attitude control gas (generally far cooler than propulsion system gases and thus more likely to contain viable organisms), outgassing from spacecraft components, and particulates broken off the craft through micrometeoroid collisions or encounters with the planet's atmosphere.[33]

For the Mars program, NASA initiated a major experimental effort to determine the sizes of the particulates likely to be ejected from spacecraft, as well as a comparison of the number of microorganisms projected to be killed by micrometeoroid impact versus those that would remain alive and continue on to impact Mars. To develop realistic projections, NASA built a micrometeoroid simulator facility to fire 5 μm particles at average velocities of 30,000 ft/sec into targets inoculated with precise numbers of microorganisms. The targets were made of typical spacecraft materials including two types of aluminum, fused silica glass similar to that covering solar cells, and fiberglass-epoxy.[34]

Another study examined the various loose particles left on the spacecraft from its manufacturing process. These might escape the spacecraft during the mission and carry microorganisms to the target planet. The study analyzed the numbers and size distributions of particles likely to be generated during spacecraft fabrication and cleaning. Still another experiment investigated the attitude control gas system as

---

32. NASA, "Glossary of Terms," NASA Planetary Protection Web site, *http://planetaryprotection. nasa.gov/pp/features/glossary.htm* (accessed 9 April 2005).
33. Craven and Wolfson, *Planetary Quarantine*, pp. 2–3.
34. Ibid.

a potential contamination source, employing a scaled-down version for the simulations. NASA staff placed filters of various sorts over gas nozzles in order to determine the extent to which microorganisms could be contained rather than ejected during firings. NASA staff also vibrated the attitude control system to simulate launch conditions. The aim was to determine whether launch environments resulted in viable organisms entering the system and getting expelled during firing operations. Some of these tests were performed using an actual Nimbus spacecraft attitude control system.[35]

Nonstandard ejecta of various types included such objects as the biological barrier on the launch rocket's shroud, whose job was to isolate the lander capsule until launch; and debris such as bolts, clamps, rings, and so on. These could accidentally separate from the spacecraft at various times in the mission, such as during spacecraft maneuvers, and fall into an impact trajectory toward the target planet.[36]

Ejecta from propulsion systems. Interplanetary spacecraft engines weighed up to several tons. Developing and applying reliable sterilization procedures for them required considerable engineering and technical effort. During the 1960s, JPL conducted a research effort on sterilizing the propulsion fuels themselves as well as the rest of the onboard propulsion systems. The testing revealed that dry-heat sterilization at temperatures up to 145°C was possible for onboard liquid and solid propellant systems. Even though the propellants themselves were heat-producing and sometimes explosive substances, their autoignition temperatures were typically 250° to 300°C, which was two to three times higher than the envisioned sterilization temperatures. The main difficulties identified in thermally sterilizing them were such effects as chemical decomposition, degradation of ballistic properties, and high vapor-pressure buildup.[37]

Developing effective sterilization procedures required that the impact of engine exhaust on microbes be understood. For the Mars Voyager mission, the "previous incarnation of Viking,"[38] General

---

35. Ibid., pp. 11–12.
36. Ibid., pp. 2–3.
37. Winston Gin, "Heat Sterilization of Pyrotechnics and Onboard Propulsion Subsystems," in *Spacecraft Sterilization Technology* (Washington, DC: NASA SP-108, 1966), pp. 433–434.
38. Stabekis e-mail comments, 21 June 2005.

	Route to Mars										
	1	2	3	4	5	6	7	8	9	10	11
Source of Contamination	initial microbe loading	Survive during trip	Ejection process	Transport process	Survive die-off	Survive vacuum	Survive ultraviolet	Survive other solar radiation	Survive entry heating	Survive martian environment	# of VOs expected to survive on Mars
Attitude control gas system											
Orbit insertion engine											
Exhaust gases											
Loose particles											
Micrometeroid ejecta											
Etc.											

Figure 3.1 Estimating the quantity of contamination reaching Mars. The matrix depicts various elements of the mathematical model used for predicting contamination carried to the Martian surface. (VO stands for viable organisms.)

Electric conducted a research project to determine the lethality of engine combustion to microorganisms. The program involved 22 test firings using three types of engines—solid fuel, monopropellant liquid fuel, and bipropellant liquid fuel—in order to estimate the survivability of microorganisms in the engine's combustion chamber during firing. Engineers injected a measured quantity of germs into the engine, fired it into a test chamber, then surveyed the chamber to establish how many of the bacteria survived the firing.[39]

Test results indicated that lethality to microorganisms derived not only from combustion, but also because of the chemical compositions of certain propellants. For instance, liquid rocket propellants such as hydrogen and nitrogen tetroxide have chemical compositions that are toxic to microorganisms. These propellants are thus self-sterilizing within a few hours. As for solid-propellant rockets, few microorganisms are able to survive the high temperatures generated during firings.[40]

Scientists believed that the environments within cold-gas propulsion systems used for spacecraft attitude control might not be lethal to microorganisms, allowing some of them to survive. To avert this

---

39. Craven and Wolfson, *Planetary Quarantine*, pp. 8–9; Craven, "Part I: Planetary Quarantine Analysis," 20.
40. Ibid.

possibility, manufacturing and assembly of attitude control systems needed to be carried out so as to minimize internal contamination. In addition, gas supplies for these systems required thorough filtration to remove any microorganisms.[41]

Mathematical Modeling of Contamination Probabilities

The model attempting to calculate Mars contamination probabilities incorporated mathematical representations of the physical phenomena associated with various mechanisms of contamination. The matrix shown in figure 3.1 lists the various elements of the mathematical model and illustrates how the task of predicting a mission's contamination probability was divided up into a collection of smaller tasks. A key factor in the probability calculations was the number of viable microbes initially present in or on each potential contamination source, several examples of which are listed in the left column of figure 3.1. Note that sources include parts of the spacecraft (i.e., "attitude control gas system"), loose particles, and exhaust gases.[42]

The model also calculated impacts on the microbes of the mechanisms that ejected them from the spacecraft and transported them to the Martian surface (see columns 3–4 in figure 3.1). Column 3 represents consequences of ejection processes. The calculations for each cell in this column drew on two types of data—the expected rate at which microbes were ejected from the spacecraft and the fraction of those microbes estimated to survive the ejection process. The environments associated with some contamination mechanisms were quite harsh, such that few microbes were expected to survive ejection. For instance, the second contamination source in the table, the "orbit insertion engine," subjected ejecta to temperatures of 6,000°F.[43]

Column 4 calculations estimated the probabilities of ejected microbes achieving Mars impact trajectories. These calculations drew

---

41. Craven, "Part I: Planetary Quarantine Analysis," 20.
42. Craven and Wolfson, *Planetary Quarantine*, pp. 4–5; Fox et al., "Development of Planetary Quarantine," 5.
43. Ibid.

from the orbital mechanics of particles leaving the spacecraft at a range of speeds and directions.[44]

Columns 5 through 9 identify calculations concerned with the impacts of various potentially lethal factors for the microbes. These included the vacuum of space, ultraviolet and other solar radiation, and atmospheric entry heating. Column 10 calculations focused on the probability that terrestrial microbes would survive and grow in the Martian environment. The Space Science Board periodically issued estimates of the ability of Mars to support terrestrial organism growth and proliferation. These were under constant review in the United States by the Planetary Quarantine Office (PQO). SSB regularly changed its probability-of-growth estimates to reflect the latest data that had been generated on planetary environments.[45]

For Column 11, the model calculated the total probability of contamination from each source and used the number in Column 1, the initial microbe loading, to estimate quantities of viable organisms from that source expected to survive on the Martian surface. Finally, summing the numbers in Column 11 yielded total numbers of viable organisms projected to survive on Mars for the particular mission under study.[46]

Developing Input Data for the Model

Determining the characteristics of the contamination mechanisms and sources in order to generate input data for the mathematical model was not a trivial task. It typically relied on subjective judgment by the field's experts combined with analysis of whatever data were available from laboratory experiments and previous space missions. For instance, determining the quantity of microbes encapsulated within various materials or located between mated surfaces was very difficult due to inaccurate sampling techniques. This resulted in a large range of the microbe estimates for the landing

---

44. Ibid.
45. Fox et al., "Development of Planetary Quarantine," 6; Craven and Wolfson, *Planetary Quarantine*, pp. 4–5.
46. Craven and Wolfson, *Planetary Quarantine*, pp. 4–6; Fox et al., "Development of Planetary Quarantine," 5.

capsule, running from less than 1 organism per cm^3 to more than 1,000 per cm^3. The variance in these densities meant that the total microbial burden in a Viking-type landing capsule might be as low as 100,000 or as high as 1 billion, implying an uncertainty of more than 20 hours in the required sterilization time.[47]

Note that each cell in the matrix corresponded to a contamination probability calculation for one aspect of the spacecraft that was considered a potential contamination source and one part of the voyage that a particle from this source might take en route to the Martian surface. For instance, the matrix cell that corresponded to "orbit insertion engine" and "survive vacuum" contained the results of probability calculations that a microbe would escape from the spacecraft's orbit insertion engine in a viable state and also endure the vacuum of space.[48]

The Craven and Wolfson NASA study initiated an experimental program to generate needed input information for the model. The largest of their endeavors focused on determining whether the spacecraft's orbit insertion engine, a bulky piece of equipment weighing five tons, presented a contamination threat if it was used near Mars. This was important information to obtain; if NASA had to sterilize the engine to protect Mars, a significant engineering effort and a considerable expenditure of funds would be required. Craven and Wolfson designed a test program that examined the exhausts from rocket engines using solid, monopropellant liquid and bipropellant liquid fuels. They inoculated these different types of engines with measured quantities of microbes, then fired the engines into a test chamber, which was bioassayed to determine bacterial survival rates.[49]

Another experimental program centered on the sizes of ejecta that would be broken loose from the spacecraft during a micrometeoroid impact, as well as how many viable organisms would be ejected, survive, and attain a Mars impact path. To answer these questions, the study used a micrometeoroid simulator facility that could fire 5 μm particles at 30,000 ft/sec. Micron-sized iron particles simulating micrometeoroids were smashed into targets made of spacecraft materials that had been

---

47. Craven and Wolfson, *Planetary Quarantine*, pp. 4–5; Fox et al., "Development of Planetary Quarantine," 5.
48. Craven and Wolfson, *Planetary Quarantine*, pp. 4–5.
49. Ibid., pp. 8–11.

inoculated with known numbers of microorganisms. Ejected particles from the targets were caught in a trap, then bioassayed.[50]

The experimental program included analysis of the extent to which manufacturing techniques deposited loose particles on spacecraft surfaces. These particles could fly off the spacecraft during the mission and possibly carry bacteria with them. The study sought to estimate the number of particles deposited on spacecraft surfaces as well as their size distributions.[51]

The attitude control gas system, another potential source of contamination, was studied by covering the exhaust nozzles of a scaled-down model with special filters, which were then bioassayed. NASA did this in conjunction with vibrating the model in a manner simulating launch vibration levels. NASA aimed to determine whether launch vibrations released viable organisms within the attitude control gas system that would be expelled when it fired.[52]

Die-off rates of unprotected microorganisms in space. NASA conducted experiments in a space simulator meant to determine the rates at which unprotected microbes die when subjected to the extreme conditions of outer space. These tests generated vital input data because many of the scenarios for possible planetary contamination relied on viable organisms traveling for some distance through space, and the model needed as accurate information as possible on their die-off rates. The tests subjected *Bacillus subtilis* variety niger (also known as *bacillus globigii*, or BG) spores to ultra-high vacuum and temperature variations that simulated thermal-vacuum relationships observed on the Mariner 4 mission to Mars. The simulator for this work was a spherical high-vacuum chamber able to achieve a 10^{-10} torr vacuum.[53] A molecular trap within the chamber captured up to 99.97 percent of the condensable molecules ejected from the test surface.[54]

Preparing input data. In developing the model, Craven and Wolfson debated whether to do the analysis using average values for input data such as the average number of microbes expected

---

50. Ibid., pp. 9–11.
51. Ibid., pp. 11–12.
52. Ibid., p. 12.
53. A torr is a unit of pressure. 1 torr is equal to the pressure exerted by a column of mercury 1 millimeter tall. 760 torr is the standard pressure at sea level on Earth.
54. Craven and Wolfson, *Planetary Quarantine*, pp. 12–13.

to reside in a contamination source. They discarded this approach because it would lead to model outputs in terms of the average, and this would *not* represent the worst cases of planetary contamination that might occur. These worst cases were of particular interest to the planetary protection community, and so Craven and Wolfson considered doing the calculations based on worst-case data input. But this approach too was dropped because the final result would likely be a very improbable one. Instead, Craven and Wolfson decided that the approach yielding the most useful results would be to formulate input data as probability distributions. For instance, probability P_1 might represent the chance that between 0 and 10 viable organisms were contained in a particular contamination source, while P_2 through P_n would represent the probabilities that successively larger numbers of viable organisms were contained in the source. The output from the model would give a distribution of probabilities representing the chances that various quantities of viable organisms would survive on Mars. Such an output is depicted in figure 3.2. This approach allowed the probabilities of both worst case scenarios and more likely occurrences to be estimated.[55]

Challenges in Sterilizing Lander Capsules

Central to NASA's planetary protection effort were its procedures to sterilize the part of a spacecraft most likely to contaminate another planet—the lander capsule. These procedures needed to be applied down to the basic component level, as well as to assemblies of these components and to the entire lander, in ways so as not to damage the spacecraft. Designing sterilization regimes to accomplish this and lander capsules that could withstand those regimes presented serious engineering challenges. In the words of Bob Polutchko, a manager in the Viking program, the rigorous sterilization requirements "made every component a brand new development. There was little or nothing off the shelf that we could buy that could meet the requirements we imposed on ourselves for planetary quarantine. From the

---

55. Ibid., pp. 4–6.

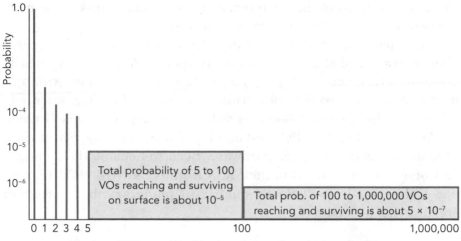

Figure 3.2 Typical probability distribution of viable organism (VO) quantities reaching and surviving on the Martian surface. In this particular distribution estimate, the probability of three organisms surviving is about 1 in 10,000, while the total probability for 100 to 1,000,000 organisms surviving is about 5 in 10 million.

smallest . . . everyday resistor or capacitor or transistor all the way to basic raw materials of thermal coatings or heat shields or parachute materials."[56] (A detailed account of the Viking experience is included in Chapter 5.)

The following sections describe some of the efforts to understand the impacts of sterilization procedures on the lander capsule and its microbial load, the development of sterilization approaches, and the identification of materials and components able to endure those procedures.

Microbial Response to Thermal Sterilization Techniques

A difficult parameter to estimate in modeling the impacts of sterilization procedures was the probability that microorganisms would survive the process. Thermal resistance is in large part dependent on

---

56. Bob Polutchko, manager of the Viking lander support office and of numerous other groups in the program, telephone interview by author, 21 February 2006.

the characteristics of the microorganism. Microorganisms of the most concern were those that were most heat-resistant.

The space science community decided to choose a bacteria representative of thermally resistant spores that would typically contaminate space hardware and use it as a standard by which to measure the effectiveness of sterilization approaches. The scientists settled on the spores of *Bacillus subtilis* variety niger, also known as *Bacillus globigii*, or BG, and employed it in the majority of their studies on sterilizing space hardware. BG also became the standard test organism mentioned in the *NASA Handbook* NHB 8020.12, that was to be used in biologically qualifying planetary flight program sterilization procedures.[57]

Other Factors Impacting Microbial Kill Rates

The level of sterilization attained is a function not only of the thermal resistance of the particular organism to be killed, but also of the characteristics and geometries of the materials on which the organisms are located. For instance, scientists have observed "at least an order of magnitude range in resistance to dry-heat sterilization for the same test species depending on whether they are on the open surface of a material, located between two mated surfaces, or encapsulated within a material."[58] BG spores encapsulated in methyl methacrylate and epoxy, a plastic, exhibited D-values (the time required to kill 90 percent of the cells)[59] of 210 to 300 minutes when exposed to flowing hot gas at 125°C. Spores from the same suspension that lay unprotected on metal surfaces exhibited D-values of only 10 to 30 minutes when exposed to the hot gas.

The physical characteristics of a spore's surroundings that inhibit or accentuate water loss have a dramatic impact on its survivability.

---

57. NASA, *Planetary Quarantine Provisions for Unmanned Planetary Missions* (Washington, DC: NASA Publication NHB 8020.12, April 1969): p. 11, folder 006697, "Quarantine/ Sterilization," NASA Historical Reference Collection; Joseph A. Stern, "Part II: Current Concepts in Sterilization," *Astronautics & Aeronautics* (August 1968): 30–31.
58. Fox et al., "Development of Planetary Quarantine," 6.
59. A fuller definition of a D-value, or decimal reduction time, is the time at a particular temperature that is required to kill 90 percent of the cells.

As an example, if water movement is restricted around spores (such as when they are encapsulated in space vehicle parts), the spores exhibit higher D-values than in situations in which water can be easily lost or gained (such as when they are thermally exposed on external surfaces of a space vehicle). Furthermore, the rate of gas flow over an exposed surface on which spores exist appears to affect their destruction rate, probably because the gas flow impacts their moisture loss.[60]

Evolution of Sterilization Technologies

A critical factor affecting spore kill rates is the efficacy of sterilization technologies, which are constantly improving. The technologies that were available when spacecraft sterilization was initiated were those that could be transferred from biological weapons applications and from pharmaceutical, food canning, surgical supply, and other industries. New methods for applying heat and gases such as ethylene oxide needed to be developed that were better suited to spacecraft sterilization.[61]

The fundamental challenge of the food processing industry—to maximize the destruction of microbes that cause food spoilage while minimizing losses to the nutritional value and taste of the food—was somewhat analogous to problems faced in spacecraft sterilization—to maximize microbe destruction but minimize damage to spacecraft components. Thus, in the first years of the space program, the food industry's analytic techniques for microbe kill rates were examined closely as models for analysis of spacecraft sterilization. An important difference, however, was that the food industry's analyses as well as supporting data were based on *moist* heat processes. Moisture present during spacecraft sterilization negatively impacted some spacecraft components and equipment. NASA thus had to develop dry-heat sterilization approaches more appropriate to its needs.[62]

---

60. Stern, "Part II: Current Concepts in Sterilization," 30–31.
61. Fox et al. , "Development of Planetary Quarantine," 7–8.
62. A. R. Hoffman and J. A. Stern, *Terminal Sterilization Process Calculation for Spacecraft*, Technical Report 32-1209 (JPL, 15 November 1967): 1, folder 006697, "Quarantine/ Sterilization," NASA Historical Reference Collection.

Identifying Mechanisms of Contamination

Developing procedures for estimating the total surface contamination of a vehicle at different points during its assembly was essential in constructing reliable decontamination technologies as well as methodologies for avoiding recontamination. The approach for estimating contamination was to extrapolate from samples taken from the vehicles during assembly and before launch. Contamination samples were, and still are, taken by swabbing vehicle surfaces, analyzing environmental settling strips, and collecting air samples. These tests helped reveal important mechanisms of contamination that had to be dealt with, such as those resulting in contamination during transport operations of the flight-ready spacecraft to the launch pad. Contamination investigations disclosed a surprisingly high estimated microorganism count for the Surveyor II spacecraft of 40 times the estimated counts for Surveyors I or III. The cause of this problem was traced back to high contamination levels on a shroud meant to cover the craft as space center staff moved it to the launch pad. Lunar Orbiter 5 (whose mission launched in August 1967) also had a high microorganism count, and its source appeared to be a shipping canister used to transport the craft across country. Key in this determination were samples taken from the canister's surfaces, which revealed large populations of microbes.[63]

The following sections examine specific types of lander capsule parts and materials, and the sterilization regimes to which they were subjected.

Electronic Components

NASA performed extensive research, notably at JPL beginning in 1962, to identify heat-sterilizable electronic parts. NASA considered a part to be sterilizable if it was "serviceable and reliable after heat

---

63. Stern, "Part II: Current Concepts in Sterilization," 31–32.

sterilization preparatory to a trip to and landing on Mars."[64] JPL developed exhaustive experimental data on the longevity and reliability characteristics of the parts able to withstand heat sterilization regimes, as well as the ability of parts to survive ethylene oxide treatment, which the Agency was then considering for use as part of its sterilization protocols. Early testing in the JPL program focused on approximately 500 capacitors constituting 15 different part types. JPL then expanded its effort, analyzing nearly 43,000 resistors, diodes, capacitors, fuses, transistors, and other parts making up 262 part types. The JPL study generated over 400 million-part test hours of data.

Electronic Packaging, Solder, Connectors, Wires, and Cables

Spacecraft electronics included not only the components mentioned in the previous section, but also the following:

- Soldered and welded joints.
- Wires and cables.
- Connectors such as for coaxial or multiconductor cables.
- Packaging materials.

All of the above needed to withstand three 36-hour, 145°C thermal sterilization treatments as well as ethylene oxide exposure.[65]

An individual spacecraft assembly operation often used a range of materials, each of which had to be thoroughly analyzed. Soldering processes provide a good illustration of this. Materials in typical soldered joints include the solder itself, various sizes of stranded or solid electric wire, terminals and connectors made from a variety of solderable materials, and flux—a substance made from rosin and acids, whose purpose is to help molten solder flow smoothly and form good joints. NASA expected that any degradation of the solder joint materials would manifest itself by degrading joint performance. To determine if this had happened, NASA conducted mechanical strength

---

64. James R. Miles, "Spacecraft Sterilization Program," in *Spacecraft Sterilization Technology* (Washington, DC: NASA SP-108, 1966), pp. 309–320.
65. Alvan G. Fitak, Leonard M. Michal, and Robert F. Holzer, "Sterilizable Electronic Packaging, Connectors, Wires, and Cabling Accessories," in *Spacecraft Sterilization Technology* (Washington, DC: NASA SP-108, 1966), p. 343.

tests, electrical resistance tests, electrical tests during vibration, and metallographic analyses of the joints.[66]

Polymers

Polymers were, and are, employed in a range of critical spacecraft parts and materials, including adhesives, electrical insulation, shrinkable tubing, circuit boards, module cases, conformal coatings, encapsulants, films, tapes, oils, and greases. Polymeric materials include many types of plastics and are typically composed of large organic molecules formed from combinations of smaller molecules (monomers) in regular patterns such as long chains. Types of polymers considered for spacecraft applications included epoxies, polyesters, polyolefins, polyurethanes, polysulfides, phenolics, polyimides, phthalates, silicones, and Teflons. Many of these materials were often found in a single electronic component of the spacecraft.

Polymers selected for use in a spacecraft needed to be compatible with NASA's dry-heat sterilization and gaseous ethylene oxide regimes. NASA analyzed the effects of these regimes on over 150 polymeric products. The testing included a cycle of heat exposure of 36 hours at 145°C and then a more rigorous exposure to three 40-hour cycles at 149°C. Combined ethylene oxide and thermal exposures were also included in the testing. NASA examined materials subjected to these conditions for indicators of property changes, which included visual alterations in the material, volume and weight changes, and alterations in tensile strength. Polymers exhibiting gross changes were dropped from further testing.[67]

One particular polymer-related project conducted by Marshall Space Flight Center (MSFC) aimed to develop a tough, heat sterilizable, transparent potting compound intended for enclosing electronic modules and printed circuit boards. MSFC required that this compound withstand a more extreme environment than many other of the materials that were tested—200°C for 24 hours—without any

---

66. Ibid., pp. 343–344.
67. Donald P. Kohorst and Herman Harvey, "Polymers for Use in Sterilized Spacecraft," in *Spacecraft Sterilization Technology* (Washington, DC: NASA SP-108, 1966), pp. 327–342.

degradation in physical or electrical properties. The potting compound had to be totally compatible with the outer space environment and demonstrate adherence to exacting strength, hardness, adhesion, thermal expansion, moisture absorption, and dielectric standards. MSFC evaluated approximately 30 materials, including silicones, epoxies, and polyesters, and also discovered that subjecting materials to low-pressure molecular distillation[68] improved their resistance to the space environment.[69]

Batteries

JPL initiated a program in 1962 to develop robust batteries capable of withstanding three heat sterilization cycles of 145°C for 36 hours each and still displaying good shelf life and energy storage characteristics. JPL deemed this battery development program necessary because, as with many other parts, no off-the-shelf battery could be found that would deliver adequate performance after being subjected to the heat sterilization regime and other stress tests. JPL conducted some in-house battery development work, although NASA contracted out much of the research work to various vendors. These contracts entailed the following projects:

- Electric Storage Battery Company (ESB) developed a copolymer of divinylbenzene and methacrylic acid that held up well to thermal sterilization procedures and was a candidate for battery separators, which were sheets of nonconducting material put between a battery's many positive and negative plates to prevent short circuits caused by plates bending and touching.
- ESB received a contract in 1965 to fabricate silver-zinc and silver-cadmium batteries that remained sound after heat

---

68. Molecular distillation is a process designed to separate polymers into fractions of different molecular weight. It is typically performed at the lowest possible temperatures in order to avoid degradation of the polymers.

69. John T. Sohell, "Development of an Improved Heat Sterilizable Potting Compound," *Semi-Annual Progress Report, Part III: OSSA Program–Supporting Research Projects*, NASA Technical Memorandum X-53069, Huntsville, AL (22 June 1964): pp. 9–11, *http://ntrs.nasa. gov/archive/nasa/casi.ntrs.nasa.gov/19740075316_1974075316.pdf* (accessed 17 February 2006).

sterilization (as well as after high impacts and vibration stresses). ESB studied all aspects of the batteries—electrodes, electrolyte solutions, separators, and case materials.

- Gulton Industries and Douglas Aircraft analyzed sterilizable battery components, but for batteries with inorganic rather than organic separators.

- Douglas Aircraft Company and The Eagle-Picher Company examined silver-zinc battery systems.

- TRW Space Technologies Laboratory had a contract with JPL to investigate battery characteristics after heat sterilization, with a focus on cells with nickel-cadmium electrodes.

- Electro-Optical Systems (EOS), a subsidiary of Xerox Corporation, developed a hydrogen-oxygen fuel cell (a device that converts the chemical energy in a fuel into electricity).[70]

Search-for-Life Instruments and Experiment Materials

Preparing instruments and experiment materials designed to search for life involved sterilizing them so as to protect the target planet and prevent contamination on the instruments from giving erroneous results. All components of the experiments had to be compatible with sterilization regimes which, in JPL's 1965 study, included dry-heat and ethylene oxide environments.[71] JPL's dry-heat regime involved sterilization of planetary-impacting hardware in an inert atmosphere at 135°C for 22 hours.

Experiment materials that needed to be sterilization-compatible included chemical reagents and nutrient media, which were formulated to help nourish and grow extraterrestrial life-forms that might be found. Many types of hardware, from hand tools up to sophisticated instrumentation, also needed to be compatible with the sterilization regime.

---

70. Ralph Lutwack, "Batteries and Spare Power Systems," and Daniel G. Soltis, "Alternate Approaches to Sterilizable Power Sources," both in *Spacecraft Sterilization Technology* (Washington, DC: NASA SP-108, 1966), pp. 361–377.

71. Carl W. Bruch, "Sterilizability of Scientific Payloads for Planetary Exploration," in *Spacecraft Sterilization Technology* (Washington, DC: NASA SP-108, 1966), pp. 503–514.

Impacts of Surface Roughness on Sterilization Operations

Spacecraft design engineers needed to know the relationships between the conditions of material surfaces throughout the lander capsule and the ease with which those surfaces could be sterilized. Tenney et al. pointed out that ordinary, machine-finished metal surfaces contained grooves and scratches that were typically "five times as deep as the diameter of streptococci or 50 times that of poliomyelitis organisms."[72] Some engineers thought that machining and polishing capsule surfaces to a high degree of smoothness would allow microbes to be more easily killed, although it would also add significantly to the program's cost and complexity of operations. NASA's biological testing showed, however, that fine machining and polishing of metal surfaces did not have the predicted reduction in the time required to sterilize the spacecraft. Marshall Space Flight Center and General Electric Company performed a series of tests using the standard BG microbe and samples of 2024 aluminum with varying degrees of surface roughness[73] ranging from 4 to 125 microinches rms.[74] Exposure of any of the surfaces to a temperature of 135°C rapidly and significantly reduced microbial spore levels. Highly machined or polished surfaces did not appear necessary for effective sterilization, given a dwell time of 24 hours at the above temperature.[75]

Terminal Sterilization Approaches

In order to sterilize an entire planetary lander capsule during the terminal sterilization procedure, the capsule and its enclosing

---

72. John B. Tenney, Erwin Fried, and R. G. Crawford, "Thermal Sterilization of Spacecraft Structures," *Journal of Spacecraft and Rockets* 3(8) (August 1966): 1239.

73. Surface roughness is a measure of the vertical deviations encountered when traversing a surface.

74. The abbreviation "rms" refers to root mean square roughness, an averaging approach. The root mean square roughness is the square root of the average of the squares of a set of roughness measurements. It gives an estimate of the average height of the bumps on a surface, typically measured in micrometers or microinches.

75. Ronald G. Crawford and Richard J. Kepple, "Design Criteria for Typical Planetary Spacecraft To Be Sterilized by Heating," in *Spacecraft Sterilization Technology* (Washington, DC: NASA SP-108, 1966), pp. 473–501.

canister would ideally spend a minimum of time at the elevated temperatures needed for resident microbes to be killed, resulting in the least impact on spacecraft materials and the least risk to reliability.[76] Terminal sterilization procedures, however, had to walk a fine line in heating the capsule, using thermal gradients that were steep enough to elevate temperatures quickly but not so steep as to create thermal stresses capable of damaging the spacecraft. A 1964 study by Marshall Space Flight Center sought to identify capsule designs and materials that could be cycled between 25°C and sterilization temperatures of 150°C as quickly as possible without warping or "adversely deflecting"[77] spacecraft structures. The study determined such things as the best types of joining methods between components that would conduct heat as rapidly as possible, thus minimizing thermal gradients and the resulting stresses from different rates of expansion.[78]

A primary issue in developing a terminal sterilization regime was to choose the best way to apply heat. A range of heater blanket, radiant heating, and oven designs were examined by NASA engineers. The landing capsule had to be heated within its canister, which was not to "be opened within any portion of the earth's atmosphere which might recontaminate the landing assembly."[79] NASA considered designs employing both evacuated sterilization canisters and those filled with gas at or near atmospheric pressures. In terminal sterilization simulations for a 1,200-pound lander, calculations showed that the times required for gas-filled canisters to reach a sterilization temperature of 150°C were almost 50 percent less than for the evacuated canisters, resulting in a 10 percent reduction of the total time required at elevated temperatures. Canisters filled with gas, which could conduct heat by convection, also reduced temperature gradients and their accompanying thermal stresses on spacecraft materials. Through the

---

76. Ibid.

77. Harry J. Coons, Jr., ed., *Semi-Annual Progress Report, Part III: OSSA Program—Supporting Research Projects,* NASA Technical Memorandum X-53069, MSFC, Huntsville, AL, 22 June 1964, pp. 6–7, *http://ntrs.nasa.gov/archive/nasa/casi.ntrs.nasa.gov/19740075316_1974075316.pdf* (accessed 17 February 2006).

78. Ibid., pp. 6–7; Crawford and Kepple, "Design Criteria," pp. 473–501.

79. J. B. Tenney, Jr., and R. G. Crawford, "Design Requirements for the Sterilization Containers of Planetary Landers" (AIAA-1965-387, proceedings of AIAA annual meeting, San Francisco, 26–29 July 1965), *http://www.aiaa.org/content.cfm?pageid=406&gTable=mtgpaper&gID=84766* (accessed 17 February 2006).

use of gas circulation fans, heat transmission between the canister wall and the landing capsule, as well as within the landing capsule itself, could be even more accelerated. To maximize heat transmission rates, the locations of equipment, cables, wire harnesses, and tubing within the capsule had to be optimized so as not to restrict gas-flow paths any more than necessary. This was especially true for components with large thermal masses, which had longer response times and were typically "the limiting items in the heating and cooling portions of the sterilization cycle."[80]

Requirements of the Canister

Fundamental to NASA's policy for planetary landings was to enclose the capsule in a bacteriological barrier (the canister) that maintained both cleanliness and sterility. The canister had to be of minimal weight but provide a reliable barrier against all types of contamination—microbes, spores, dirt, and dust. This canister had to be compatible with the thermal sterilization regime. It could not be opened from the time of terminal sterilization until the capsule had ascended beyond Earth's atmosphere and had separated from unsterile components of the spacecraft. Finally, the canister, whose exterior might not be sterile, needed to include the capability for being opened in space and jettisoned in a manner so as not to hit the target planet.[81]

Capsule Recontamination Issues

The aim of terminal sterilization procedures was to ensure that the landing capsule bioload was low enough so as to meet planetary protection requirements. Other parts of the spacecraft, however, were generally not sterilized and could potentially recontaminate the landing capsule at some point in the mission after terminal sterilization had been carried out. As mentioned above, the canister completely

---

80. Crawford and Kepple, "Design Criteria," pp. 479–480.
81. Tenney and Crawford, "Design Requirements"; Crawford and Kepple, "Design Criteria," pp. 490–491.

enclosed the landing capsule, providing an effective biobarrier until such point during the mission that it was safe to remove the capsule. The chance of recontaminating it was minimized if the capsule remained within its biobarrier canister as long as possible—preferably until planetary encounter.

Recontamination probabilities were also reduced if mission control separated the capsule from the spacecraft engines *while the capsule was still contained within its canister.* Once the capsule and canister attained a distance from the spacecraft bus beyond its potential to recontaminate (which might occur, for instance, should its attitude-control jets spew out viable bacteria along with propulsive gases), mission control could then safely separate the landing capsule from its canister.[82]

A potential difficulty with the above separation strategy was that the ratio of canister mass to capsule mass was about 1 to 10, while the ratio of canister mass to that of the capsule and bus was roughly 1 to 100. If the capsule separated from its canister *after* separating from the bus, the capsule could experience much greater perturbations to its velocity and orientation than if it had stayed anchored to the heavy bus while the canister lid was removed and discarded.[83]

This second strategy, removing the canister while the capsule stayed attached to the bus, had a downside as well. Unsterilized exhaust, dislodged dust, or other ejecta from the bus might contact the unprotected capsule, possibly resulting in microorganisms settling on its exterior and reaching the target planet. A way to prevent this was for NASA to program the capsule to carry out a slow rolling maneuver on its approach to the target planet, in such a manner as to expose all external surfaces to the Sun's ultraviolet flux for 10 minutes or more. Typical bacteria cells cannot survive such radiation for more than one-half minute, while 10 minutes is sufficient to kill even the most resistant microbes.[84]

---

82. Dwain F. Spencer, "Effects of Sterilization on Separation, Entry, Descent, and Landing Phases of a Capsule Mission from an Engineering Mechanics Perspective," in *Spacecraft Sterilization Technology* (Washington, DC: NASA SP-108, 1966), pp. 461–471.
83. Ibid., pp. 461–465.
84. Ibid., pp. 461–465.

Soviet vs. U.S. Planetary Protection Approaches

The nature of planetary protection is such that it cannot be accomplished by only one nation. Any country sending probes to other planets can contaminate those bodies and compromise the scientific investigations of all other countries. This was why U.S. and European scientists received the accounts of Soviet exploration of Venus with serious concern. On 1 March 1966, in a hearing of the U.S. Senate Committee on Aeronautical and Space Sciences, George Mueller, head of NASA's Office of Manned Space Flight, reported an account from the Soviet Tass news agency announcing the landing of a spacecraft on Venus. NASA's best interpretation of the event, however, was that this was not a soft landing, but an impact with the planet.[85]

Press reports from the USSR indicated that the payload capsule and bus from its Venera 4 spacecraft entered Venus's atmosphere in a state described as sterilized. But there was no evidence at that time that the capsule had received anything beyond surface sterilization using gaseous germicides and high-energy irradiation of certain components, and the Soviets made no claims whatsoever about sterilizing the bus, which U.S. scientists believed to contain a circulating coolant. Although NASA gained some confidence in the 1970s in Soviet planetary protection approaches (see the discussion in Chapter 2 regarding USSR planetary protection approaches), the United States was quite worried at the time of Venera 4 that Soviet spacecraft might seriously contaminate other planets.

Considerable data collected prior to the Venera 4 mission indicated that Venus surface temperatures were very high—perhaps too high to support life—but there were cooler environments above the surface that space scientists thought possibly capable of sustaining life. A worst case scenario: the spacecraft bus was internally contaminated with terrestrial microbes and had broken open in the dense Venusian

---

85. U.S. Senate Committee on Aeronautical and Space Sciences, "NASA Authorization for Fiscal Year 1967," Hearings before the Committee on Aeronautical and Space Sciences, United States Senate, 89th Cong., 2nd Sess., on S. 2909 (1 March 1966): pp. 101–102.

atmosphere, releasing those microbes into an environment in which they might be able to replicate.[86]

Protecting Against Unplanned Impacts

Studies by the U.S. space science community revealed important differences between our country's planetary protection precautions and those of the Soviet Union. One of these was the variance in strategies for preventing an unwanted impact of the bus with the target planet. A spacecraft bus's interior, and sometimes its exterior as well, are typically left unsterilized. One method for preventing a collision with the target planet is called a *bus-deflection maneuver,* in which mission control staff initially place the bus-payload capsule assembly on a collision course with the target planet. Days before arrival at the planet, a mechanism in the spacecraft separates the bus and capsule. The bus then uses its own propulsion system to deflect it onto a new course that will avoid planetary impact. If all systems operate as planned, this strategy effectively protects the planet from an unwanted impact. But if the bus propulsion system does *not* function correctly and does not change the bus's course sufficiently, then it will crash into the planet and possibly contaminate it with terrestrial organisms.[87]

Payload capsule deflection is an alternate strategy that uses trajectory biasing and does not present the contamination dangers of bus deflection. In this approach, mission staff bias (aim) the capsule-bus assembly on a near-encounter trajectory, then send commands to separate the capsule from the bus and give the capsule a velocity increment that will change its path to an impact trajectory with the planet. Meanwhile, the bus will continue on its nonimpact course. The importance of the payload capsule deflection strategy is that even if spacecraft systems fail to operate correctly, the bus assembly will not end up impacting or contaminating the planet. If the bus and capsule cannot be separated, then neither will hit the planet.

---

86. Carl Sagan, Elliott C. Levinthal, and Joshua Lederberg, "Contamination of Mars," *Science* 159 (15 March 1968): 1191–1195.
87. Bruce C. Murray, Merton E. Davies, and Phillip K. Eckman, "Planetary Contamination II: Soviet and U.S. Practices and Policies," *Science* 155 (24 March 1967): 1505–1511.

Payload capsule deflection is a more fail-safe strategy from a planetary protection point of view, but it is also a more complex and expensive strategy than bus deflection. To redirect the payload capsule onto just the right trajectory as to encounter the planet, the capsule needs to contain its own small propulsion system and propellant supply. Also, if the capsule is not equipped with its own internal stabilization and orientation system, then it must be spin-stabilized, which necessitates spinning it up before separating it from the bus. These requirements make capsule-deflection a significantly more difficult operation from an engineering point of view and more expensive due to the additional equipment costs.[88]

The cheaper bus-deflection approach can be employed in a manner that offers additional protection for the target planet by mounting an abort rocket on the bus as a backup system. The abort rocket would contain its own timing system that would initiate firing unless Earth sent an override command. Thus, if mission control lost contact with the bus or lost the ability to operate it before carrying out the deflection maneuver, the backup abort rocket would still fire and propel the bus away from an impact trajectory. The space science community analyzed this abort rocket strategy in great depth during plans for a possible 1969 Mars mission. The scientists came to the conclusion that this approach would sufficiently reduce the probability of accidental bus impact with the planet so as to meet the COSPAR recommendations discussed earlier in this chapter.[89]

The limitations of 1960s USSR planetary protection procedures can be illuminated by comparing Soviet planetary impact prevention approaches with those of the United States. It is telling that Soviet mission planners chose to use bus-deflection rather than capsule-deflection strategies, didn't employ abort rocket capabilities on the buses, and did not heat-sterilize the buses, although they may have received other types of sterilization treatments. In addition, the Soviets may not have completely heat-sterilized the capsules that were intended to impact the planet itself. The USSR might have assumed that partial heat sterilization, possibly combined with gaseous sterilization and other nondestructive methods, adequately

---

88. Murray et al., "Planetary Contamination II," 1506–1507.
89. Ibid.

addressed COSPAR's recommendations. Alternatively, the Soviets may have decided that following the COSPAR recommendations would present such onerous restrictions as to prevent serious solar system exploration.[90] Analysis of data from USSR and U.S. Venus and Mars missions revealed additional important differences between the Soviet and U.S. approaches to planetary protection. The Soviet Venus 2 mission launched on 12 November 1965 with the objective of passing as close as 40,000 kilometers (25,000 miles) from the Venusian planetary surface. In actuality, through calculations made from trajectory measurements once the spacecraft entered interplanetary space, scientists determined that the craft may have passed only 24,000 kilometers from the planet's surface. In contrast, the United States gave its Mariner 4 spacecraft an initial trajectory that would take it no closer than 600,000 kilometers from Mars, providing a considerably higher measure of planetary protection.[91]

The USSR's Venus 3 space vehicle carried a landing craft that the Soviets claimed was thoroughly sterilized, although they never specified what procedures were used. No such claim was made for the spacecraft bus. Because the Soviets were not able to reach Venus 3 for the last scheduled radio contact, U.S. space scientists assumed that the planned bus-deflection maneuver was not executed. As a result of this loss of communication with the spacecraft, U.S. scientists believe that the unsterilized Venus 3 bus as well as the payload capsule, which also may not have been adequately sterilized, crashed into Venus on 1 March 1966.[92] Regarding this likely crash, British scientist Sir Bernard Lovell charged that the USSR "endangered the future of biological assessment of Venus."[93]

In addition to the Venus 3 incident, the USSR also may have crashed a spacecraft into, and contaminated, Mars. Space scientists Bruce Murray, Merton Davies, and Phillip Eckman argued that the Soviet Zond 2 mission to Mars, which launched on 30 November 1964, had a high likelihood of ending its life in an impact with Mars.

---

90. Ibid., 1507.
91. Ibid.
92. Ibid., 1508.
93. James J. Haggerty, Jr., "Space Contamination Worries Scientists," *Journal of the Armed Forces* (9 April 1966): 9.

England's Jodrell Bank Observatory for radio astronomy successfully tracked the course of Zond 2 until at least the middle of February 1965. Tracking data suggested that the probe was traveling on a collision course with Mars. When Bernard Lovell asked a high-level Soviet scientist who was visiting the United Kingdom, M. V. Keldysh, if this was indeed the case, Keldysh responded that the probe would miss Mars by the slim margin of 1,500 km, within the range of uncertainty of the Jodrell Bank calculation. A characteristic of the Zond 2 trajectory suggested, however, that the Soviets intended to send the payload capsule on an impact with Mars, and it was only the bus, after a bus-deflection maneuver, that was supposed to miss Mars by 1,500 km.

When Western scientists analyzed the Zond 2 trajectory, they found that it was set up to minimize the relative velocity of approach at Mars. This gave an important clue as to the USSR's intent for this mission. To achieve maximum scientific value, a payload on an impact trajectory needed to have adequate time to transmit all the data it could back to Earth before the moment that it crashed into the surface of the planet. The extremely thin Martian atmosphere did not provide the potential for atmospheric braking of the spacecraft that Earth's or Venus's atmosphere would have. Thus, a mission to Mars intending to impact with the planet's surface needed a trajectory such as Zond 2's that would reduce the relative velocity between spacecraft and planet as much as possible. Furthermore, the Soviet Union indicated that Zond 2 was the same type of spacecraft as Venus 3, which had a payload capsule intended for planetary impact.[94]

The above accounts indicate the possibility that Soviet missions have already contaminated both Venus and Mars with viable terrestrial microorganisms. If this has indeed happened, then the basis for the COSPAR Mars recommendations has been compromised, since the number of microorganisms already released on Mars would have considerably exceeded the total that COSPAR expected from *all* robotic exploration during the period of biological exploration of this planet.

U.S. space scientists expressed concern that the "expensive prophylactic measures"[95] our country was investing in for spacecraft

---

94. Murray et al., "Planetary Contamination II," 1508–1509.
95. Washington Post Foreign Service, "U.S. Scientist Says Soviet Probes May Contaminate Planets," *Washington Post* (23 May 1969): A15.

sterilization might have to be reexamined if Soviet craft were already contaminating other planets. At the 10th annual COSPAR meeting in Prague, Czechoslovakia (now the Czech Republic), Richard W. Porter of the U.S. National Academy of Sciences (NAS) Space Science Board (SSB) stated that our planetary protection policies might need review, should the USSR land spacecraft on Mars having used the same sterilization techniques and precautions as in their Venus shots. Porter complained that the USSR had given no indication to the space science community of the extent of its spacecraft sterilization program and stressed that the "introduction of earthly bacteria into the environments of other planets . . . could spoil man's first chance to make a pure investigation of biological evolution elsewhere in the solar system."[96] His strong statements did not only fall on the ears of those who agreed with him. Forty-seven Soviet scientists also attended the meeting, which was the first large COSPAR gathering to host both Russian and U.S. scientists since the USSR had invaded Czechoslovakia the previous August.

---

96. Ibid.

BACK CONTAMINATION
The Apollo Approach

4

Apollo program objectives:
- To make the United States preeminent in space by building a broad-based capability for manned space flight.
- To demonstrate this capability by landing men on the Moon, and returning them safely to Earth, within this decade.

—*Hearings before the Committee on Science and Astronautics, U.S. House of Representatives*[1]

On 25 May 1961, President John F. Kennedy challenged the nation to send a man to the surface of the Moon and bring him back him safely. He also asked that this be accomplished by the end of the 1960s. President Kennedy considered such an endeavor to be a vital action in "the battle that is now going on around the world between freedom and tyranny."[2]

Flying a human to the Moon and back would have tremendous scientific as well as political returns, but many in the space science community realized the potential danger that such a mission posed to Earth. In 1960, the National Academy of Sciences' Space Science Board (SSB) had warned that life could theoretically exist on other

---

1. U.S. House of Representatives, *1967 NASA Authorization*, Hearings before the Committee on Science and Astronautics, 89th Cong., 2nd Sess., on HR 12718 (10 March 1966).
2. John F. Kennedy, "Man on the Moon" Address—A "Special Message to Congress on Urgent National Needs," 25 May 1961, Home of Heroes Web site, *http://www.homeofheroes.com/presidents/speeches/kennedy_space.html* (accessed 27 June 2005).

bodies of the solar system and advised NASA and other interested government organizations, including the Public Health Service (PHS), to establish an interagency committee on interplanetary quarantine that would develop safe policies for handling spacecraft and samples returning from planetary missions.[3]

The SSB, whose membership included 16 highly respected U.S. scientists, issued an important cautionary statement that at some "not very great distance below the lunar surface,"[4] an environment of low temperature, high vacuum, and protection against destructive radiation existed that was ideal for preserving microbes. NASA management thanked the SSB for identifying this concern but did not take any immediate action. This lack of response was possibly due to the immediate organizational and technical problems that the new Agency's management was struggling with that precluded any serious attention being given to possible infections from the Moon. In the initial years of NASA, the life sciences did not have the strong voice in Agency policy decisions that space physicists and astronomers had. Plus, few biologists of the day had much research interest in space; the focus of new research was on Earth. In addition, most space scientists believed that the risk to Earth from lunar contamination was extremely small.[5]

Policy Development for Preventing Back Contamination

During 1962, as NASA struggled with the task of actualizing President Kennedy's Moon challenge, some space scientists continued to raise concerns about the unlikely but possible prospect of back contamination and its potentially catastrophic impacts. In particular, Carl

3. William David Compton, *Where No Man Has Gone Before: A History of Apollo Lunar Exploration Missions* (Washington, DC: NASA SP-4214, 1989): p. 45; Kent Carter, "Moon Rocks and Moon Germs: A History of the Lunar Receiving Laboratory," *Prologue: Quarterly of the National Archives and Records Administration* 33 (winter 2001): 236.
4. Kent C. Carter, *The Prevention of Back Contamination from the First Manned Lunar Landing: A Study in Organization*, M.A. thesis, University of Cincinnati (1972): p. 7.
5. Homer E. Newell, "Space Science Board," in *Beyond the Atmosphere: Early Years of Space Science* (Washington, DC: NASA SP-4211, 1980), Chap. 12-2, *http://www.hq.nasa.gov/office/pao/History/SP-4211/ch12-2.htm* (accessed 26 January 2011).

Sagan warned that "there was a remote possibility that lunar explorers might bring deadly organisms back with them that could destroy life on Earth."[6] Sagan had not yet become well known through his books, articles, and public television series, and most scientists did not seriously think that toxic microorganisms, or life of any sort, were likely to exist on our barren, airless Moon. Sagan's point, however, could not be ignored. The consequences of back contamination, while improbable, could be so dire that serious prevention measures needed to be implemented. This view was shared by others in the space science community. The SSB issued a warning in the June 1962 report of its Iowa City Summer Conference, advising that the introduction of alien organisms into the biosphere of our planet could cause "a disaster of enormous significance to mankind"[7] and recommended that "NASA do everything possible to minimize the risk of back contamination."[8] The SSB went on to suggest using quarantine and other procedures to manage returning astronauts, samples, and spacecraft. These warnings still had little impact on NASA plans.[9]

This situation changed as plans for Apollo missions progressed, and it became increasingly likely that shipments of Moon rocks and soil samples would soon appear on Earth. Finally, the voices of concerned politicians as well as biologists began to be heard. In April 1963, during budget hearings, Senator Margaret Chase Smith (R) of Maine raised questions about the potential biological threat posed by the Moon. NASA could hardly afford to ignore congressional concerns regarding its priorities. The Agency's management quickly assured the senator that studies of back contamination issues would be made and appropriate actions taken to address her concerns.[10] Nevertheless, NASA did not follow through with its promise, viewing the chances of back contamination harmful to Earth to be "too remote to warrant further attention."[11] NASA also decided to ignore a warning from the commercial sector. The Lockheed Missile and Space Company was interested in the business opportunities of establish-

6. Carter, "Moon Rocks and Moon Germs," 235.
7. Carter, *The Prevention of Back Contamination*, p. 7.
8. John R. Bagby, "Rocks in Quarantine: Apollo's Lessons for Mars," *Planetary Report* 14 (July–August 1994): 6.
9. Compton, *Where No Man Has Gone Before*, p. 45.
10. Carter, "Moon Rocks and Moon Germs," 236.
11. Carter, *The Prevention of Back Contamination*, pp. 8–9.

ing a permanent U.S. presence on the Moon, but its director of spacecraft sterilization also recognized the potential dangers of doing so. For many millions of years, extraterrestrial material that conceivably contained organic and even living bacterial components had been impacting the Moon. Isolating Earth from such potentially disease-carrying matter was thus a prudent course of action.[12]

What NASA was interested in was protecting Moon samples from getting contaminated by Earth's environment, which would severely impair their scientific value. Space scientists in academia, outside of NASA, also needed the Moon rocks to be protected. Their own research agendas demanded that they obtain unspoiled lunar samples whose utility had not been destroyed by exposure to terrestrial contamination. In addition, these scientists understood that "reputations could be enhanced and prizes won from research on pristine lunar material"[13] and vigorously supported sample handling procedures that would not compromise their analyses of lunar material.

In late 1963, geoscientists Elbert A. King and Donald A. Flory from NASA's Manned Spacecraft Center (MSC) in Houston (which was renamed the Lyndon B. Johnson Space Center [JSC] in 1973)[14] proposed that the Agency take action to handle lunar samples under properly controlled conditions. In February 1964, King (who would, during the Apollo mission, become NASA's first Lunar Sample Curator)[15] and Flory constructed conceptual plans for a lunar sample receiving laboratory (LSRL) to appropriately handle samples from the Moon, and they presented the plans to Max Faget, MSC's director of engineering and development.[16]

---

12. David S. F. Portree, "Romance to Reality: Moon & Mars Mission Plans," Mars Institute, 18 May 2005, http://www.marsinstitute.info/rd/faculty/dportree/rtr/ft06.html (accessed 9 December 2005); Carter, "Moon Rocks and Moon Germs," 236–237.

13. Carter, "Moon Rocks and Moon Germs," 236.

14. John M. Logsdon (moderator), "Managing the Moon Program: Lessons Learned from Project Apollo," Proceedings of an Oral History Workshop, 21 July 1989, NASA History Division, Monographs in Oral History, No. 14, July 1999, http://klabs.org/history/monographs/no_14/monograph_14.pdf (accessed 26 January 2011); JSC, "Manned Space Flight Laboratory Location," NASA Release 61–207, 20 August 2002, http://www.jsc.nasa.gov/history/jsc40/memo_1a.htm (accessed 11 December 2005); JSC, "Space Center Houston: There's Always Something New," 2002, http://www.spacecenter.org/about2.html (accessed 11 December 2005).

15. Judy Allton, "25 Years of Curating Moon Rocks," http://www-curator.jsc.nasa.gov/curator/lunar/lnews/lnjul94/hist25.htm, JSC Astromaterials Curation Web site, http://www-curator.jsc.nasa.gov/curator/ (last updated 28 September 2005, accessed 13 December 2005).

16. Compton, Where No Man Has Gone Before, p. 41.

By February 1964, NASA's Office of Space Science and Applications (OSSA) was putting together a committee of prominent scientists to choose the researchers and organizations "best qualified to carry out individual and different phases of the sample investigations."[17] NASA thus needed the capability to open sample return containers, analyze the contents, then repackage and distribute them to eager scientists around the country for more in-depth study, while maintaining the material in as near-pristine condition as when they were collected on the Moon.[18]

John Eggleston, who served as the chief of the Lunar and Earth Sciences Division at NASA's Manned Spacecraft Center in Houston during the Apollo mission, stated in a February 1964 memo to Max Faget that appropriately managing the samples required construction of a central facility capable of handling lunar material "under precisely controlled, uncontaminated, sterile conditions."[19] This was the first formal statement by a NASA manager that the need existed for such a resource.[20] The facility he envisioned would require a sterile vacuum chamber approximately 10 by 10 by 7 feet that would be pumped down to a pressure of about one ten-millionth of Earth's atmospheric pressure for extended periods and maintained at temperatures varying from –30°C to +25°C. The reason for vacuum processing was to preserve any loosely bonded lunar gases contained in the samples, and also to protect against the possibility that lunar samples might react with Earth's atmosphere.[21] The vacuum chamber would be kept as sterile as possible from microorganisms, as well as chemically clean, and would be

17. John M. Eggleston to M. A. Faget, "Initial Handling of Geological and Biological Samples Returned From the Apollo Missions," 25 February 1964, LBJ SCHC, "LRL Chronological Files 1964" folder, box 76-11, LRL Chron. Files 1964–1965.
18. Carter, *The Prevention of Back Contamination*, p. 13.
19. Eggleston to Faget, "Initial Handling of Geological and Biological Samples Returned From the Apollo Missions."
20. Committee on Planetary and Lunar Exploration, SSB, National Research Council, "A History of the Lunar Receiving Laboratory," in *The Quarantine and Certification of Martian Samples* (Washington, DC: National Academies Press, 2002), Appendix B; Carter, *The Prevention of Back Contamination*, p. 10.
21. Committee on Planetary and Lunar Exploration, SSB, National Research Council, "A History of the Lunar Receiving Laboratory," in *The Quarantine and Certification of Martian Samples* (Washington, DC: National Academies Press, 2002), Appendix B.

dedicated only to handling lunar samples.[22] This chamber would have to contain remote-controlled equipment and manipulators able to open the sealed sample containers, divide up and prepare the samples as necessary, analyze them, then repack them in vacuum-tight containers suitable for transport to investigators. Eggleston believed that no existing facility even came close to fulfilling these requirements.

A more detailed vision of the required lunar sample facility was incorporated into an in-house MSC study entitled "Sample Transfer Facility," which Aleck Bond, an MSC systems test and evaluation manager, enclosed in a 14 April 1964 memo. The envisioned facility needed the capabilities to check lunar samples for the presence of any viable organisms, perform some control testing of the material, and then repackage portions of each sample in accordance with the analytical techniques being used by each investigator. The facility also required the following:

- 2,500 square feet of area, with an overhead 5- to 10-ton crane.
- Class 100 cleanroom conditions throughout.
- Approximately 200 square feet area containing no viable organisms (to be used for a virological laboratory).[23]

While design concepts for a central receiving facility continued to be examined by MSC engineers, several Apollo science planning teams, which had been established under the Space Science Steering Committee of NASA's OSSA, were developing the technical requirements for such a facility and identifying issues involved in handling materials from the Moon. On 16 June 1964 at a MSC Lunar Sampling Summer Symposium, these teams gave their recommendations. The science teams envisioned a facility whose responsibilities would be much broader than the repackaging function originally conceived by MSC engineers. They recommended that the facility provide for biocontainment and that "all operations be conducted <u>behind a biological barrier</u> [underlining from cited document] to prevent any harmful

---

22. Compton, *Where No Man Has Gone Before*, pp. 41–42.
23. JSC, "NASA Johnson Space Center Oral History Project Biographical Data Sheet" for Aleck Constantine Bond, *http://www.jsc.nasa.gov/history/oral_histories/BondAC/ACB_Bio. pdf* (accessed 21 March 2006); Aleck C. Bond to Chief, Office of Technical and Engineering Services, "Sample Transfer Facility," 14 April 1964, LBJ SCHC, "LRL Chronological Files 1964" folder, box 76-11, LRL Chron. Files 1964–1965.

organisms which might be contained in the lunar material from contaminating the surrounding environment."[24] They also recommended analyzing all lunar materials before release for the presence of pathogenic organisms. The facility would thus serve receiving and distribution as well as quarantine and bioanalysis functions.

Since certain experiments had to begin immediately after samples reached the receiving facility in order to ensure the generation of useful data, these experiments would need to be initiated within the containment facility while the samples were still being quarantined. Thus the necessary equipment would have to be built into the receiving facility. Examples of time-sensitive experiments included low-level radiation counting and investigations of gas contained in the samples, because such characteristics of the samples could decay quickly with time.[25]

By the summer of 1964, NASA realized that it was facing the complex task of constructing a multipurpose facility—one that could simultaneously keep terrestrial contamination out (to avoid degrading the lunar samples) and Moon pathogens in. Furthermore, NASA needed to do this while allowing scientists and technicians to enter the facility; conduct detailed, in-depth research on the lunar samples; and then leave without taking contaminants with them. Although biocontainment facilities such as Fort Detrick were in operation in the United States, no organization "had ever built or managed a facility which could do scientific research and biological containment and testing simultaneously."[26] Building and operating such a facility constituted a daunting technical and management task that NASA had to work through before the United States could fulfill President Kennedy's challenge to land a person on the Moon. Adding to the challenge was the need to reconcile the demands of two powerful interest groups. Many in the scientific community considered the lunar samples invaluable as research material but also believed that quarantining them was unnecessary and expensive and could compromise the science analyses. The biological community, on the other hand,

24. Carter, *The Prevention of Back Contamination*, p. 13.
25. Ibid.
26. Carter, *The Prevention of Back Contamination*, p. 14.

strongly cautioned that lunar material was a potential source of deadly pathogens and needed to be contained until proven safe to release.

On 29–30 July 1964, SSB hosted a Washington, DC, conference titled "Potential Hazards of Back Contamination from the Planets"[27] in which rigorous discussions of back contamination issues took place. SSB invited agencies representing both the biological and physical science communities, including the Public Health Service, the U.S. Department of Agriculture (USDA), the National Institutes of Health, the National Academy of Sciences, the U.S. Army Biological Laboratories at Fort Detrick, and NASA. Representatives from the academic community also attended, including professionals from the University of Minnesota School of Public Health, the University of Pennsylvania Hospital Department of Medicine, and the University of Michigan Department of Zoology.

The assembled scientists recognized that the Moon was an extremely unlikely place to harbor living organisms, but that if the Apollo spacecraft did bring back lunar microbes, they could harm or even kill life on our planet or flourish in the terrestrial environment to such an extent as to overrun our biosphere. The conference thus supported quarantine procedures "no matter how slight the possibility of life might appear to be."[28] The conference made the following recommendations, which eventually became policy guides for the management of samples, vehicles, and personnel returning from the Moon:

- Alien life-forms, astronauts, spacecraft, and lunar samples returning from the Moon should immediately be placed in isolation.
- Astronauts should be quarantined for at least 3 weeks (some conference participants thought that 4–5 weeks would be preferable)
- Spacecraft, suits, and equipment should not be decontaminated until thorough biological studies have been conducted on them.

27. Allan H. Brown (chairman), "Potential Hazards of Back Contamination from the Planets—29–30 July 1964," Space Science Board of the National Academy of Sciences (19 February 1965), JSC Historical Archives, "1 January–31 March 1965" folder, box 76-11, LRL Chron. Files, 1964–1965.
28. Bagby, "Rocks in Quarantine: Apollo's Lessons for Mars," 6.

- Preliminary sample analysis should be conducted behind "absolute biological barriers, under rigid bacterial and chemical isolation."[29]
- NASA should immediately develop the detailed operational procedures that would be necessary to carry out these recommendations.
- In preparation for missions, trial runs using nonpathogenic microbes should be made in order to test equipment and develop the best methods for minimizing contamination of the landing capsule.
- Postflight testing should include experiments designed to identify microbes harmful to terrestrial plants and animals. Protecting Earth's biosphere must be a priority.

The conference also identified a moral and practical imperative to implement protective measures against back contamination and recommended that the United States take a leadership role in that effort. Since our country recognized the potential hazard of back contamination, and since we had the power to incur this hazard, we had to acknowledge the responsibility for doing what we could to avoid it.

NASA decided to accept the recommendations of its science planning teams, the SSB conference, and various interest groups, and it instructed the MSC engineering team to revise its plans and include a biocontainment function in the receiving facility.[30]

Incorporating Planetary Protection into the Lunar Sample Receiving Laboratory Design

King and Flory now envisioned a far more elaborate facility than they had before, exceeding 8,000 square feet (740 square meters) in

---

29. Committee on Planetary and Lunar Exploration, SSB, National Research Council, "A History of the Lunar Receiving Laboratory," in *The Quarantine and Certification of Martian Samples* (Washington, DC: National Academies Press, 2002), Appendix B; Compton, *Where No Man Has Gone Before*, pp. 45–46; Carter, *The Prevention of Back Contamination*, pp. 14–15.
30. Brown, "Potential Hazards of Back Contamination from the Planets"; Carter, *The Prevention of Back Contamination*, pp. 15–16.

area and including offices for 30 visiting scientists and laboratories for chemical, biological, mineralogical, and petrological analyses as well as radioactive material measurements. The facility would have a vacuum chamber and also a cabinet filled with purified, inert nitrogen gas for processing lunar samples. In addition, the facility would contain a separate, sterile laboratory for biological investigations and measurement instruments for monitoring the atmospheric composition in the vicinity of the laboratory, so that investigators who received samples would know what contaminants were likely to be in them.[31]

The cost estimate for constructing the expanded facility reached $14 million as of 12 August 1964.[32] King and Flory presented their conceptual plans to Bob Gilruth, MSC's Director, on 13 August 1964, and Gilruth approved them. Max Faget then submitted MSC's preliminary plans for the facility to Willis B. Foster, who headed NASA OSSA's Manned Space Flight Division, asking for $300,000 to cover the costs of a detailed conceptual study.[33]

Foster did not favor MSC's expanded laboratory concept, but thought that building a simple receiving laboratory without extensive analytical capabilities was the more appropriate action.[34] Foster had been expecting a total cost estimate for the receiving facility of about $2 million and was "understandably shaken when he realized the extent to which MSC plans had grown."[35] Foster supplied only $150,000 for the concept study and recommended that MSC meet with OSSA to resolve the two parties' hugely different visions of what the facility's mission should be and what it should cost.

31. Committee on Planetary and Lunar Exploration, "A History of the Lunar Receiving Laboratory."
32. Q. G. Robb to Chief, Test Facilities Branch, "Background Material on the Lunar Sample Facility," 6 April 1965, LBJ SCHC, "1 April–30 September 1965" folder, box 76-11, LRL Chron. Files 1964–1965.
33. Carter, *The Prevention of Back Contamination*, p. 15; Compton, *Where No Man Has Gone Before*, p. 42.
34. Compton, *Where No Man Has Gone Before*, p. 42.
35. Carter, *The Prevention of Back Contamination*, pp. 15–16.

Edward Chao's Committee and the Prioritization of Planetary Protection

MSC and OSSA corresponded often over the next two months on appropriate models for the facility. Several statements of work for the conceptual study were also drafted and revised. On 23 October 1964, Willis Foster sent MSC Director Gilruth a letter informing him of the formation of the OSSA Ad Hoc Committee on the Lunar Sample Receiving Laboratory.[36] OSSA gave this committee the charter to investigate LSRL needs and establish general facility concepts and requirements. OSSA did not, however, specify the level of importance that planetary protection needed to have in the design.

Edward Ching Te Chao, an eminent U.S. Geological Survey (USGS) geochemist on loan to NASA, chaired the ad hoc committee, which was often referred to as the Chao Committee. It met four times and functioned from 9 November 1964 until 15 March 1965. Membership included representatives from academia (the University of California, Harvard University, and the University of Minnesota), the Department of Energy (DOE)'s Oak Ridge National Laboratory, the National Bureau of Standards, the Army Biological Laboratories, and NASA.[37] In the view of James McLane, MSC's engineer in charge of the early planning of LSRL, Ed Chao seemed at the committee's first meeting to have a preset agenda—to "keep the size and cost of the proposed lab to an absolute minimum."[38]

The next meeting, held 11 December 1964, focused on LSRL performance parameters as well as a topic of relevance to planetary protection—the procedures for biological decontamination and testing. The third meeting on 15 January 1965 also involved discussion of issues related to potential living organisms in the lunar material, with the Chao Committee's biological subcommittee giving its presentation.[39]

The wrap-up meeting took place on 26 February 1965, but by the beginning of March 1965, the Chao Committee had not yet completed

---

36. Robb, "Background Material."
37. Carter, *The Prevention of Back Contamination*, pp. 16, 137.
38. Compton, *Where No Man Has Gone Before*, p. 42.
39. Q. G. Robb to Chief, Test Facilities Branch, "Background Material on the Lunar Sample Facility," 6 April 1965, LBJ SCHC, "1 April–30 September 1965" folder, box 76–11, LRL Chron. Files 1964–1965.

its final report. MSC personnel knew that adding quarantine capabilities to LSRL would significantly complicate its construction. Cognizant of the tight schedule that had to be followed to construct LSRL, MSC grew impatient to find out what Chao's group recommended.[40]

Although the committee did finish its report on 15 March 1965, it did not immediately issue it. On 22 March, James McLane called Chao, who explained that a primary reason the report had not yet been released was that the committee had begun new discussions with various U.S. agencies and important issues had been raised. In particular, the U.S. Public Health Service (PHS) had aired its concerns regarding appropriate quarantine requirements and its envisioned role in planetary protection matters. These discussions with PHS, which eventually led to a long-term partnership between it and NASA, are examined in some detail later in this chapter.[41]

By 8 April 1965, McLane had finally received a draft of the Chao Committee's report and had begun reviewing and circulating it. Planetary protection concerns had been very visibly addressed in the report's first recommendation: Quarantine all lunar material while testing it for the presence of biological life or other substances that might be infectious, toxic, or harmful to humans, animals, or plants. The Chao Committee understood the importance of lunar sample biocontainment and defined a cardinal criterion for LSRL "to provide an absolute barrier between the lunar material and the world at large until the material is certified as biologically harmless."[42] The Chao Committee thus recognized planetary protection as a primary goal of the facility.

Other recommendations included the following:
- Store, control, and package lunar materials and distribute them to scientists around the world.

---

40. Susan Mangus and William Larsen, *Lunar Receiving Laboratory Project History* (NASA/CR-2004-208938, June 2004), p. 14; Robb to Chief, "Background Material."

41. Mangus and Larsen, *Lunar Receiving Laboratory Project History*.

42. Quotes in the sentence are from Carter, *The Prevention of Back Contamination*, p. 17, and Ad Hoc Committee on the Lunar Receiving Laboratory, *Concepts, Functional Requirements, Specifications and Recommended Plan of Operation for the Lunar Sample Receiving Laboratory* (draft), NASA OSSA, 15 March 1965, pp. 4 and 6, an attachment to J. C. McLane, Jr., to J. G. Griffith, "Lunar Receiving Laboratory," 8 April 1965, LBJ SCHC, "1 January–31 March 1965" folder, box 76-11, LRL Chron. Files 1964–1965.

- Conduct physical, chemical, biological, and radiological analyses that cannot wait until after the quarantine period.
- Protect all lunar material from terrestrial contamination in order to ensure the validity of later scientific investigations.[43]

The Chao Committee did not have the unquestioned support of MSC personnel, some of whom had reservations about the need for the committee to even exist. In a November 1964 meeting that took place between NASA OSSA's Foster and MSC staff members Elbert King, James McLane, and Donald Flory, the MSC personnel expressed their belief that the ad hoc committee should function only in an advisory capacity and should definitely not have the final word on LSRL design criteria. MSC needed to own that function. In addition, they felt that the ad hoc committee should not determine where LSRL would be built.[44]

Various organizations tried to position themselves to host LSRL. Goddard Space Flight Center (GSFC) declared that it should be in the running, since other "data bank"[45] facilities would possibly be constructed at GSFC as well. Several university laboratories expressed interest in having LSRL at their facilities and, in the opinion of John Eggleston, MSC's Assistant Chief for Space Environment, Ed Chao envisioned locating LSRL at the U.S. Geological Survey facility in Flagstaff, Arizona. The Chao Committee, however, eventually recommended MSC as the best site for LSRL.[46] The Chao Committee also reported the intense concern in the U.S. scientific community regarding NASA's presumed intention to use LSRL "as the basis for a major in-house research program."[47] Such an action would have directly threatened the domains of scientists outside of NASA, who strongly wanted the detailed lunar sample analyses to be assigned to them.

---

43. Ad Hoc Committee, *Concepts.*
44. Mangus and Larsen, *Lunar Receiving Laboratory Project History,* p. 9.
45. Ibid., p. 8.
46. Ibid., p. 14.
47. Carter, *The Prevention of Back Contamination,* pp. 16–17.

The Hess Committee's Vision for a Receiving Facility

Homer Newell, NASA's Associate Administrator for Space Science and Applications, decided that in order to design a suitable LSRL, an independent, non-NASA assessment was needed of the appropriate types of analyses to conduct on lunar samples upon their return to Earth as well as of the facilities and staffing required. In December 1964, Newell wrote to NAS SSB chairperson Harry H. Hess, requesting SSB's judgment on these issues. Hess formed a committee composed of three SSB members and two scientists from academia. One of the scientists, Clifford Frondel of Harvard, also served on the Chao Committee. The Hess Committee (also known as the Ad Hoc Committee on the Lunar Sample Handling Facility) met in Washington, DC, on 14 January 1965 to examine Newell's concerns and considered four key questions:

- What types of analyses needed to be performed immediately upon return of samples from the Moon?
- What types of research were more appropriate to postpone until samples could be sent to the best available research laboratory?
- What types of LSRL capabilities would be needed for its analyses?
- What personnel would LSRL require?[48]

The Hess Committee gave recommendations in its 2 February 1965 report on fulfilling both scientific and planetary protection objectives. The report concluded that the only critical scientific analysis that needed to be performed by LSRL, rather than by investigators to whom the samples would eventually be sent, was of the short-lived radioactivity induced in lunar samples by cosmic rays. This parameter would have to be measured soon after the samples arrived on Earth because induced radioactivity rapidly drops to a minimal level. No other measurements were as urgent as this from the point of view of fulfilling science objectives. From the planetary protection perspective, however, public health authorities needed to immediately examine the lunar material in order to identify any harmful biological properties.

---

48. SSB, *Report of the Ad Hoc Committee on Lunar Sample Handling Facility*, 2 February 1965, an attachment to H. H. Hess to Homer E. Newell, 2 February 1965, LBJ SCHC, "1 January–31 March 1965" folder, box 76-11, LRL Chron. Files 1964–1965; Compton, *Where No Man Has Gone Before*, p. 43.

These biological examinations could be performed at LSRL, but they could also be performed at Army Biological Warfare Laboratories or U.S. Public Health Service facilities.[49]

The Hess Committee recognized a need that Newell had not asked it to evaluate: the requirement to quarantine lunar material until biological investigations could show them to be biologically harmless. In a letter written to Newell on 2 February 1965, Hess stated that the members of his committee had indeed determined that such isolation procedures were necessary. Notable in its absence was any recommendation that the quarantining be carried out at MSC. The report of the Hess Committee, in fact, stated that "we can see no inherently compelling reason why"[50] LSRL should be in Houston.

The report of the Hess Committee also estimated facility costs. Its recommended "relatively restricted-mission sample receiving laboratory for quarantine and biological measurement"[51] coupled with the necessary low-level counting facility would, the committee envisioned, cost a comparatively modest $2.5 million. For all other sample analyses, the Hess Committee recommended using existing laboratories rather than building new ones. Besides the budget savings that could accrue from such a plan, potential political benefits may also have played a part in the committee's recommendation. Hess chaired the SSB, a National Academy of Sciences organization concerned with the well-being and advancement of the U.S. scientific community. The SSB worried that NASA LSRL staff members might overstep their bounds and develop competing research programs with space scientists outside of the Agency. Thus, it was not surprising that the Hess Committee advocated a sample receiving laboratory with a limited mission.[52]

The committee held that the extensive analyses MSC was proposing would best be done by "our national scientific community"[53] rather than by LSRL or any other single group. While the committee did not recognize any compelling reason for locating LSRL at MSC, it did

49. SSB, *Report of the Ad Hoc Committee*; Compton, *Where No Man Has Gone Before*, p. 43.
50. SSB, *Report of the Ad Hoc Committee*.
51. H. H. Hess to Homer E. Newell, 2 February 1965, LBJ SCHC, "1 January–31 March 1965" folder, box 76-11, LRL Chron. Files 1964–1965; SSB, *Report of the Ad Hoc Committee*.
52. Mangus and Larsen, *Lunar Receiving Laboratory Project History*, p. 15.
53. Hess to Newell, 2 February 1965.

understand that doing so would add considerably to the capabilities of MSC, and this would not be a bad thing since MSC was to play a critical part in the Apollo program. But construction of a radiation-counting facility at MSC might prove to be more expensive than at other sites, due to the waterlogged soil and the need to locate such a facility deep underground to shield it from naturally occurring background radiation.

The committee recommended that LSRL, wherever it was located, be operated continuously rather than just during Apollo missions, so that it could conduct maximum research in as cost-effective a manner as possible. This, too, did not favor locating LSRL at MSC, because the committee suspected that MSC, with its "minimal scientific capability,"[54] would likely have to shut down LSRL between lunar voyages.

The Hess Committee's recommendation to NASA to quarantine all lunar samples until biological analyses showed that they contained no harmful living organisms surprised staff at both NASA Headquarters and MSC, who had not totally accepted that preventing back contamination was such a serious concern. Although NASA's Director of the Biosciences Program in the OSSA had been communicating with NAS and PHS, he had not been expecting the importance that the Hess Committee gave to potential back contamination dangers presented by lunar samples. His surprise was in large part due to the generally held belief in the scientific community that the Moon was sterile.[55]

●●●

The U.S. Public Health Service Involvement in Back Contamination Issues

On 18 March 1965, a report published by PHS's Communicable Disease Center (later renamed the Centers for Disease Control and

---

54. Compton, *Where No Man Has Gone Before*, p. 43.
55. Compton, *Where No Man Has Gone Before*, pp. 43–46.

Prevention [CDC]) in Atlanta, Georgia,[56] added its opinion to the back contamination debate by recommending specific protective procedures for handling and quarantining lunar samples.[57] In addition, the Surgeon General of the United States, whose office oversees PHS,[58] met casually with NASA's Administrator to discuss the two agencies' relationship in preventing back contamination and had proposed the following actions:

- Continue to expand PHS's space biology program.
- Assign a PHS liaison office to NASA.
- Develop, through a NASA-PHS partnership, an interagency committee to study and give guidance on matters of "outbound and inbound contamination problems."[59]

The Administrator and Surgeon General reportedly agreed that negotiations between the two agencies needed to continue at the staff level. In order to kick off this process, Lawrence B. Hall, whom PHS sent on assignment to NASA's Bioscience Program in the OSSA to serve as Special Assistant for Planetary Quarantine, drafted a proposal embodying the three items above as well as a NASA organizational structure that could meet the goals of both agencies.[60]

In April 1965, Hall examined another aspect of lunar exploration— the jurisdictional issues of bringing extraterrestrial material into the country. In particular, Hall asked counsel for NASA's Office of Space Medicine to advise him on the statutory authority of PHS, the Department of Agriculture, the Department of the Interior, and NASA itself to manage and control the import of extraterrestrial material into the United

56. National Archives, "Records of the Centers for Disease Control and Prevention," *http://www.archives.gov/research/guide-fed-records/groups/442.html* (accessed 26 January 2011). Online version is based on *Guide to Federal Records in the National Archives of the United States* (Washington, DC: National Archives and Records Administration, 1995).
57. Committee on Planetary and Lunar Exploration, "A History of the Lunar Receiving Laboratory."
58. U.S. Dept. of Health and Human Services, "Office of the Surgeon General," U.S. Dept. of Health and Human Services Web site, *http://www.surgeongeneral.gov/aboutoffice.html* (accessed 12 December 2005).
59. O. E. Reynolds to Associate Administrator for Space Science and Applications, "Status of the Public Health Service—National Aeronautics and Space Administration negotiations on back contamination," 10 May 1965, folder 19.4.1.2, "Lunar Receiving Lab 4715," NASA Historical Reference Collection.
60. Benny B. Hall for Orr E. Reynolds, to Director of Personnel Division, "Detail of Sanitary Engineer Officer, U.S. Public Health Service, to Office of Space Science and Applications, National Aeronautics and Space Administration, renewal of," 14 July 1965, folder 19.4.1.2, "Lunar Receiving Lab 4715," NASA Historical Reference Collection; O. E. Reynolds, "Status."

States.[61] This was an important matter to establish, because if an agency had such authority, it could refuse to admit lunar samples, the spacecraft, and possibly even astronauts into the country until they were subjected to required examinations and quarantines. During a meeting on 17 August 1965, NASA and PHS agreed that PHS had the responsibility "for the health of the nation and for any potential threat to that health from extraterrestrial life, particularly from back contamination."[62] Furthermore, the Department of Agriculture had a similar responsibility in matters concerning U.S. crops and animals of economic importance. The Department of the Interior's (DOI's) Fish and Wildlife Service also had a potential concern with the matter but was not, NASA and PHS believed, interested in pressing for a part in planetary protection activities.

PHS's Office of the Surgeon General agreed to submit a proposal to the Administrator of NASA outlining the planetary protection actions it deemed necessary to protect against extraterrestrial health threats. Furthermore, the chief of PHS's CDC, James L. Goddard, affirmed that he "was prepared to staff any required quarantine activity"[63] at LSRL. NASA initiated actions to bring PHS representatives into regular contact with the Headquarters Office of Manned Space Flight and Houston's Manned Spacecraft Center, so as to provide needed planning input for the Apollo mission as soon as possible.

One of these contacts took place on 27 September 1965, when representatives from PHS as well as the Department of Agriculture met with those from the NASA Headquarters Office of Manned Space Flight and Houston's MSC for "an informal exchange of views . . . on the subject of back contamination."[64] Elbert King of MSC's Space Environment Division, after discussing quarantine issues with medical specialists

61. Lawrence B. Hall to Counsel to OMSF, "Statutory Authority for Planetary Quarantine on the Part of the Public Health Service, Department of Agriculture, Department of Interior, and the National Aeronautics and Space Administration," 22 April 1965, folder 19.4.1.2, "Lunar Receiving Lab 4715," NASA Historical Reference Collection.
62. Orr E. Reynolds to Record, "Summary of Meeting Between Representatives of the National Aeronautics and Space Administration and the Public Health Service, July 31, 1965," 17 August 1965, JSC Historical Archives, "1 April–30 September 1965" folder, box 76-11, LRL Chron. Files 1964–1965.
63. Reynolds to Record, "Summary of Meeting."
64. W. W. Kemmerer, Jr., to Record, "Summary of Meeting Between Representatives of the National Aeronautics and Space Administration, Public Health Service and the Department of Agriculture, MSC, Houston, Texas, September 27, 1965," 30 September 1965. This is an

from MSC, put forward a view that certain conditions on the Moon probably sterilized samples taken from its surface and rendered them less hazardous to life on Earth than samples taken from below the surface. These sterilizing conditions included the following:

- Intense ultraviolet radiation.
- High-energy protons from the solar wind.
- High surface temperatures.
- Low atmospheric temperatures.
- Absence of free water.[65]
- Hypervelocity impacts of small meteoroids.[66]

If the surface of the Moon was indeed sterile, then back contamination precautions could be altered. Astronauts and samples that had come into contact only with the lunar surface could be handled "less stringently than subsurface samples."[67] This would have placed fewer restrictions on the scientific research performed on the samples, which many scientists both within and outside of NASA would have appreciated. The cost of building LSRL might also have been lowered. But MSC's opinion that samples collected on the lunar surface presented low risks was not shared by representatives from the other agencies. PHS representative and chief of the CDC James Goddard[68] questioned King's assumptions and suggested that there might be places on the Moon's surface, such as in sheltered areas, where microorganisms could survive. Since no one at the meeting could refute this, Goddard pushed for strict quarantining of *all* lunar material. The importance of the issues, Goddard said, justified this action "even if it cost $50 million to implement an effective quarantine."[69] Goddard as well as other non-NASA personnel at the meeting agreed

---

attachment to Aleck C. Bond to Deputy Director, "Lunar Sample Receiving Laboratory," 30 September 1965, JSC Historical Archives, "1 April–30 September 1965" folder, box 76-11, LRL Chron. Files 1964–1965.

65. Kemmerer to Record, "Summary of Meeting"; Compton, *Where No Man Has Gone Before*, pp. 46–47.

66. W. W. Kemmerer, Jr., and E. A. King to Maxime A. Faget, "Proposed MSC Quarantine Policy for the Apollo Program," 23 September 1965, JSC Historical Archives, "1 April–30 September 1965" folder, box 76–11, LRL Chron. Files 1964–1965.

67. Kemmerer to Record, "Summary of Meeting."

68. J. H. Allton, J. R. Bagby, Jr., and P. D. Stabekis, "Lessons Learned During Apollo Lunar Sample Quarantine and Sample Curation," *Advances in Space Research* 22 (1998): 374.

69. Compton, *Where No Man Has Gone Before*, p. 47.

that the general quarantine period for all samples brought back from the Moon needed to be at least 30 days, and more if LSRL found any biologic evidence.

Other points that non-NASA personnel agreed on included the following:

- Potential contamination of astronauts presented as great a hazard as lunar samples. Appropriate quarantine periods discussed were from 21 to 30 days.
- Astronauts needed to be trained in using sterile techniques and in the principles of microbiology.
- A high-hazard terrestrial pathogen was an appropriate model to use in designing LSRL's protective systems.
- The LSRL quarantine area had to be capable of housing all personnel entering it should they become exposed to lunar material. LSRL also needed the capability to function as a hospital in the event personnel fell ill.
- Biologic findings were likely to result from accidental contamination with Earth microbes, rather than from detection of actual lunar microbes.[70]

The then-current concepts for an LSRL appeared adequate for conducting initial plant and animal testing involving exposure to lunar material, and the Department of Agriculture representatives thought that more extensive testing would not be necessary unless evidence was found of the presence of unusual organisms. If this happened, additional investigations could be carried out at other locations.[71] In particular, Department of Agriculture representatives reached a consensus that LSRL did not need to provide large animal testing facilities, but that if life was detected in the Moon samples, a portion of them should be sent to the USDA Plum Island Animal Disease Center, which did have such facilities.[72]

---

70. Kemmerer to Record, "Summary of Meeting."
71. Kemmerer to Record, "Summary of Meeting."
72. Lawrence B. Hall to Deputy Administrator, "Informal Conference on Back Contamination Problems," 15 October 1965, JSC Historical Archives, "1 October–31 December 1965" folder, box 76-11, LRL Chron. Files 1964–1965; USDA, "USDA and DHS Working Together," http://www.ars.usda.gov/main/site_main.htm?modecode=19-40-00-00&pf=1&docid=3833&cg_id=0 (accessed 27 March 2006).

Participants at the 27 September 1965 meeting came to a consensus that, in spite of precautions that might be taken, astronauts would not be able to "completely escape respiratory and dermal exposure to lunar contaminants and . . . the inside of the command module would receive a low but significant level of contamination."[73] A heated discussion ensued over the lack of plans up until that time for handling the astronauts after their return to Earth. Meeting participants examined in some depth the types of facilities that would be needed on the several recovery ships. The conferees also defined to a certain extent what was unacceptable treatment of returning space travelers by raising the hypothetical question: If Apollo astronauts were handled in the same way as the Gemini astronauts had been, would PHS permit them to enter the United States? Gemini 9 astronauts, for instance, who had not made a landing on another celestial body, did not wear biological isolation garments (BIGs) when they came aboard the prime recovery ship, the aircraft carrier USS *Wasp*.[74] This would not be acceptable for Apollo astronauts who had landed and walked on the Moon. PHS, after considering the question, "emphatically replied that they would refuse such entry"[75] to the United States for Apollo astronauts if they only followed the precautions that Gemini astronauts had.

Those at the 27 September 1965 meeting also discussed the problem of contaminating the Moon with terrestrial microbes. NASA needed to avoid such an occurrence if possible because, in the view of the meeting's participants, "this could lead to confusion in the examination of returned samples and possibly prolong the length of sample and astronaut quarantine."[76]

Lawrence Hall, NASA's Planetary Quarantine Officer, thought that one particularly useful result of the 27 September 1965 meeting was that it informed MSC staff of the points of view of U.S. regulatory agencies on planetary protection matters. He felt, however,

73. Hall to Deputy Administrator, "Informal Conference on Back Contamination Problems."
74. Apollo Explorer, "Gemini 9-A Astronauts Welcomed Aboard USS Wasp," Gemini Project, Apollo Explorer Web site, *http://www.apolloexplorer.co.uk/photo/html/gt9/10074392.htm* (accessed 27 April 2006). Piloted Gemini missions did not touch down on any other planet than Earth, so their astronauts did not need to be quarantined as did the Apollo astronauts.
75. Hall to Deputy Administrator, "Informal Conference on Back Contamination Problems"; Compton, *Where No Man Has Gone Before*, p. 47.
76. Kemmerer to Record, "Summary of Meeting."

that those agencies still did not understand the reasons for MSC's reluctance to address back contamination issues to the extent that the agencies wanted. One of the reasons was that MSC considered lunar surface material to most likely be sterile and thus not a source of back contamination. Another reason was that MSC's and NASA's main objective was to put an astronaut on the Moon before the end of the decade. They were much more focused on planetary exploration than on planetary protection and had concerns that excessively rigid planetary protection requirements could interfere with attaining their main objective. PHS's chief purpose, on the other hand, was to protect health.[77] Hall recommended that further informal exchanges between NASA and the regulatory agencies be scheduled to work out policies that were hopefully acceptable to both parties.[78] Such meetings were held, but throughout much of the Apollo mission, MSC and PHS continued to differ on how extensive the quarantine needed to be.

MSC staff came away from the 27 September 1965 meeting realizing that LSRL would need to be considerably larger and more expensive than had been projected. Although MSC questioned PHS's strict position regarding planetary protection, MSC did not have the data at that time to justify a lower level of quarantine. Facilities would have to be constructed in which astronauts and support staff could be quarantined for at least three weeks. Postmission briefings, which would be numerous and extensive, would have to be conducted through a biological barrier. In addition, recovery operations for the spacecraft, the astronauts, and samples would have to be far more complex than MSC and NASA had hoped. PHS officials believed that from the moment the Command Module (CM) splashed down in the ocean, exposure of its interior to the atmosphere of Earth had to be prevented. Astronaut crews would need to be immediately isolated, even in the rubber rafts that the recovery crews used. Once the astronauts reached the recovery ship, they would have to go straight into a mobile quarantine chamber.[79]

George Mueller, NASA Associate Administrator for Manned Space Flight, paid a visit to MSC on 13 October 1965 and received a briefing

---

77. Compton, *Where No Man Has Gone Before*, pp. 46–47.
78. Hall to Deputy Administrator., "Informal Conference on Back Contamination Problems."
79. Compton, *Where No Man Has Gone Before*, p. 47.

on the 27 September meeting. He directed MSC to propose a budget to Headquarters that incorporated the above-mentioned features of LSRL, including "isolation of astronauts and returned hardware during the quarantine period."[80]

MSC's Systems Engineering Division Chief, Owen Maynard, passed the word to all branches of his division that the Apollo missions had to implement a much stricter quarantine of spacecraft, astronauts, and samples than originally envisioned. It was unacceptable for his personnel to take the view that no changes could be made in Apollo mission plans, given present weight, cost, and schedule limitations. In Maynard's words, all Apollo staff were "morally obligated to prevent any possible contamination of the Earth"[81] and thus would have to find ways of conforming to PHS quarantine guidelines and incorporating additional planetary protection requirements into the mission. Preventing back contamination had become an *ethical* issue. It had also become a compelling political matter. Failure to take adequate planetary protection measures could seriously damage the public's image of the Apollo program as well as of NASA in general, hampering the Agency's future actions. These were additional reasons why NASA had to develop an effective quarantine program.[82]

PHS and NASA Roles and Actions

During the fall of 1965, United States Surgeon General William H. Stewart exercised his authority, demanding that Apollo mission astronauts and samples be quarantined. NASA decided that PHS needed to be given some responsibility in the matter and help design LSRL to effectively prevent back contamination. NASA and PHS initiated efforts

---

80. Walter W. Kemmerer and James C. McLane, Jr., "Lunar Sample Receiving Laboratory Office Staff Paper," undated, p. 3, in a package beginning with "Lunar Sample Receiving Laboratory Siting Factors," ca. May 1966, JSC Historical Archives, "May 1966" folder, box 76-12, LRL Chron. Files.

81. Owen E. Maynard to PS Branches, "Earth Contamination from Lunar Surface Organisms," PS8/M-194/65, 29 October 1965, JSC Historical Archives, "1 October–31 December 1965" folder, box 76-11, LRL Chron. Files 1964–1965.

82. Compton, *Where No Man Has Gone Before*, p. 47.

to "reach a formal agreement on cooperation."[83] On 15 November 1965, NASA Deputy Administrator Hugh L. Dryden wrote to the Surgeon General, expressing NASA's wish that PHS, whose mission was to protect the health of the United States, "should assume a leadership role in the anticipation of problems that man-in-space flight may bring to the earth."[84] Dryden proposed assigning a PHS officer as a liaison with NASA and establishing an interdepartmental advisory committee that understood the potential impact of extraterrestrial contamination of our planet's biosphere. Dryden's hope was that the committee would be chaired by a senior PHS scientist-administrator. Dryden asked Stewart for recommendations concerning the characteristics of the facilities, equipment, and staff needed. He also asked Stewart for help "in justifying to the Bureau of the Budget the urgency and necessity of this program."[85]

On 22 December 1965, Stewart responded directly to NASA Administrator James E. Webb, agreeing that a senior PHS officer be made a liaison between the two agencies and that an interagency committee be established to "guide NASA in the development and conduct of a program to avoid possible contamination of the earth biosphere by extraterrestrial life."[86] Stewart recommended that this committee be chaired by James Goddard, chief of the CDC. He also mentioned that PHS personnel were working with MSC to develop a Preliminary Engineering Report and budget materials for the envisioned LSRL and that he would be pleased to assist in justifying the envisioned planetary protection program to the Bureau of the Budget.

83. Lawrence B. Hall to Director, Manned Space Flight Program Control, "Quarantine Requirements—Lunar Landing Program," 4 November 1965, in packet of material beginning with Lawrence B. Hall to Director, Space Medicine, "Proposal from the Public Health Service to the National Aeronautics and Space Administration," 2 November 1965, JSC Historical Archives, "1 October–31 December 1965" folder, box 76-11, LRL Chron. Files 1964–1965.
84. Hugh L. Dryden to William H. Stewart, 15 November 1965, in package of material beginning with Frank A. Bogart to Deputy Director, Space Medicine, "Formulation of PHS-NASA Working Relationships Re Lunar Sample Receiving," 11 January 1966, JSC Historical Archives, "January 66" folder, box 76-12, LRL Chron. Files.
85. Ibid.
86. William H. Stewart to James E. Webb, 22 December 1965, in package of material beginning with: Frank A. Bogart to Deputy Director, Space Medicine, "Formulation of PHS-NASA Working Relationships Re Lunar Sample Receiving," 11 January 1966, JSC Historical Archives, "January 66" folder, box 76-12, LRL Chron. Files.

The Preliminary Engineering Report that Stewart mentioned had been initiated in late May of 1965. NASA had selected a Detroit architecture and engineering firm, Smith, Hynchman, and Grylls, to review MSC's preliminary plans for LSRL and identify whether any alternatives existed to constructing a new facility. On 1 October 1965, this firm submitted a draft two-volume report, which concluded that modifying one or more existing facilities to satisfy both scientific and quarantine requirements would cost as much as constructing a new facility and take as long a time. Also, after evaluating 29 possible sites for LSRL, the report recommended MSC as the best location. This was the same conclusion that the Chao Committee and the Apollo science planning teams had come to almost nine months earlier.[87]

A Revised Plan for LSRL

The MSC Technical Working Committee for the Design of the LSRL, which had been established on 14 June 1965, was made up primarily of engineers who had been involved in LSRL planning since early 1964. This committee updated the Preliminary Engineering Report to reflect new requirements that had been implemented since the firm of Smith et al. had begun work on it and submitted it to NASA on 21 October 1965 for approval as the Revised Preliminary Engineering Report. This report called for construction of an 86,000-square-foot facility at a cost of $9.1 million. An additional $7 million, approximately, would be needed to buy the equipment for the individual laboratories. The facility would require a staff of about 80 scientists and technicians and would cost $1.6 million per year to operate.[88]

Transport of Mission Components from Splashdown to LSRL

NASA awarded a contract to the MelPar Company of Falls Church, Virginia, to examine issues involved in transporting astronauts, spacecraft, and Moon samples from the planned splashdown point in

---

87. Carter, *The Prevention of Back Contamination*, pp. 22–24.
88. Ibid.

the Pacific Ocean to LSRL in a manner that would not contaminate Earth or the samples.[89] MelPar also designed and oversaw construction of the Mobile Quarantine Facility that housed the personnel, equipment, and material returning from the Moon during their trip to LSRL.

●●●

Formation of the Interagency Committee on Back Contamination

In January 1966, NASA began setting up the multiagency board that Surgeon General Stewart had recommended, calling it the Interagency Committee on Back Contamination (ICBC). During the months that followed, NASA invited a range of federal agencies and organizations to bring their own points of view and expertise to the new board and eventually brought the organizations and individuals listed in table 4.1 onto it. In a letter from NASA Administrator Webb to Secretary of Agriculture Orville L. Freeman, Webb cited the importance of each group's specialized knowledge and experience in helping to protect Earth's living resources as well as the integrity of lunar samples and scientific experiments.[90]

---

89. Carter, *The Prevention of Back Contamination*, p. 23.
90. NASA, Management Instruction NMI 1052.90, regarding protection of Earth's biosphere from lunar sources of contamination, 24 August 1967, folder 009901, "Lunar Quarantine and Back Contamination," NASA Historical Reference Collection; James E. Webb to Orville L. Freeman, 28 February 1966, folder 006696, "Sterilization/Decontamination," NASA Historical Reference Collection; NASA memo for record, "Interagency Committee on Back-Contamination," 1 March 1966, folder 006696, "Sterilization/Decontamination," NASA Historical Reference Collection; Gerald E. Meloy, memo for record, "Interagency Committee on Back-Contamination," 1 March 1966, folder 006696, "Sterilization/Decontamination," NASA Historical Reference Collection.

Table 4.1 Membership of the Interagency Committee on Back Contamination.[91]

Department of Health, Education, and Welfare—Public Health Service
David Sencer, ICBC Chairman and Chief of CDC
John Bagby, Assistant Chief of CDC
Department of Agriculture
Ernest E. Saulman, Deputy Director of Science and Education
Archibald J. Park (alternate), Assistant to the Administrator, Agricultural Research Service
Department of the Interior
John Buckley, Office of Science Adviser
Howard H. Eckles (alternate), Assistant to Science Adviser
National Academy of Sciences
Wolf Vishniac, Department of Biology, University of Rochester
Allen Brown (alternate), Department of Physiology, University of Pennsylvania
NASA Ames Research Center
Harold P. "Chuck" Klein, Assistant Director for Life Sciences
Richard S. Young (alternate), Chief, Exobiology Division
NASA Manned Spacecraft Center (MSC)
Charles A. Berry, Director, Medical Research and Operations
Walter Kemmerer (alternate), Chief, Biomedical Specialties Branch
Aleck C. Bond, Manager, Systems Test and Evaluation
James C. McLane (alternate), Chief, Facilities Requirements Office of the Lunar Receiving Laboratory Program Office
G. Briggs Phillips, Liaison, Lunar Receiving Laboratory Program Office
NASA Headquarters
Leonard Reiffel, Apollo Program Office
Col. John E. Pickering, Director, Lunar Receiving Operations, Office of Space Medicine

James Goddard, the Surgeon General's choice for chairperson of ICBC, got the opportunity to serve as Commissioner of the Food

---

91. NASA memo for record, "Interagency Committee on Back Contamination," JSC Historical Archives, "May 1966" folder, box 76-12; Margaret S. Race, "Anticipating the Reaction: Public Concern About Sample Return Missions," part of the Mars Library, California Space Institute, Univ. of California at San Diego (UCSD), 9500 Gilman Drive, Dept. 0524, La Jolla, CA 92093-0524, 1999–2000 (copyright date), Mars Now Web site, *http://calspace. ucsd.edu/marsnow/library/mars_exploration/robotic_missions/landers/sample_return/ public_information1.html*; Committee on Planetary and Lunar Exploration, "A History of the Lunar Receiving Laboratory"; Mangus and Larsen, *Lunar Receiving Laboratory Project History*, p. 12; Compton, *Where No Man Has Gone Before*, p. 16.

and Drug Administration,[92] and thus on 4 February 1966, Stewart selected the new CDC Chief, David Sencer, for the position. At that meeting, Sencer chose John Bagby, the Assistant Chief of CDC and the other PHS delegate on ICBC, to represent PHS's views in support of LSRL during upcoming congressional hearings.[93] Bagby's argument to Congress to support construction of LSRL will be reviewed in the next section.

MSC assigned two regular delegates to ICBC—Charles Berry and Aleck Bond. Berry was concerned primarily with the biomedical and quarantine aspects of LSRL, while Bond focused mainly on the impact of quarantine measures on LSRL design, development, and operation.[94]

The prime objective of ICBC was to advise and guide the Administrator of NASA on planetary protection matters, with a charter to protect public health, agriculture, and other resources "against the possibility of contamination by hypothetical lunar organisms conveyed in returned sample material or other material exposed to the lunar surface (including astronauts) and to preserve the biological and chemical integrity of lunar samples."[95] ICBC was also to aid in protecting the validity of scientific experiments and ensure against compromising the operational aspects of the Apollo program. ICBC was formed to advise on policy, not on technical details, although the complexities of the Apollo mission required that it examine many technical matters. As plans for the Apollo mission developed, ICBC became intimately involved in many of the decisions regarding controls for preventing back contamination, quarantine protocols, and isolation facilities. It helped establish specific criteria for quarantine operations, determine the lunar material analyses needed in order to

---

92. Letter to Dr. William H. Stewart, prepared by H. S. Brownstein (name of signatory unclear), January 1966 (day unclear), in the package of materials beginning with NASA, "Interagency Committee on Back Contamination," JSC Historical Archives, "May 1966" folder, box 76-12.

93. Jack Bollerud to Director, MSF Field Center Development, "Public Health Service Proposed Congressional Statement in Support of the NASA Lunar Sample Receiving Facility," 14 February 1966, JSC Historical Archives, "February 1966" folder, box 76-12.

94. George M. Low to NASA Headquarters, "Manned Spacecraft Center Representation on the Interdepartmental Advisory Committee for Back Contamination Problems," 31 January 1966, JSC Historical Archives, "January 1966" folder, box 76-12.

95. Commission on Physical Sciences, Mathematics, and Applications (CPSMA) and SSB, "Lessons Learned from Apollo," in "Program Oversight," in *Mars Sample Return: Issues and Recommendations* (National Academies Press: Washington, DC, 1997), Chap. 8, http://www.nap.edu/books/0309057337/html/34.html (accessed 26 January 2011).

evaluate planetary protection risks, and design contingency plans to handle unintended releases from the quarantine facility.[96]

The MSC-PHS-Army "Apollo Personnel Quarantine Facilities Review"

Building quarantine capabilities into LSRL would, as the Preliminary Engineering Report drafts indicated, increase the facility's cost by millions of dollars. George Mueller, head of NASA's Office of Manned Space Flight, was concerned about the rising costs of LSRL and how to justify them to Congress. He recognized the need for quarantining, but still questioned whether a new, dedicated quarantine facility needed to be built, or whether a more cost-effective approach could be found.

On 22 November 1965, Mueller sent a teletype to James C. McLane of MSC's LSRL Office requesting a survey of "existing US facilities with a potential for related personnel quarantine"[97] that might possibly accommodate returning Apollo astronauts and their support crews. McLane responded the next day that MSC had already initiated the study, was conducting it in cooperation with PHS and the U.S. Army,

96. NASA, "Apollo History," *Proceedings of the Planetary Protection Workshop Held in Washington, DC, June 23–24, 1988*, p. 1, folder 006695, "Sterilization/Decontamination," NASA Historical Reference Collection; Richard S. Johnson, John A. Mason, Bennie C. Wooley, Gary W. McCollum, and Bernard J. Mieszkuc, "The Lunar Quarantine Program," in *Biomedical Results of Apollo* (Washington, DC: NASA SP-368, 1975), Chap. 1, http://history.nasa.gov/SP-368/sp368.htm (accessed 26 January 2011); Race, "Anticipating the Reaction"; Committee on Planetary and Lunar Exploration, "A History of the Lunar Receiving Laboratory"; Mangus and Larsen, *Lunar Receiving Laboratory Project History*, p. 12; Compton, *Where No Man Has Gone Before*, p. 16; Robert Seamans draft document, in package beginning with NASA, "Interagency Committee on Back Contamination," JSC Historical Archives, "May 1966" folder, box 76-12.

97. Deputy Director, Space Medicine, to James C. McLane, 22 November 1965, in a collection of papers beginning with Maxime A. Faget to NASA Headquarters (Attn: Col. Jack Bollerud), "Study of Existing Personnel Quarantine Facilities," 8 December 1965, JSC Historical Archives, "1 October–31 December 1965" folder, box 76-11, LRL Chron. Files 1964–1965.

and would produce a final report by 3 December 1965.[98] The title of the report was "Apollo Personnel Quarantine Facilities Review."[99]

Other organizations assisting in the study—some of which owned facilities that were examined for their quarantine capabilities—included representatives from the military (the U.S. Air Force and Navy), the academic community (the University of Minnesota and the University of Notre Dame), industry (Pitman-Moore Division of the Dow Chemical Company), the American Medical Association, and the National Institutes of Health. Some of the facilities examined, including large environmental chambers, pharmaceutical industrial plants, and aerospace life science laboratories, were quickly rejected due to "grossly evident space or containment inadequacies."[100] The study committee selected for more careful evaluation 12 facilities that it felt had the best potential for personnel biocontainment. The committee analyzed these 12 facilities using the criteria listed below as measures of suitability for quarantining Apollo astronauts, samples, and equipment:

- Presence of a barrier wall effective for biocontainment.
- Pass-through locks and autoclaves.
- Sufficient filtration capability for incoming air.
- Negative pressure maintenance of facility.
- Effluent heat treatment capability.
- Habitability for up to 15 people for 60 days.
- Sufficient area for medical examinations and specialized medical equipment.
- Layout compatible with security requirement.
- Observation provisions for debriefings and medical surveillance.
- Capability for handling and culturing high-hazard, pathogenic fungi, bacteria, and viruses.

---

98. James C. McLane to NASA Headquarters (Attn: H. S. Brownstein), 23 November 1965, JSC Historical Archives, "1 October–31 December 1965" folder, box 76-11, LRL Chron. Files 1964–1965.
99. W. W. Kemmerer, "Apollo Personnel Quarantine Facilities Review," December 1965, in a collection of papers beginning with Maxime A. Faget to NASA Headquarters (Attn: Col. Jack Bollerud), "Study of Existing Personnel Quarantine Facilities," 8 December 1965, JSC Historical Archives, "1 October–31 December 1965" folder, box 76-11, LRL Chron. Files 1964–1965.
100. Ibid.

Table 4.2 summarizes the capabilities of the 12 facilities analyzed. The three hospitals in the table were those that provided the highest level of bio-isolation available in the country.[101]

Table 4.2 Candidate facilities for lunar crew quarantine.[102]

Facility	Location	Cost To Modify Facility for Apollo Use	Comments on Suitability
U.S. Army medical facility	Ft. Detrick, MD	1	Care facility for personnel working with high-hazard pathogens. Met most suitability criteria, but its use in the Apollo program would severely compromise its DOD mission.
Naval Biological Laboratory	Oakland, CA	2	Did not meet any suitability criteria.
Naval Institute of Health Clinical Center	Bethesda, MD	2	Primarily a hospital research facility. Had no biobarrier, pass-through locks; air filtration systems; negative-pressure capabilities; or isolated waste treatment.
U.S. PHS Hospital	Staten Island, NY	2	Facilities for three people only.
Cook County Hospital	Chicago, IL	2	General hospital facility. Had no biobarrier, pass-through locks, air filtration systems, negative-pressure capabilities, or isolated waste treatment.
Belmont Hospital	Worcester, MA	2	General hospital facility. Had no biobarrier, pass-through locks, air filtration systems, negative-pressure capabilities, or isolated waste treatment.
U.S. PHS Hospital	Carville, LA	2	Leprosy facility that was not designed for highly infectious agents.
University of Notre Dame (Lobund germ-free laboratory)	South Bend, IN	2	Animal facilities only.

101. Ibid.
102. Ibid.

Facility	Location	Cost To Modify Facility for Apollo Use	Comments on Suitability
National Institutes of Health National Center	Bethesda, MD	2	Designed to isolate single patients from environment; *not* designed to protect environment from patients.
Ames Research Center	Moffett Field, CA	2	Did not meet any suitability criterion.
U.S. Air Force School of Aerospace Medicine	San Antonio, TX	2	Animal facilities only.
U.S. Dept. of Agriculture	Washington, DC	2	Designed for animals only.

Legend
1 = Major cost
2 = Cost approaching that to build a new facility

Note from the table that some of the facilities had been designed for housing animals rather than people. Only one facility came close to the level of biocontainment and accommodations that were needed for Apollo: the Fort Detrick, Maryland, Army hospital. It was equipped to isolate and care for personnel who had handled highly risky microorganisms such as in Fort Detrick's biological warfare facilities (which became the U.S. Army Medical Research Institute of Infectious Diseases [USAMRIID] in 1969). Converting the Fort Detrick hospital to accommodate the needs of the Apollo program would have required major expense and significant new construction, but this would have been cheaper than converting the other facilities listed in the table, which would have required expenses nearly equal to those for constructing a brand new facility. A downside of using Fort Detrick for the Apollo program, however, was that this would seriously impede the Army's own occupational medicine research programs that it was conducting there. Thus, none of the existing sites in the MSC study could be used for LSRL without serious barriers to overcome.[103]

---

103. Norman M. Covert, "A National Resource," in *Cutting Edge: A History of Fort Detrick, Maryland* (4th ed.) (Fort Detrick, MD: Fort Detrick Public Affairs Office, October 2000), Chap. 12, *http://www.detrick.army.mil/cutting_edge/chapter12.cfm* (accessed 26 January 2011); Compton, *Where No Man Has Gone Before*, p. 48; Kemmerer, "Apollo Personnel Quarantine."

Different functions of LSRL could have been distributed among various facilities around the country, such as those examined above. But the "Facilities Review" pointed out that proximity of astronauts, samples, and spacecraft would make security and operational requirements far easier to attain and would minimize travel requirements for medical, engineering, and scientific staff. The "Facilities Review" also made a case for locating LSRL at MSC, arguing that this would allow monitoring and operation of the quarantine areas for samples, astronauts, and spacecraft by a relatively small resident group of NASA, PHS, and Department of Agriculture personnel and would eliminate considerable duplication of effort. This same argument could have been used, however, to justify locating all LSRL functions at another site. Nevertheless, the study's findings gave Mueller some justification for seeking a budget from Congress to build a new LSRL facility rather than trying to use an existing site.

●●●

Winning Congressional Support

In February 1966, John Bagby, representing the points of view of PHS's CDC and ICBC, wrote a strong statement to Congress providing reasons for including biocontainment capabilities in LSRL. This statement was included in the record of the 24 February 1966 hearing of the House Subcommittee on Manned Space Flight. Bagby acknowledged that the chances were extremely low that the lunar environment could support life as we know it, but he also stressed that we had no data assuring the absence of such life. His argument for biocontainment highlighted the dangerous lack of knowledge that space scientists had regarding the impacts of potentially life-bearing lunar material on our planet. There was the chance that if lunar organisms were destructive to Earth life and were set free in our ecosystem, they could cause a disaster of enormous proportions. Thus, the United States needed a facility that could reliably keep any extant life in the lunar samples from contacting our biosphere.[104]

Bagby made the point that "the single most important scientific objective of the Apollo missions is the return of the lunar samples"[105] and that the success of the experiments that would be performed on them depended heavily on preventing their contamination by terrestrial material. Thus, a quarantine facility was needed to prevent the flow of contamination in two directions—from the terrestrial environment to the samples and from the samples to Earth's ecosystem. Bagby also summarized for Congress the recommendations of the July 1964 SSB conference for handling returning astronauts, spacecraft, equipment, and samples, as well as the 15 March 1965 report of the NASA OSSA Ad Hoc Committee on the Lunar Sample Receiving Laboratory (the Chao Committee report),[106] which covered the concepts, requirements, specifications, and operation of LSRL. Finally, Bagby's presentation focused on the Preliminary Engineering Report that PHS as well as NASA MSC staff had worked on. This report identified necessary capabilities of LSRL, including the following:

- Microbiology testing of lunar samples.
- Isolation facilities for astronauts and potentially exposed personnel.
- Isolation capabilities for samples during all analyses.
- Isolated facilities for processing onboard camera film and data tape.

---

104. John R. Bagby, "Statement by the Assistant Chief, Communicable Disease Center, Public Health Service, DHEW, On the Containment of Lunar Samples, Astronauts, and Support Personnel," 11 February 1966, attached to Jack Bollerud to Director, MSF Field Center Development, "Public Health Service Proposed Congressional Statement in Support of the NASA Lunar Sample Receiving Facility," 14 February 1966, JSC Historical Archives, "February 1966" folder, box 76-12. Bagby's statement was included in the record of the U.S. House of Representatives, "Questions Submitted for the Record by the Subcommittee on Manned Space Flight During the Fiscal Year 1967 NASA Authorization Hearings," 24 February 1966 hearings, in Hearings before the Committee on Science and Astronautics, *1967 NASA Authorization*, 89th Cong., 2nd Sess., on HR 12718, pp. 418–421.

105. Ibid.

106. Ad Hoc Committee on the Lunar Receiving Laboratory, *Concepts, Functional Requirements, Specifications and Recommended Plan of Operation for the Lunar Sample Receiving Laboratory* (draft), NASA OSSA, 15 March 1965, an attachment to J. C. McLane, Jr., to J. G. Griffith, "Lunar Receiving Laboratory," 8 April 1965, LBJ SCHC, "1 January–31 March 1965" folder, box 76-11, LRL Chron. Files 1964–1965.

- Capability to perform time-critical analyses (such as radiation counting).[107]

Reduced Presidential and Congressional Support for the Space Program

NASA sought funding for LSRL during a time when support for human spaceflight was becoming increasingly difficult to acquire from Congress. This was in part due to the fact that President Lyndon Johnson was sinking large amounts of money into his Great Society social programs while at the same time trying to keep the U.S. budget under $100 billion. Meanwhile, the situation in Southeast Asia, and our country's involvement in it, was heating up. As a result, the budgets of other U.S. programs, including our space program, were getting slashed. NASA had requested $5.556 billion for fiscal year (FY) 1967; the Office of Management and Budget (OMB) recommended that it be cut to $5.012 billion, and President Johnson accepted the opinion of his budget advisers. As NASA prepared for its annual presentation to Congress, the chances of receiving even the reduced budget recommended by OMB seemed questionable.[108]

The basis of President Johnson's support for the Apollo mission had been strong but was steadily eroding. During the period from 1961 through 1963, while he was Vice President under Kennedy, Johnson's support for the space program rested heavily on issues of U.S. security. He made this position very clear in such statements as one he gave in May 1963, "I do not think this generation of Americans is willing to go to bed each night by the light of a Communist moon."[109] Johnson's eagerness to surpass the USSR in the race for space and his support for the U.S. space exploration effort remained solid during the first year of his presidency—after Kennedy's assassination in November

---

107. Ibid.
108. Michael Meltzer, *Mission to Jupiter: A History of The Galileo Project* (Washington, DC: NASA SP-2007-4231, 2007); Carter, *The Prevention of Back Contamination*, p. 26; Carter, "Moon Rocks and Moon Germs," 239; Compton, *Where No Man Has Gone Before*, p. 48.
109. Robert Dallek, "Johnson, Project Apollo, and the Politics of Space Program Planning," in *Spaceflight and the Myth of Presidential Leadership*, ed. Roger D. Launius and Howard E. McCurdy (Urbana and Chicago: University of Illinois Press, 1997), p. 75.

1963 until November 1964. Furthermore, in January 1964, Johnson told Congress that U.S. efforts to put a man on the Moon during the 1960s continued unchanged. But various U.S. security-related events eventually reduced his eagerness to back NASA's current space program. President Kennedy had emerged victorious from the Cuban missile crisis with the USSR in 1962; the United States steadily built up its own impressive arsenal of missiles; several nuclear powers—the USSR, the United Kingdom, and the United States—signed the *Limited Test Ban Treaty* in 1963, which prohibited nuclear tests in the atmosphere, underwater, and in space; and the USSR decided to de-emphasize its lunar landing program in favor of developing its near-Earth orbital capabilities.[110]

One of the main drivers for the race to the Moon had been the perceived threat from the USSR and its aggressive space program. But Johnson no longer worried that the Soviets were beating us in military might, especially missile capabilities, and the USSR was no longer threatening to land on and control the Moon.[111] President Kennedy had concurred as well with Johnson's view that the major danger had passed, especially regarding the space race. In fact, on the day that he was assassinated, Kennedy had prepared a speech that he planned to give at the Dallas Trade Mart, asserting that "we have regained the initiative in the exploration of outer space . . . the United States of America has no intention of finishing second in space," and that "there is no longer any fear in the free world that a Communist lead in space will become a permanent assertion of supremacy and the basis of military superiority."[112]

Congress, too, was changing its opinion of the magnitude of the Soviet threat. In the view of astronaut and later Senator John Glenn, appeals to the dangers presented by the USSR no longer swayed

---

110. U.S. Department of State, "Treaty Banning Nuclear Weapon Tests in the Atmosphere, in Outer Space and Under Water," *http://www.state.gov/t/ac/trt/4797.htm* (accessed 8 July 2005).
111. Carter, *The Prevention of Back Contamination*, pp. 25–26.
112. John F. Kennedy, "Remarks Prepared for Delivery at the Trade Mart in Dallas," Selected Speeches, John Fitzgerald Kennedy Library-Columbia Point, Boston, MA, 22 November 1963, *http://www.jfklibrary.net/j112263b.htm* (accessed 3 April 2006); Dallek, "Johnson, Project Apollo, and the Politics of Space Program Planning," pp. 74–75.

Congress, as they had in the past, to increase space spending. In Glenn's words, "the anti-Russian theme had worn out."[113]

The Competition for Limited Funds: NASA's Struggle To Win Congressional Approval

In 1965, the war in Vietnam and its costs began to escalate sharply. Those costs as well as President Johnson's antipoverty and Great Society programs had significant impacts on the 1965 and subsequent federal budgets.[114] This made it increasingly difficult for NASA to get its budget requests passed. NASA had its first congressional hearing for its fiscal year 1967 budget request on 18 February 1966 in front of the Subcommittee on Manned Space Flight of the House Committee on Science and Astronautics. This committee had been authorized in 1958 primarily to provide oversight of NASA and the nonmilitary space program.[115]

When NASA Administrator James Webb presented the Agency's FY 1967 budget authorization request, 60 percent of which was for human spaceflight, he found that the House subcommittee was actually sympathetic to most of the Apollo program's funding requests. In spite of tight budgets, many seasoned politicians in Congress appeared to be "completely swept up in the pure romanticism of landing a man on the moon."[116] Nevertheless, the House subcommittee gave the proposed LSRL budget an exceptionally thorough examination. It was in this subcommittee that "the majority of the battle over funding for the receiving lab would be fought."[117]

---

113. Dallek, "Johnson, Project Apollo, and the Politics of Space Program Planning," p. 75.

114. Ibid., p. 81.

115. Charles E. Schamel, Mary Rephlo, Rodney Ross, David Kepley, Robert W. Coren, and James Gregory Bradsher, "Records of the Committee on Science and Astronautics," in *Guide to the Records of the U.S. House of Representatives at the National Archives, 1789–1989* (Record Group 233), Legislative Branch—Center for Legislative Archives (Washington, DC: National Archives and Records Administration, 1989), Chap. 19, *http://www.archives. gov/legislative/guide/house/chapter-19.html* (accessed 6 April 2006); Carter, "Moon Rocks and Moon Germs," 239.

116. John Shedletsky, "Then We Must Be Bold: The Rhetoric of Putting a Man on the Moon," John Shedletsky's Program in Writing and Rhetoric, 30 May 2002, *http://www.stanford. edu/~jjshed/portfolio/PWRpaper1-7.htm*.

117. Carter, *The Prevention of Back Contamination*, p. 28.

NASA fortunately had a strong ally on the House subcommittee—its chairman, Olin "Tiger" Teague, whose Texas district included MSC. Teague was a conservative Democrat and an ex-Marine fighter pilot who had been highly decorated in World War II. He was known for his directness, determination, and "heavy emphasis on educating the Congress and the public on the practical value of space."[118] He strongly supported human spaceflight and had been a major influence in Congress's selection of Houston as the location of MSC. But the House subcommittee also included representatives not nearly so friendly to NASA's endeavors and, in particular, to its plans for LSRL.[119]

The House Subcommittee on Manned Space Flight—18 and 24 February 1966 Hearings

At the 18 February hearings, NASA gave a prepared statement on its budget request. This statement mentioned LSRL funding only once, indicating that this facility was meant to "house all activities and functions pertinent to receipt, processing and distribution of samples of lunar materials, as well as the quarantine of lunar exploration teams and spacecraft . . . "[120]

The subcommittee briefly discussed back contamination issues, with NASA Deputy Administrator Robert C. Seamans assuring those at the hearing that NASA would take all precautions to prevent it. NASA reiterated and expanded on this position in response to questions that the subcommittee submitted for the record as part of its 24 February 1966 hearings. NASA drew from the recommendations of the February 1965 NAS SSB conference, "Potential Hazards of Back Contamination,"[121] which stated that although the probability of find-

---

118. House Committee on Science, "A History of the Committee on Science," http://www.house.gov/science/committeeinfo/history/index.htm (accessed 4 April 2006).

119. Mangus and Larsen, Lunar Receiving Laboratory Project History, p. 18.

120. U.S. House of Representatives, 18 February 1966 hearings in Hearings before the Committee on Science and Astronautics, 1967 NASA Authorization, 89th Cong., 2nd Sess., on HR 12718, p. 248.

121. Allan H. Brown, (chairman), "Potential Hazards of Back Contamination from the Planets—29–30 July 1964," Space Science Board of the National Academy of Sciences (19 February 1965), JSC Historical Archives, "1 January–31 March 1965" folder, box 76-11, LRL Chron. Files 1964–1965.

ing viable organisms on the Moon was extremely low, the ecological balance of Earth was at risk, and so it was mandatory to take the necessary steps to prevent back contamination.[122]

No serious opposition to NASA's planetary protection plans were presented at the February hearings. But that situation changed the next month.

The House Subcommittee—1 March 1966 Hearings

The House Subcommittee on Manned Space Flight hearings continued through March. At the 1 March 1966 hearing, NASA received its first direct challenge on constructing LSRL. This challenge was led by James G. Fulton of Pennsylvania, the ranking Republican on the subcommittee and a longstanding critic of the way that NASA managed its programs. A continuing complaint of Fulton was that NASA tended to concentrate its operations geographically around its Centers, rather than spread out the work through more areas of the country. Fulton mentioned several reasons why NASA was implementing the wrong strategy. One aspect was that of national defense. He worried that if the Agency centered too many capabilities in one place, they could all be wiped out with a single weapon. He also held that centering scientific talent in too few places would drain talent from every other area. He wanted the talent distributed in order to raise the cultural level throughout the country. He thus questioned the Agency's intent to locate LSRL in Houston, already a Center for space programs. In addition, he suspected that LSRL would constitute an unnecessary duplication of existing facilities and actively tried to block its construction.

Fulton's position, as well as that of his Republican colleague on the subcommittee, Donald Rumsfeld of Illinois, was that NASA should make use of the already existing quarantine and scientific research

---

122. U.S. House of Representatives, "Questions Submitted for the Record by the Subcommittee on Manned Space Flight During the Fiscal Year 1967 NASA Authorization Hearings," 24 February 1966 hearings, in Hearings before the Committee on Science and Astronautics, *1967 NASA Authorization*, 89th Cong., 2nd Sess., on HR 12718, pp. 417–421; Carter, *The Prevention of Back Contamination*, pp. 28–29.

facilities in other parts of the United States. Fulton gave the example of the University of Pittsburgh, which had already established a department of Moon geology.[123]

Both congressmen were worried that by constructing LSRL in Houston, NASA was trying to use planetary protection as a springboard for building an extensive, costly in-house research program rather than relying on the talents of scientists outside of NASA. Rumsfeld in fact stated his concern that "next year and [in] succeeding years NASA will come before this committee and request funds totaling millions of dollars to beef up this capability which we are being asked to begin this year. The argument will be that we have to keep it an integrated program, and the capability exists in Houston. NASA will say: 'We have this $9 million facility, therefore we need to add to it, add another $30, $40, or $100 million.'"[124]

Republicans also opposed Texas Democrats' channeling still more federal funds into their construction projects, such as would be the case if LSRL was built at MSC.[125] Fulton expressed the view that if an integrated LSRL was needed (and he was skeptical that it was), then a better location for it would be in a place that was not receiving its fair share of space exploration contracts, such as regions in the Midwest. Even though Fulton's home state of Pennsylvania was not in the Midwest, his arguments highlighted the fact that members of Congress were motivated to obtain contracts and income for their home states and that NASA had to be aware of these motivations and perhaps play to them if they were to win congressional support.[126]

George Low, Deputy Director of MSC, and William Lilly, Director of NASA Manned Space Flight Program Control and a budget specialist, attempted to address the concerns raised by subcommittee members, but with little effect. Neither Low nor Lilly had come prepared with a detailed defense against the attacks. Low did point out that the Preliminary Engineering Report as well as the Chao Committee

123. U.S. House of Representatives, 1 March 1966 hearings in Hearings before the Committee on Science and Astronautics, *1967 NASA Authorization*, 89th Cong., 2nd Sess., on HR 12718, pp. 477–478.
124. U.S. House of Representatives, 1 March 1966 hearings, p. 479.
125. Carter, *The Prevention of Back Contamination*, p. 30.
126. Compton, *Where No Man Has Gone Before*, pp. 48–49; Mangus and Larsen, *Lunar Receiving Laboratory Project History*, p. 18; Carter, "Moon Rocks and Moon Germs," 240.

report both recommended MSC as the best location for LSRL. He also stressed that "there will be more than 50 scientific principal investigators who will get portions of the sample for their own investigations, and these will be mostly at universities throughout the country."[127]

NASA's arguments did not sway Fulton or Rumsfeld. The latter congressman expressed interest in a scheme that would send the lunar samples to one location and the astronauts to another. He believed that the astronauts were not going to be deeply involved in the packaging, distribution, or radiation counting of the lunar material, so why not use two locations? Low didn't agree, insisting that the total cost of two facilities would far exceed the cost of the one planned LSRL. He gave one supporting example: If personnel working with lunar material got exposed to it (such as through a breach in the biological barrier), then at a single integrated facility, they could simply go into quarantine along with the astronauts. At two widely separated facilities, however, the situation would be more problematic.

Rumsfeld responded to this by asking if NASA had available a more quantitative analysis of the cost of developing two receiving labs using existing facilities. He wanted to see statistics and cost calculations in order to determine how the cost of enhancing two geographically separated, existing facilities—one for astronauts and the other for lunar samples—compared to the cost of building one integrated facility.[128]

Low stated that NASA might not have such an analysis in the form that Rumsfeld wanted to see, but the Agency did have information on the various existing sites that were investigated. Acting subcommittee chair Emilio Daddario of Connecticut requested that NASA provide such information for the record, and Low agreed. NASA hastily put together a document titled "Additional Data Submitted Regarding the Lunar Receiving Laboratory,"[129] which was added to the 1 March 1966 hearing record. This document employed a shortened version of the term for the facility: Lunar Receiving Laboratory, or LRL, which was the name being increasingly used by NASA and other organizations. The "Additional Data" document included a written justification for NASA's LRL plans, stressing that they outlined necessary actions

---

127. U.S. House of Representatives, 1 March 1966 hearings, p. 478.
128. Ibid., pp. 443, 478–480; Carter, *The Prevention of Back Contamination*, pp. 31–32.
129. U.S. House of Representatives, 1 March 1966 hearings, pp. 480–483.

to meet the concerns of public health authorities in the retrieval of sample materials from the Moon as well as the time dependency requirements for certain types of testing. The "Additional Data" document made several important points, including the following:

- No existing U.S. laboratory was capable of performing all the required functions of LRL.
- The containment of all hazardous material and personnel within one facility significantly decreased the risk of contamination when transferring material from one functional area to another.
- The potential for contaminating surrounding areas was minimized if all sources of contamination were contained at one geographical location.
- Cost savings would result from using a single location for the operating and maintenance functions of quarantine areas.

Nowhere did the "Additional Data" document quantitatively compare the costs of two existing versus one new receiving facility, although reference was made to the Preliminary Engineering Report, which did compare several facility alternatives and showed that a cost saving would be realized by integrating all of the functions in one facility.[130]

Not all of the criticism during the 1 March 1966 hearings came from Republicans. Some came from the Democratic side of the subcommittee. Texas Democrat Robert R. Casey, for instance, asked for more explanation of LRL's utility, complaining that "the ordinary layman is uninformed as to the necessity of this lab and what its purpose is going to be."[131] Casey's comment indicated that if NASA expected subcommittee members to approve LRL funding as part of the Agency's budget, then it had to do a better job of justifying the facility in ways that would be supported by the committee's constituents. Up until the hearings, debates over LRL had taken place primarily within NASA and the scientific community as well as with other agencies. Little effort had been made to sell LRL to the public or to Congress. For instance, NASA's 256-page prepared statement on its overall budget request mentioned LRL only once. NASA needed to conduct a far-reaching campaign to

---

130. Ibid., pp. 480–483.
131. Ibid., p. 473.

communicate the necessity of back contamination protection and the critical part that LRL would play in accomplishing this.[132]

The House's Rejection of NASA's Plans

The week following Lilly's and Low's hearing, the subcommittee expressed its position regarding LRL funding. Unsatisfied that NASA had adequately defended its plans for LRL, the subcommittee removed its funding from the authorization bill. One of the contributing factors to this action may have been the perception that NASA's arguments were not objective, but were "after the fact"[133] justifications for constructing an LRL and locating it at MSC. This opinion was supported by the fact that MSC staff had already developed conceptual studies for LRL, always assuming that it would be built in Houston. Critics believed that the focus of MSC's analyses had been on identifying reasons why no other facility was suitable for LRL, rather than on logically determining what the best location within the United States actually was.[134]

Loss of LRL would have been a huge blow to the Apollo program and to MSC. Without the facility, the Apollo program could not meet all the requirements that had been imposed on it by the scientific community and PHS. Low stated the situation very bluntly: without the laboratory, NASA could not bring samples back from the Moon and, in fact, could not even complete the Apollo mission. This was not an exaggeration. Quarantine issues had become so important that a failure to have LRL ready for Apollo 11 would postpone the entire Apollo program.

In response to this dire situation, Mueller called for a more detailed study of existing venues than he had ordered just a few months earlier, demanding that the study be ready in time to help secure FY 1967 funding for LRL. This new study was to identify and examine additional existing facilities that could either provide quarantine

---

132. Mangus and Larsen, *Lunar Receiving Laboratory Project History*, p. 18; Carter, "Moon Rocks and Moon Germs," pp. 239–240.
133. Mangus and Larsen, *Lunar Receiving Laboratory Project History*, p. 19.
134. Compton, *Where No Man Has Gone Before*, pp. 48–49.

capabilities for the Apollo crews and lunar samples or carry out the necessary scientific studies.[135]

The Pickering Committee Study

To fulfill Mueller's orders, MSC formed a site survey board led by Col. John Pickering, Special Assistant to the Director of Space Medicine in NASA's OSSA. The Pickering Committee screened 27 existing facilities and selected the following eight of them for in-depth assessment during the week of 16–23 March 1966:

- PHS's CDC in Atlanta, Georgia.
- Army Biological Center at Fort Detrick, Maryland.
- National Institutes of Health in Bethesda, Maryland.
- Oak Ridge National Laboratory in Oak Ridge, Tennessee.
- School of Aerospace Medicine, Brooks Air Force Base in San Antonio, Texas.
- Ames Research Center at Moffett Field, California.
- Navy Biological Laboratory in Oakland, California.
- Los Alamos Scientific Laboratories in Los Alamos, New Mexico.[136]

The committee then developed criteria for an acceptable facility, based on the Apollo mission's requirements. The facility needed to have the following capabilities:

Quarantine Capabilities
- Two-way biological containment.
- Treatment for liquid effluents.
- Sterilization for gaseous effluents.

Sample Handling Capabilities
- High vacuum sample handling with special chamber seals.
- Remote manipulators for samples and containers.
- Exceptionally clean pumping system.
- Low-level radiation measurement capability.

---

135. Compton, *Where No Man Has Gone Before*, pp. 48–49; Carter, *The Prevention of Back Contamination*, p. 33.
136. U.S. House of Representatives, Hearings before the Committee on Science and Astronautics, *1967 NASA Authorization*, 89th Cong., 2nd Sess., on HR 12718, 31 March 1966, Appendix M, pp. 1217–1218.

An acceptable facility also had to have available space, administrative and technical support services, ample utilities, and good logistics of travel to and from the site. In addition, the facility's management had to be willing to accept the project and integrate it into the existing work schedule.[137]

On 24 March 1966, the Pickering Committee presented its results in a draft *Site Investigation Study, Lunar Receiving Laboratory.* This study was eventually made a part of House Subcommittee on Manned Space Flight hearings record. The study presented important findings regarding each of the eight sites, which helped the Pickering Committee and Congress to make decisions on LRL. These findings are summarized below:

Site 1: U.S. PHS Communicable Disease Center (CDC), Atlanta, Georgia. The CDC did not have adequate radiation counting facilities, quarantine laboratories with two-way containment capabilities, or crew reception facilities, all of which were vital to the envisioned LRL. In addition, laboratory support facilities at the CDC were being fully used and would not be available to LRL if it located there. No facilities were available that could be adapted to meet LRL needs. The conclusion was that all components of LRL would have to be provided by new construction at an estimated cost of $9.4 million, as compared to the $9.1 million estimated for building LRL at MSC. The additional cost for building it at the CDC was due in part to rock formations close to the surface that would have to be blasted or drilled away.

Site 2: U.S. Army Biological Center, Fort Detrick, Maryland. No acceptable radiation counting facilities existed. No laboratory facilities were available for LRL use, and no two-way quarantine facilities existed. Laboratory support capabilities were widely scattered and not adaptable to LRL use. A small Army hospital unit (20 beds) was at the site but was fully utilized in a respiratory disease research program and so was not available for a Crew Reception Area (CRA). Land was available to build an LRL, but doing so would require an extension of utilities into the area. The estimated cost of building an LRL at this site was $11 million.

---

137. U.S. House of Representatives, Appendix M, pp. 1217–1218; Mangus and Larsen, *Lunar Receiving Laboratory Project History*, pp. 18–19; Compton, *Where No Man Has Gone Before*, pp. 48–49; Carter, *The Prevention of Back Contamination*, p. 33.

Site 3: National Institutes of Health (NIH), Bethesda, Maryland.
Excellent radiation counting capabilities existed, but there were no quarantine facilities with two-way containment. Laboratory support capabilities were there but were committed to other use. The large clinical center on site could conduct tests on returning astronauts, but it was located 25 miles away from where LRL would be constructed. Because LRL would not be able to be integrated into the existing NIH campus, LRL would require completely new construction at an estimated cost of $11.8 million. The land acquisition necessary would involve complex negotiations with city, state, and district planning commissions and would conflict with the development of a pathogen-free, large-animal farm at the site.

Site 4: Atomic Energy Commission (AEC) Oak Ridge National Laboratory, Oak Ridge, Tennessee. Low-level radiation counting facilities existed, but not those with the specifications needed by LRL. No two-way containment facilities were at the site. An animal-virus laboratory was under construction, but use of it by NASA would impair its planned mission. Laboratory support facilities were present but were not adjacent to the potential sites of LRL. The region was not near a jet aircraft landing field. Conflicts with the weapons-development mission of the site were also envisioned. A necessary capital investment of about $9 million was estimated.

Site 5: USAF School of Aerospace Medicine, Brooks Air Force Base, San Antonio, Texas. Radiation counting facilities existed, although not underground (which would have made it easier to reduce background radiation). No quarantine facility was present. No laboratory support facilities were available for use. Space was available for new facility construction, which would cost an estimated $11 million. The high cost was due in part to unusual soil conditions, which would require special foundations.

Site 6: NASA Ames Research Center, Moffett Field, California.
An above-ground radiation counting laboratory existed, although shielding would have to be greatly increased and a radon-removal system installed in order to reduce background radiation, which was necessary for the counting measurements to be as sensitive as possible. No two-way quarantine facilities were available. Existing facilities could be modified to provide some of the laboratory support capability needed. No crew reception facilities existed at the site. Land was available for construction of an LRL at an estimated cost of $10.3 million, with ready access to utilities and air transportation.

Site 7: Naval Biological Laboratory, Oakland, California. No radiation counting or crew reception facilities existed, and laboratory support was not available. Considerable one-way quarantine capabilities existed, but not two-way. The availability and acceptability of commercially owned land adjacent to the site was questionable. A capital investment of $10.5 million was needed if the land could be obtained.

Site 8: AEC Los Alamos Scientific Laboratories, Los Alamos, New Mexico. Radiation counting facilities existed and could possibly have been used by an LRL for a period of several weeks, if scheduled in advance. Laboratory support capabilities were present as well. No suitable quarantine facilities existed. Land was available for new facility construction at an estimated cost of $10.3 million. The remote location of the site added to construction cost estimates. Residential housing in the area was severely limited. The site had limited air and road access.[138]

At the end of its intensive survey, the Pickering Committee reached conclusions similar to ones in previous studies:

- Without extensive modifications, none of the examined facilities could meet the criteria for even one of LRL's main functionalities.
- None of the facilities could be economically adapted to meet all criteria of LRL.
- Using any one of the sites examined for LRL would result in the reduction or modification of another nationally significant research activity.
- To minimize public health hazards, one facility integrating several functionalities (rather than multiple specialized facilities) was advisable.
- The optimal location for LRL was MSC in order to attain the best operational characteristics and minimum operating costs. This was because efficient management of the Apollo mission would best be achieved by locating astronauts, spacecraft, and Moon samples in proximity to MSC engineers and physicians, especially in the weeks following recovery.

The Pickering Committee's draft *Site Investigation Study, Lunar Receiving Laboratory* that was released on 24 March 1966 was

---

138. U.S. House of Representatives, Appendix M, pp. 1238–1251.

Table 4.3 Lunar Receiving Laboratory conceptual studies and site evaluations.

Date	Study	Summary	Source
Apr-1964	MSC "Sample Transfer Facility" in-house study	Identified need for the facility to check lunar samples for viable organisms.	Aleck C. Bond to Chief, Office of Technical and Engineering Services, "Sample Transfer Facility."
16-Jun-1964	NASA Lunar Sampling Summer Symposium	Identified need for biocontainment to protect terrestrial environment, and for analyzing all lunar materials before release for the presence of pathogenic organisms.	Carter, *The Prevention of Back Contamination*, p. 13.
29–30-July-1964	SSB "Conference on Potential Hazards of Back Contamination from the Planets"	Advocated quarantine of astronauts, spacecraft, and samples. Called for astronaut quarantine of at least three weeks. Recommended that the United States take a leadership role in planetary protection.	Brown, "Potential Hazards of Back Contamination From the Planets."
2-Feb-1965	Hess Committee report (NAS)	Recognized need for quarantining lunar material, but did not identify any reason why the isolation facility should be at MSC.	SSB, *Report of the Ad Hoc Committee on Lunar Sample Handling Facility*.
15-Mar-1965	Chao Committee report	Supported quarantining and testing lunar material for living organisms. Eventually recommended that the LRL be located at MSC.	Ad Hoc Committee on the Lunar Receiving Laboratory.

Date	Study	Summary	Source
18-Mar-1965	PHS CDC report	Recommended specific protective procedures for handling and quarantining lunar samples.	Mangus and Larsen, *Lunar Receiving Laboratory Project History*, p. 14.
1-Oct-1965	Preliminary Engineering Report	Concluded that modifying an existing facility to satisfy both scientific and quarantine requirements of the LRL would cost as much as constructing a new facility and take as long a time. Recommended MSC as best place for LRL.	Carter, *The Prevention of Back Contamination*, pp. 22–24.
8-Dec-1965	MSC "Study of Existing Personnel Quarantine Facilities"	Concluded that no existing site was suitable for LRL without first overcoming serious cost, land availability, and other barriers. Study recommended one site rather than several distributed sites for LRL and also made case for locating it at MSC.	Deputy Director, Space Medicine, to James C. McLane, 22 November 1965.
7-Apr-1966	Pickering Committee's *Site Investigation Study, Lunar Receiving Laboratory*	Visited eight sites with existing facilities. None acceptable without expensive modifications. Recommended that LRL be built at MSC.	Carter, *The Prevention of Back Contamination*, p. 33.

immediately read by at least some members of Congress, including Representative James Fulton of Pennsylvania. MSC distributed the final version of its study on 7 April 1966. The study underlined the point that from a project management perspective, *Houston was the best location for LRL.*[139]

The House Subcommittee Rehearing—31 March 1966

Mueller appeared before the House subcommittee for a rehearing on 31 March 1966 devoted exclusively to LRL. The stated purpose of the rehearing was to "give NASA an additional opportunity to present other information in a complete way,"[140] given the dissatisfaction that the subcommittee had with NASA's justifications for LRL up until that time. Mueller, who was NASA's Associate Administrator, Office of Manned Space Flight, brought several NASA witnesses with him to talk on the need for LRL. These included Lt. Gen. Frank A. Bogart, Deputy Associate Administrator (Management); Col. John Pickering, Special Assistant to the Director of Space Medicine; and Angelo P. Picillo, Facilities Engineer. Mueller began the proceedings by summarizing the envisioned LRL's objectives and the evolution of LRL concepts and requirements, focusing in particular on the need for quarantine. He emphasized how identification of this need arose not only out of NASA's ranks, but also from working in close cooperation with NAS Space Science Board members.[141]

Representative Fulton of Pennsylvania continued his attack from the 1 March hearing, pointedly criticizing the fairness of NASA's intent to locate LRL at MSC. Fulton called into question the makeup of the Pickering Committee, noting that seven of the nine members were from NASA and five of those were from MSC. He commented that "If I were going to set up a site selection committee that would select

---

139. U.S. House of Representatives, Appendix M, pp. 1207, 1212; Compton, *Where No Man Has Gone Before*, pp. 49–50, 279.
140. U.S. House of Representatives, Appendix M, p. 1207.
141. U.S. House of Representatives, Appendix M, pp. 1208–1212; Carter, *The Prevention of Back Contamination*, p. 33.

Houston as the site, I would then, of course, pick five out of the nine members from Houston."[142]

Mueller was better prepared for Fulton's attacks than Lilly and Low had been on 1 March. Mueller noted that the Chao Committee, which recommended at an earlier date than the Pickering Committee that LRL be located in Houston, was composed of members from a whole spectrum of organizations, including universities, Department of Defense facilities, a national laboratory, and the National Bureau of Standards, and was definitely not dominated by NASA. (A summary of the Chao and Pickering Committees' findings, as well as those of other task forces, is included in table 4.3.) In fact the chairman himself, Edward Chao, was from the U.S. Geological Survey (USGS) and on loan to NASA.

Fulton then brought up many other questions, such as why the envisioned site for LRL was not an "isolated land mass or island"[143] near the ocean pickup point for the spacecraft. Would not that have been a safer location that did not require the transport of potentially harmful material halfway around the world? Fulton also questioned why all functions of LRL should be located in one place rather than letting other parts of the country in on the action. At this point, acting subcommittee chair Emilio Daddario stopped any further questions and asked Mueller to complete his testimony on LRL plans, believing that many of Fulton's questions would be answered during the course of the testimony.[144]

One of Fulton's chief concerns was that the selection of the LRL site be handled fairly and objectively, and he questioned whether the heavy NASA makeup of the Pickering Committee would further this aim. Mueller responded by saying that the Pickering Committee was formed to make sure that the results of the earlier investigation conducted by the Chao Committee (which recommended Houston as the best LRL site) were sound and to investigate whether any existing facilities that could potentially serve as LRL had been overlooked. Mueller's point was that the Pickering Committee was not meant to ensure that Houston got LRL, but rather to do what Fulton wanted done—conduct

---

142. U.S. House of Representatives, Appendix M, pp. 1208–1212; Carter, *The Prevention of Back Contamination*, p. 33.
143. U.S. House of Representatives, Appendix M, p. 1213.
144. Ibid., pp. 1213–1214.

an objective study to find the best site. The Pickering Committee was able to use results from NASA's Preliminary Engineering Report, which gave it "a much more definite set of requirements that could be used to measure both the cost and the advantages and disadvantages of various locations."[145] The Pickering Committee thus employed the latest data to narrow down its investigation to the eight sites discussed earlier in this chapter as possible candidates for LRL.

Mueller also defended the selection of Pickering Committee members by declaring that "These are the people who have been interested, intimately involved in trying to design the lunar receiving laboratory as a facility, so they were our experts as to what was required to carry out this program."[146] These people approached the task with the goal of finding another site that could more economically fulfill the necessary planetary protection requirements that were established by other agencies. They found that this could not be done.

Mueller said that NASA's original belief had been that every effort should be made to use existing facilities wherever they could be found in the United States. He stated that as little as a year earlier, he believed "there must be an existing facility somewhere in the United States . . . so that we would not have to . . . build a new facility. We therefore started with a bias against new facilities, and it was only after very careful study and very careful analysis that we came very reluctantly to the conclusion that a new facility had to be built . . . ."[147] These remarks were borne out by a NASA internal document entitled the "Lunar Sample Receiving Laboratory Office Staff Paper,"[148] which recorded NASA's original intent to use existing or modified existing facilities. The shortcomings of existing vacuum systems, biological laboratories, containment capabilities, crew isolation facilities, and radiation counting laboratories forced the conclusion that new facilities were, in fact, required.

---

145. Ibid., p. 1217.
146. Ibid., p. 1229.
147. U.S. House of Representatives, Appendix M, p. 1227.
148. Walter W. Kemmerer and James C. McLane, Jr., "Lunar Sample Receiving Laboratory Office Staff Paper," undated, pp. 2–13, in a package beginning with "Lunar Sample Receiving Laboratory Siting Factors," ca. May 1966, JSC Historical Archives, "May 1966" folder, box 76-12, LRL Chron. Files.

After Mueller was finished, Col. John Pickering, the head of the site survey team, explained in considerable detail to the House subcommittee the assessments that his team had done and presented its findings. These findings, which were summarized earlier in the chapter, identified the many inadequacies of existing facilities and clearly showed that money would not be saved by modifying them into an LRL rather than constructing a new such facility.[149]

Fulton and Rumsfeld continued to oppose NASA's plan for LRL. Fulton questioned the actual danger back contamination presented to Earth, commenting that it had been repeatedly hit by meteoroids containing "traces of life-giving substances in them."[150] Mueller agreed, but pointed out that the high temperatures those bodies reached while plunging through Earth's atmosphere would have destroyed any complex molecules and viable organisms.

Richard L. Roudebush, a Republican representative from Indiana, raised a question related to one brought up by Congressman Fulton earlier in the hearings, asking if there would be a savings in locating LRL in an isolated area such as an island or a ship at sea. Mueller responded that NASA had carefully considered this and believed that a ship installation would markedly increase the cost, simply because a typical ship would by itself cost far more than LRL. As to locating the facility on an island, it was true that this could place it near one of NASA's landing points, but the Apollo program required two such sites—a primary one in the Pacific Ocean and another in the Atlantic. If proximity to the landing site needed to be attained, two facilities would be necessary.[151]

Representative Roudebush questioned whether construction of LRL could be delayed a year.[152] Roudebush's inquiry was in line with recent concerns of President Johnson who, in order to curb rising inflation, wanted a slowdown in all federal construction. Mueller, however, made clear that NASA was up against a hard deadline. In order to meet the late President Kennedy's goal of getting a man on the Moon within the decade of the 1960s, LRL needed to first be completed and fully operational, given the importance of preventing back contamina-

---

149. U.S. House of Representatives, Appendix M, pp. 1228–1233.
150. U.S. House of Representatives, Appendix M, p. 1231.
151. Ibid., p. 1234.
152. Ibid., pp. II, 1234.

tion from returning astronauts and lunar material. Mueller admitted that NASA should have recognized the need for an LRL far earlier than it had, in 1961 rather than in 1964. But if construction did not start immediately on the facility, it would prove extremely difficult to get it built and tested in time for the first lunar landing, and this would delay the entire Apollo program.[153]

Roudebush then asked if LRL could be scaled down in scope. He inquired as to why unprocessed lunar samples could not be sent directly to the university research laboratories that would examine them in depth, rather than first examining and testing the samples at LRL. He wondered if NASA was "going overboard"[154] on the criteria for LRL. Mueller replied that LRL was to be the "minimum possible facility"[155] to meet planetary protection requirements and that the equipment required by the facility did not at present exist anywhere else. He also pointed out that the earliest estimates for LRL added up to $23 million. Since that time, he and many others had done as much as they could to reduce those costs to the present estimate of $9.1 million.

NASA's 31 March presentations to the House subcommittee convinced most of its members that constructing a new LRL facility at MSC was justified, and the funds to begin this process were restored to the Agency's authorization bill. The *1967 NASA Authorization Bill* left the full Committee on Science and Astronautics on 4 April 1966 and was sent to the House floor recommending a total NASA budget of $4,986,864,150 for 1967 operations, including the entire $9.1 million for constructing LRL. Fulton filed a Dissenting View to the House committee's report, saying that not enough data had come to light on which to responsibly design and build the laboratory and restating all of the objections he had brought up during subcommittee hearings. He also tried to push through an amendment that would have gotten rid of LRL funding. Nevertheless, the authorization bill passed the House vote on 3 May 1966 by an overwhelming 349 to 10, with the LRL budget of $9.1 million entirely intact.[156]

---

153. Ibid., pp. II, 1234–1235; Carter, "Moon Rocks and Moon Germs," 240.
154. U.S. House of Representatives, Appendix M, p. 1234.
155. Ibid., p. 1234.
156. Compton, *Where No Man Has Gone Before*, p. 50; Margaret S. Race, "Anticipating the Reaction: Public Concern About Sample Return Missions"; Carter, "Moon Rocks and Moon Germs," 240; Carter, *Prevention of Back Contamination*, pp. 35–36.

The Senate's Review of LRL

The U.S. Senate Committee on Aeronautical and Space Sciences raised some similar concerns to those of the House subcommittee, but George Mueller, head of NASA's Office of Manned Space Flight, had far less trouble convincing the Senate committee of the need for LRL. The Senate committee, in fact, discussed LRL only once during its 11 days of hearings. Mueller explained briefly that LRL was an "essential part of the quarantine process"[157] and necessary to meet the requirements of PHS. When asked by James J. Gehrig, the staff director of the committee, whether the requirement for LRL was placed on NASA in written form, Mueller referred to the statement written by John Bagby, Assistant Chief of the CDC, on the containment of lunar samples, astronauts, and personnel. As with the House subcommittee, this document was made part of the record.[158]

The Senate committee's concern was not that LRL was needed, but that NASA had not done sufficient analysis to justify its estimated $9.1 million cost for constructing the facility. The committee took an action that was typical of the way Congress often responded to NASA budget requests: it recommended a reduction in the budget (of $1 million) in order to ensure that LRL planning and construction expenses would be minutely reviewed and carefully controlled. According to Carter,[159] Congress frequently assumed that budget requests from NASA, as well those from other agencies with large Research and Development (R&D) programs such as the Department of Defense, had been expanded from what was actually needed. Most members of Congress did not have the capabilities to determine by just how much, and so arbitrary budget cuts were recommended.

NASA and other agencies, aware of this congressional tendency, sometimes tried to circumvent the cuts by adding extra funding

---

157. U.S. Senate, Hearings before the Committee on Aeronautical and Space Sciences, 89th Cong., 2nd Sess., on S. 2909, *NASA Authorization for Fiscal Year 1967*, p. 137.
158. John R. Bagby, "Statement by the Assistant Chief, Communicable Disease Center, Public Health Service, DHEW, On the Containment of Lunar Samples, Astronauts, and Support Personnel," U.S. Senate, Hearings before the Committee on Aeronautical and Space Sciences, 89th Cong., 2nd Sess., on S. 2909," *NASA Authorization for Fiscal Year 1967*, pp. 137–139.
159. Carter, *The Prevention of Back Contamination*, pp. 36–38.

to its initial budget requests. NASA may have done just that in the case of LRL, since nearly 11 percent of the requested budget was included in a rather nondescriptive category entitled "Acceleration of Construction."[160] Nevertheless, when NASA realized in May 1966 that the Senate meant to reduce LRL funding, the Agency took steps for lowering facility costs. On 9 May 1966, NASA approved two important groups to manage design and construction of LRL—a policy board and a program office, one of whose important tasks was to discover ways of meeting the reduced cost constraints that the Senate was attempting to impose while making sure that neither the biocontainment mission nor the research capabilities of LRL got compromised.[161] In order to ensure that construction costs would not exceed $8.1 million, NASA eventually compressed the LRL area by 3,800 square feet, eliminated one of its two vacuum systems, and amalgamated all activities into one building rather than having a separate structure for personnel and Apollo crew quarantine.[162]

The Joint House-Senate Conference Committees

The Senate passed a NASA authorization bill on 24 May 1966 for a total amount of $5.008 billion, including the reduced amount of $8.1 million for LRL. Since LRL budgets recommended by the House and Senate differed with each other, Congress formed a joint conference committee in order to work out an agreement. The joint committee came to a meeting of the minds on 19 July 1966, which was 19 days past the beginning of FY 1967. The joint committee agreed on a NASA budget of $5,000,419,000, with LRL receiving $8.1 million. The House accepted this budget on 21 July 1966 and the Senate on 22 July.

The budget process was not yet completed, however. Congressional funding required a second step: the passing of a bill to actually appropriate funds for NASA. The House Subcommittee on Independent Offices of the Committee on Appropriations had begun this process

---

160. Ibid., p. 38.
161. Carter, "Moon Rocks and Moon Germs," 241; Compton, *Where No Man Has Gone Before*, p. 51.
162. Compton, *Where No Man Has Gone Before*, p. 52.

on 5 April 1966, before the first bill had passed. The subcommittee agreed on a reduced appropriation, for a total NASA budget of only $4.95 billion, but with the LRL budget restored to $9.1 million. This appropriation bill passed the full House on 10 May 1966.

The amounts approved in the corresponding Senate appropriations bill, which passed on 2 August 1966, again differed from the amounts in the House appropriations bill, and the matter once more went to a joint conference committee. The two houses of Congress finally reached an agreement on 18 August 1966 to give NASA a total of $4.968 billion. The compromise bill, which once again included only $8.1 million for LRL, was immediately accepted by the House; the Senate approved it on 24 August, nearly two months after the start of FY 1967.[163]

Designing, Constructing, and Outfitting LRL

Defining LRL Scientific Requirements and Physical Characteristics

During 1965, NASA Headquarters's Manned Space Science Division worked on defining scientific requirements for LRL, while MSC in Houston focused on developing the facility's architectural and engineering design requirements. By the end of the year, as mentioned earlier, MSC completed its Preliminary Engineering Report as well as quarantine requirements for crew and support personnel. The results of these efforts, which were presented to Congress in 1966, included plans for an 86,000-square-foot (8,000-square-meter) facility with a personnel isolation area, sample laboratories, and a radiation-counting capability, part of which was to be built 50 feet (15 meters) below ground and heavily shielded from outside radiation.[164]

---

163. Carter, *The Prevention of Back Contamination*, pp. 38–39; Compton, *Where No Man Has Gone Before*, pp. 50–51; Kent Carter, "Moon Rocks and Moon Germs: A History of NASA's Lunar Receiving Laboratory," *Prologue: Quarterly of the National Archives and Records Administration* 33 (winter 2001): 240.
164. Compton, *Where No Man Has Gone Before*, p. 51.

LRL had three major objectives:

1. To prevent the back contamination of Earth from any lunar organisms that might have been brought back from the mission.
2. To prevent lunar sample contamination from Earth's atmosphere and microorganisms in order to protect the integrity of the scientific investigations on those samples.
3. To determine if and when it was safe for lunar material to come into contact with the terrestrial environment.[165]

To achieve these objectives, LRL required a complex design and integration of many functional areas:

- Living and examination quarters for the astronauts, including the following:
  - Medical and dental examination rooms.
  - Operating room.
  - Biomedical laboratory.
  - Astronaut exercise, debriefing, office, and sleeping area.
  - Support staff office and dormitory areas.
  - Kitchen, dining, and lounging areas.
  - Facilities for sterilizing and passing food and laundry in and out.
- Sample analysis laboratories of various types:
  - Radiation counting laboratory, a section of which was 50 feet below ground.
  - Gas analysis laboratory.
  - Physical-chemical test laboratory for performing mineralogical, petrological, and geochemical analyses.
  - Spectrographic laboratory and darkroom.
  - Biopreparation laboratory for sample handling, preparing, weighing, and packaging for distribution.
  - Bioanalysis laboratory for blood and other tests on mice.
  - Holding laboratory for germ-free mice.
  - Holding laboratory for normal mice.
  - Lunar microbiology laboratory to isolate, identify, and possibly grow lunar microorganisms.

---

165. Dora Jane Hamblin, "After the Moonflight, A Wall Against the Unknown—A Close Watch on the Men and Their Prizes," *Life* (4 July 1969): 54, folder 012990, "Lunar Receiving Lab," NASA Historical Reference Collection.

- Bird, fish, and invertebrate laboratory for exposing animals, including shrimp, quail, cockroaches, and oysters, to lunar samples.
- Plant laboratory for exposing plants such as germ-free algae, spores, seeds, and seedlings to lunar samples.
- Microbiological laboratory for culturing lunar and astronaut samples.
- Crew virology laboratory for postflight astronaut examinations.
- Biosafety laboratory to monitor all systems.
- Microbiology laboratory for clinical tests of quarantined personnel.
- X-ray room.
- Vacuum system for receiving and processing Moon samples.
- Air conditioning system to sterilize air both entering and leaving LRL.
- Two-way biological barrier for quarantining astronauts for more than two weeks and lunar samples for over a month.
- Apollo Command Module storage.
- Support facilities.
- Staff offices.[166]

Oak Ridge National Laboratory, which was owned at the time by the Atomic Energy Commission (and later by the Department of Energy), designed and constructed LRL's vacuum system and its related equipment.[167] AEC also developed LRL's low-level radiation counting equipment and the software (or "computer programs," as they were referred to at that time) to analyze the radiation data.[168]

Contracting for LRL Construction and Equipment Installation

MSC envisioned that construction of LRL would be completed and the facility ready for practice-testing by the end of December 1967. This testing would be completed by January 1969, which would hopefully be in

166. Hamblin, "After the Moonflight," 52–53.
167. Compton, *Where No Man Has Gone Before*, p. 51.
168. Carter, *The Prevention of Back Contamination*, p. 54.

time for the facility to analyze the first lunar samples. The delay in congressional approval of the LRL budget until nearly two months after the start of FY 1967, however, seriously threatened this tight schedule, especially since Congress insisted that MSC could only solicit construction bids but could not even open them until passage of NASA's authorization bill.

MSC was able to work around this problem by initiating some tasks early, before actually receiving congressional funding. Procurement rules during this time were more flexible than they are today, allowing a job to be initiated before official congressional authorization.[169] Some of the correspondence between NASA managers in 1966 gave hints as to how preofficial work on a contract was defended by MSC. In a 16 May 1966 letter from NASA Headquarters, for instance, George Mueller advised MSC Director Robert Gilruth that LRL "must be operational by November 15, 1967,"[170] and thus to "proceed with your plan toward the early award of a construction contract for the substructure, utilities and building shell."[171] This gave Gilruth potential license to get a head start on the work before congressional approval.[172] He was under enormous pressure to construct LRL in time to support the first Moon landing, which NASA thought might be as soon as late 1968. He acknowledged this pressure in a 9 June 1966 letter to Mueller, mentioning that meeting the 15 November 1967 deadline to have LRL fully operational was "a very difficult task to accomplish"[173] with a cost ceiling of $9.1 million, which ended up being cut to only $8.1 million.

In an 8 July 1966 letter, Col. John Pickering, Executive Secretary of ICBC, also hinted at initiation of work before receipt of congressional funding. He wrote that the final LRL design effort was already under way and that "procurement actions for incremental construction are being readied for award on July 15,"[174] even though he must

---

169. Kent Carter, telephone conversation with author, 21 April 1966; Carter, *The Prevention of Back Contamination*, pp. 49–50; Compton, *Where No Man Has Gone Before*, p. 51.
170. George E. Mueller to Robert R. Gilruth, 16 May 1966, JSC Historical Archives, "May 1966" folder, box 76-12, LRL Chron. Files.
171. Ibid.
172. Kent Carter, e-mail to author, 24 April 2006.
173. George M. Low for Robert R. Gilruth, to George E. Mueller, 9 June 1966, JSC Historical Archives, "June 1966" folder, box 76-12, LRL Chron. Files.
174. John E. Pickering letter (no recipient), 8 July 1966, in the package, "Interagency Committee on Back Contamination—Dr. Berry," JSC Historical Archives, "May 1966" folder, box 76-12, LRL Chron. Files.

have known that congressional authorization and appropriation bills were very unlikely to be approved and signed by that date.

In June 1966, two months before congressional approval, MSC did start work, contracting with S. H. Barrett and Associates of Houston for $15,000 to develop shop drawings and fabricate some of the structural steel that it would need for LRL. MSC purchased the steel for this task by borrowing the funds from another project.[175]

On 5 July 1966, MSC chose Warrior Constructors, another Houston firm, to perform Phase I construction of LRL, which involved site preparation, excavation, pouring the foundation, and erecting the building's steel framework. Although Congress had not yet approved NASA's budget and the official contract with Warrior was not yet signed, the company agreed to meet a 1 January 1967 deadline for Phase I and initiated work on 11 July 1966.

On 28 July 1966, George Mueller sent MSC approval from NASA Headquarters to open bids for Phase I construction on 1 August 1966—somewhat before the FY 1967 authorization bill passed Congress, but later than the Agency had hoped. NASA officially signed a contract for Phase I construction with Warrior Constructors on 12 August 1966 for $1.69 million. The Mueller letter also instructed MSC to conduct final negotiations with a contractor selected for Phase II construction—finishing the LRL structure and installing and testing all laboratory equipment—but to execute no contract until after NASA Headquarters had authorized funding (which didn't occur until 24 August). MSC complied with this guidance, selecting Warrior Constructors on 1 August for final Phase II contract negotiations.[176] On 19 August, NASA picked WNN Contractors, a consortium of Warrior, National Electronics Corporation of Houston, and Notkin & Company of Kansas City, to complete Phase II. Phase I and Phase II contracts totaled $7.8 million.[177]

---

175. Carter, *The Prevention of Back Contamination*, pp. 49–50.

176. NASA, "Texas Firm Selected to Complete Lunar Receiving Lab," release no. 66-200, 1 August 1966, in a package beginning with a membership list of the Interagency Committee on Back Contamination, JSC Historical Archives, "May 1966" folder, box 76-12.

177. George E. Mueller teletype to MSC, "Lunar Receiving Laboratory," 28 July 1966. JSC Historical Archives, "July 1966" folder, box 76-12, LRL Chron. Files; Henry C. Dethloff, "A Contractual Relationship," in *Suddenly, Tomorrow Came: A History of the Johnson Space Center* (Washington, DC: NASA SP-4307, Johnson Space Center, 1993), Chap. 8, pp. 153–154; Compton, *Where No Man Has Gone Before*, p. 51; Carter, *The Prevention of Back Contamination*, pp. 49–50; Carter, "Moon Rocks and Moon Germs," 241.

During LRL construction, all communications between NASA personnel and their contractors were officially supposed to go through a NASA contracting officer, but because of the time-sensitive nature of the work, LRL Program Office staff began issuing assignments and guidance directly to the contractors. This informal line of communication was eventually accepted by all parties and supplemented with guidance through official NASA channels. The arrangement allowed contractors to respond rapidly to new issues and necessary changes in requirements. The open communication procedures helped Phase I construction activities to proceed smoothly, notwithstanding technical problems that included construction site flooding and a building material shortage. Warrior Constructors completed Phase I on schedule by January 1967 for $1.71 million—about 1 percent over the initial contract amount, due to NASA-requested changes.

Phase II proved to be more challenging to complete on time. MSC staff had to reduce construction costs in order to stay within Congress's condensed budget, but no time was available for a major redesign of the facility, for this would have meant a significant delay that the Apollo program could not afford. The solution that MSC came up with, as mentioned above, was to reduce the total LRL area by 3,800 square feet, consolidate all operations in one building rather than constructing a separate crew quarantine facility, and reduce the number of vacuum systems from two to one in the area where Apollo staff would open sample containers.[178]

Phase II construction ran into other challenges as well, including labor disputes, late equipment delivery, and unceasing visits by congressional VIPs, which became so disruptive that program manager Joseph V. Piland eventually ordered the staff giving VIP tours to "remain out of the immediate areas of construction and refrain from any conversation with the construction personnel."[179] NASA had to extend its 1 August 1967 deadline for completion of the LRL building by 40 days. This was not quite enough because the contractors did not entirely finish the building until the end of

---

178. Carter, *The Prevention of Back Contamination*, pp. 49–50; Carter, "Moon Rocks and Moon Germs," 241.
179. Carter, "Moon Rocks and Moon Germs," 241.

September, although Wilmot Hess held a press conference on 29 June 1967 to officially open it.[180]

Dealing with late equipment problems had its own set of complexities. The root cause of the late equipment situation was that the scientists who were developing specialized, cutting-edge equipment for LRL were intent on making products of the highest quality and reliability, regardless of the cost or time spent. Although Piland repeatedly communicated the need to deliver the equipment on schedule and within budget, most of it ended up getting to LRL late and overrunning cost targets. WNN Contractors could not complete its installation of equipment until late 1967, and even then, technical problems with the equipment had to be worked through in order to render it operational.[181]

Preparing and Implementing Planetary Protection Operational Procedures

Beginning months before Congress officially appropriated funds for LRL, the Interagency Committee on Back Contamination worked intensively with NASA to define procedures for the receiving facility that would minimize back contamination probabilities. These included various means of biologically isolating astronauts, samples, spacecraft, and personnel in contact with them during residence at LRL. In its 13 April 1966 meeting, ICBC approved a plan to contract with the Baylor medical school in Texas, or a similarly qualified institution, to develop a biological analysis protocol for lunar samples that would include planetary protection considerations.[182]

---

180. Compton, *Where No Man Has Gone Before*, p. 119; Carter, *The Prevention of Back Contamination*, p. 53.

181. Carter, "Moon Rocks and Moon Germs," 241.

182. ICBC, "Operational Requirements for an Integrated Lunar Receiving Laboratory," in a package beginning with ICBC member list, JSC Historical Archives, "May 1966" folder, box 76-12; ICBC, "Interagency Committee on Back Contamination: Views and Recommendations," attachment to George M. Low for Robert R. Gilruth letter to George E. Mueller, 8 June 1966, JSC Historical Archives, "June 1966" folder, box 76-12, LRL Chron. Files.

Federal laws required that "all precautionary steps be taken to prevent the introduction of pathogens [to the Earth's biosphere] that are harmful or destructive to human, animal, or plant life."[183] Implementing such steps was complex and multifaceted because it involved spacecraft hardware, crew and support personnel, and many strict procedures and training requirements. NASA gave Charles A. Berry, a medical doctor and MSC's Director of Medical Research and Operations, the responsibility for implementing effective measures at LRL and the rest of his Center to prevent back contamination as a result of lunar landing missions. He was charged with meeting regulatory Agency requirements and with approval of all protocols and specifications for back contamination prevention, including those employed on return flights and during recovery operations as well as those carried out at LRL. Berry worked closely with Walter W. Kemmerer, another medical doctor and the Assistant for Biomedical Operations in MSC's Science and Applications Directorate. MSC would be under intense scrutiny when lunar samples were received, with regulatory Agency representatives expected to be in residence. PHS, for instance, appointed G. Briggs Phillips as its MSC agent. Berry and Kemmerer had the job of certifying to these personnel and their agencies that all quarantine measures were being executed according to the agreed-upon procedures.[184]

Charles Berry also worked closely with ICBC which, as discussed above, included representatives from several regulatory agencies in its membership. Berry and the chair of ICBC, David Sencer, agreed that ICBC needed to approve the quarantine protocols implemented at MSC as well as to develop schemes for the release of lunar samples, spacecraft, and personnel after analysis for potential back contamination risks. This could be interpreted as an expansion of ICBC responsibilities from the original one discussed by NASA and PHS management, which envisioned the committee to be only an *advisory body* to NASA that would, as mentioned earlier in the chapter, "guide NASA in the

183. MSC, "Management Instruction: Assignment of Responsibility for the Prevention of Contamination of the Biosphere by Extraterrestrial Life," MSCI 8030.1, 9 January 1967, p. 1, in a package beginning with John E. Pickering to Walter W. Kemmerer, 25 January 1967, JSC Historical Archives, "January 13–31, 1967" folder, box 76-14.
184. Ibid., pp. 1–2.

development and conduct of a program to avoid possible contamination of the earth biosphere by extraterrestrial life."[185]

ICBC's original purpose was to give counsel on policy more than technical details. Now ICBC needed to approve the detailed and fairly technical quarantine actions that would be carried out at LRL as well as to formulate the decision trees to be followed for the release of people, equipment, and lunar material from quarantine.[186] In addition, as a Memorandum of Understanding between NASA and ICBC member agencies stated, ICBC needed to certify that the established release protocols and criteria had been met before the Administrator of NASA authorized their release.[187]

In February 1967, ICBC developed guidelines for analyzing lunar samples during their period of quarantine. ICBC greatly simplified the envisioned analysis by not requiring a search for living microorganisms in general, but only for infectious microorganisms. The suggested analyses included aerobic and anaerobic culturing; inoculating tissue cultures, eggs, plants, amphibia, invertebrates, and other animals; and conducting biochemical analyses.[188]

The Baylor University Operating Procedures

On 16 June 1967, a team of biologists and physicians at Baylor University College of Medicine published a formal, in-depth set of procedures for operating LRL and quarantining lunar samples and

185. William H. Stewart to James E. Webb, 22 December 1965, in package of material beginning with Frank A. Bogart to Deputy Director, Space Medicine, "Formulation of PHS-NASA Working Relationships Re Lunar Sample Receiving," 11 January 1966, JSC Historical Archives, "January 1966" folder, box 76-12, LRL Chron. Files.
186. James E. Webb, "Management Instruction: Interagency Advisory Committee on Back Contamination," in the package, "Reference Material for Meeting of Interagency Committee on Back Contamination," 2 March 1967, JSC Historical Archives, "March 1967" folder, box 76-15, Lunar Receiving Lab.
187. George M. Low to Lt. Gen. Frank A. Bogart (Ret.), 9 February 1967, and its attachment, "Memorandum of Understanding Between U.S. Public Health Service, Department of Agriculture, Department of the Interior and the National Aeronautics and Space Administration," JSC Historical Archives, "February 1967" folder, box 76-14.
188. Committee on Planetary and Lunar Exploration, "A History of the Lunar Receiving Laboratory."

astronauts.[189] Writing these procedures to test for lunar pathogenic organisms took far more time and effort than expected, because the personnel who envisioned the possible existence of Moon organisms could not agree on approaches for identifying them. Baylor alumnus Walter W. Kemmerer, a medical doctor who had been helping in the LRL planning process and, as mentioned above, was the Assistant for Biomedical Operations in MSC's Science and Applications Directorate, oversaw work on the procedures. PHS exerted its influence on the development and testing of the procedures as well as on the selection of the staff that would carry them out. PHS wishes needed to be respected because the organization could refuse to certify the procedures, which might delay the entire Apollo program.[190]

The Baylor protocol attempted to identify the biological studies that could detect the presence of agents that might be infectious or toxic to humans, animals, and plants. Furthermore, the protocol sought to provide safety clearance for the Moon samples, if possible, within 30 days. The protocol included three elements:

- Crew microbiology that compared preflight microbiology profiles with profiles following a return to Earth in order to establish any new microbial burden carried by the astronauts and whether they were free from communicable disease.
- In vitro attempts to culture microorganisms from Moon samples.
- Placement of Moon samples into terrestrial living systems for biohazard testing.[191]

Lunar material could conceivably harm terrestrial organisms due to any inherent toxicity or to a capability to propagate within Earth environments. The types of possible toxic materials envisioned included the following:

- Radioactive materials.
- Inorganic polymers.

---

189. Ibid.
190. Carter, "Moon Rocks and Moon Germs," 241; C. A. Berry to W. W. Kemmerer, 28 December 1965, JSC Historical Archives, "1 October–31 December" folder, box 76-11, LRL Chron. Files 1964–1965.
191. Baylor University College of Medicine, *Comprehensive Biological Protocol for the Lunar Sample Receiving Laboratory*, NASA CR9-2209 (16 June 1967): p. iii; Summary of protocol by John Rummel, 26 February 2004, *http://planetaryprotection.nasa.gov/pp/summaries/ nr9_2209.htm* (accessed 19 July 2005).

- Low-molecular-weight compounds that could act as cellular and metabolic poisons, mutagens, irritants, antimetabolites, or antivitamins.
- Metallo-organic compounds.

Harmful replicative materials might take the form of the following:

- Viral, bacterial, or fungal microorganisms carried from Earth to the Moon and returned in a mutated form.
- Plant materials of lunar origin capable of reproducing on Earth.
- Xerophilic fungi (molds that do not require free water for growth) of lunar origin.
- Living matter on the Moon at organizational levels above that of small metazoa (multicellular animals) or metaphytes (multicellular green land plants). The Baylor protocol excluded these from consideration because their probability of existence was estimated to be extremely low.[192]

The Maximum Biotesting Fraction

During the LRL planning process, controversies arose over the maximum fraction of material brought back from the Moon that the facility would be allowed to use to test for signs of life and potential hazards to Earth's ecology. LRL was competing for access to lunar material with the rest of the space science community. There were disagreements between various stakeholders of the Apollo mission, including opposing groups within NASA, as to which types of analyses would be prioritized. In particular, questions were raised regarding the relative importance of planetary protection studies as compared to other scientific inquiries.

The 1965 NASA Summer Conference on Lunar Exploration and Science,[193] known as the Falmouth Conference, had been attended largely by the scientific community outside of NASA, and it sorely

---

192. Baylor University College of Medicine, *Comprehensive Biological Protocol*, pp. iv–v.
193. W. David Compton, "Appendix 3b: Recommendations of the 1965 Summer Conference on Lunar Exploration and Science, Falmouth, MA, 19–31 July 1965," in *Where No Man Has Gone Before: A History of Apollo Lunar Exploration Missions* (Washington, DC: NASA SP-4214, 1989).

wanted maximal opportunity to analyze lunar material in depth. The Falmouth Conference recognized that a small amount of lunar sample should be used for quarantine testing but believed it important to limit this quantity so as to not unduly restrict the amount of rock available for a broad range of other experiments. Both the Falmouth Conference's Bioscience and Geochemistry Group Reports recommended that no more than 5 percent of the total lunar sample (an expected quantity of 1.2 kg or 2.6 lb) be used for quarantine testing. A 9 September 1966 NASA memo from Homer Newell to George Mueller strongly supported these recommendations, formally setting forth 5 percent as a "general guideline for the collection and handling of Lunar materials."[194]

A memo written several months later on 18 January 1967 from Charles Berry, MSC Director of Medical Research and Operations, suggested a different limit for biomedical assessment—10 percent of the lunar sample. ICBC was involved in developing this new limit, which Berry explained was based on an analysis conducted by MSC's Computations and Analysis Branch and "took into consideration the concentrations of organisms anticipated on the lunar surface, and the required probability of having an adequate number of organisms in each sample."[195]

Scientists inside and outside of NASA took exception to this change in sample allotment. A memo from MSC's Geology and Geochemistry Branch, for instance, asked that Berry's Medical Research and Operations Directorate reduce the new fraction to remain within the established guidelines, arguing that "The great demand for these samples for approved scientific experiments requires . . . the maximum possible sample."[196] The memo also asserted that Berry's claim of needing 10 percent of the lunar sample to furnish an adequate quantity of organisms for quarantine testing was highly speculative and

---

194. Chief, Geology and Geochemistry Branch to Deputy Director of Science and Applications, "Allotment of Lunar Sample for Quarantine Testing," 25 January 1967, in the package, "Reference Material for Meeting of Interagency Committee on Back Contamination," 2 March 1967, JSC Historical Archives, "March 1967" folder, box 76-15, Lunar Receiving Lab.
195. Charles A. Berry to Experiments Program Office, "Required Lunar Sample for Biomedical Assessment," 18 January 1967, in the package, "Reference Material for Meeting of Interagency Committee on Back Contamination," 2 March 1967, JSC Historical Archives, "March 1967" folder, box 76-15, Lunar Receiving Lab.
196. Chief, Geology and Geochemistry Branch, 25 January 1967.

the computations he referred to were not sufficient justification for increasing the allotment.

G. Briggs Phillips, PHS-CDC's liaison to NASA, responded to scientists' concerns with the assurance that regulatory agencies on ICBC were committed to "not using any greater amount of sample for quarantine than is absolutely necessary."[197] He also presented a document summarizing his and ICBC's position on the matter. This document, "Guidelines for the Selection of Protocol Tests," did not include any quantitative sample requirements, although it did list the general types of testing recommended, such as *in vitro* analyses versus *in vivo* exposures to living systems, and broad spectrum rather than selective culture media tests. The document appeared to be written at least in part to provide additional assurance that the smallest possible quantity of lunar sample would be consumed for quarantine assessments.[198]

The controversy over required sample size continued throughout the year. At the 7 June 1967 ICBC meeting, Walter Kemmerer proposed an approach that focused on the actual mass of lunar rock needed, rather than some percentage of the total lunar sample. He came up with a mathematical method for determining this, based on the desired reliability of 0.999 in the bioassessment tests, the number of test systems being planned, and assumptions concerning the concentrations of viable microbes that might be contained in a unit of sample. Kemmerer arrived at an estimate of 1.2 kg as the minimum amount of lunar material that would meet the quarantine test requirements and mentioned that 5 percent of a 22-kg sample return from the Moon would just about supply this amount. ICBC discussions followed on developing a *minimum,* rather than desired, quarantine testing protocol for a situation in which a lesser mass of sample was all that the Apollo mission could return from the Moon.[199]

---

197. G. Briggs Phillips to Clark Goodman, 2 February 1967, in the package, "Reference Material for Meeting of Interagency Committee on Back Contamination," 2 March 1967, JSC Historical Archives, "March 1967" folder, box 76-15, Lunar Receiving Lab.

198. G. Briggs Phillips, 31 January 1967, and attachment, "Guidelines for the Selection of Protocol Tests," in the package, "Reference Material for Meeting of Interagency Committee on Back Contamination," 2 March 1967, JSC Historical Archives, "March 1967" folder, box 76-15, Lunar Receiving Lab.

199. John E. Pickering, "Minutes: Interagency Committee on Back Contamination," 7 June 1967, JSC Historical Archives, "June 1967" folder, box 76-15, Lunar Receiving Lab.

In an 18 July 1967 letter to MSC Science and Applications Director, John Pickering also proposed resolving the argument over quarantine sample size by using mass rather than percentage of total sample as a measure, noting that the "desirable sample quantity for accomplishing the complete biological protocol of quarantine tests"[200] was about 1.2 kg of loose lunar material. He reiterated this proposal in a 15 September 1967 communication, indicating that the matter of sample size had still not been resolved and needed ICBC's early attention for "some rather serious deliberations"[201] on his proposed resolution. Nevertheless, the issue was still under debate in January 1968, when Charles Berry sent a memo to MSC Director of Science and Applications, arguing that from a quarantine testing point of view, a minimum sample size was absolutely necessary in order to "conduct an adequate protocol allowing unconditional release of the sample."[202] Berry held that at least this minimum amount of sample should be unconditionally dedicated to biomedical testing protocols, even if the total amount of sample obtained from the Moon was less than what was envisioned. Berry was, in essence, giving the highest priority to planetary protection considerations, arguing that they must not be ignored just because the quantity of returned lunar material was not enough to fulfill the needs of all the scientists who had planned experiments. Sufficient lunar material had to be tested in order to determine the following:

1. Whether there was valid concern about viable organisms coming from the Moon and potentially contaminating Earth.
2. What care and treatment the returning astronauts needed prior to their release from quarantine.

Although Kemmerer's and Pickering's proposals provided paths for resolving the controversy, the issue of a sufficient sample for quarantine testing was still on the table a year later in April 1969, just months before the launch of Apollo 11. In a letter to NASA Administrator

---

200. John E. Pickering to Wilmot N. Hess, 18 July 1967, JSC Historical Archives, "July 1967" folder, box 76-15, Lunar Receiving Lab.
201. John E. Pickering to Wilmot N. Hess, 15 September 1967, and attachment, "Interagency Committee on Back Contamination—Activities Summary," in a package beginning with John E. Pickering to Wilmot N. Hess, 28 September 1967, JSC Historical Archives, "October 1967" folder, box 76-15, Lunar Receiving Lab.
202. Charles A. Berry to Director of Science and Applications, "Minutes of the 6th meeting of the Interagency Committee," doc. no. DA-68-M623, 3 January 1968, JSC Historical Archives, "1–9 January 1968" folder, box 76-16, Lunar Receiving Lab.

Thomas Paine, David Sencer listed "Adequate . . . sample for quarantine testing"[203] as one of the requirements that needed to be met before LRL could be certified as a biologically safe containment facility and before any release of lunar materials could be effected.

Sencer added a caveat that made his statement into a threat: PHS was currently proceeding with a fallback plan to use other facilities than LRL for the lunar quarantine, "in the event that the LRL cannot be certified."[204] Sencer signed his letter not as the chair of ICBC, but with more weighty titles—the Assistant Surgeon General and the Director of the CDC. A letter of 2 July 1969 added additional clout, identifying an agreement between regulatory agencies to designate the MSC Director of Medical Research and Operations—Charles Berry, who had strongly advocated the use of up to 10 percent of the total lunar sample for quarantine testing—as their agent to "impose a quarantine during any phase of the Apollo 11 mission as it applies to the crew, spacecraft and/or the returned lunar materials."[205] The regulatory agencies and ICBC could thus, through Berry, hold up the threat of additional quarantine if they felt that an insufficient portion of lunar sample had been made available for biotesting.

Procedures for Release from Quarantine

ICBC developed a fundamental framework for deciding when quarantine requirements had been met and when the release of astronauts and their support personnel, lunar samples, and spacecraft was warranted. Although the decision procedures took pains to envision all likely quarantine occurrences, they also included contingency provisions that gave ICBC as well as individual regulatory agencies "adequate opportunity to provide requirements and suggestions for

---

203. David J. Sencer to Thomas O. Paine, 7 April 1969, and attachment entitled "Requirements," JSC Historical Archives, "April 1969" folder, box 76-24.
204. Ibid.
205. T. E. Jenkins for George H. Hage, letter to NASA Headquarters, "Back Contamination and Quarantine—Apollo 11," 2 July 1969, JSC Historical Archives, "2 July 1969" folder, box 71-45/46.

situations not covered in the formal plans."[206] For instance, the astronaut release scheme was based on the most likely occurrence, that Apollo crew and support personnel would exit the Crew Reception Area after 21 days if there were "no alterations in the general health of the quarantined people and no other indications of infectious diseases due to lunar exposure."[207] But if there were such occurrences, different branches of the decision tree would be followed (subject to review by ICBC and approval by NASA). Should an alteration in personal health occur, quarantine release would still not be affected if the medical team could show that it was noninfectious or of terrestrial origin. But if the incident could not be readily diagnosed, quarantine would likely be prolonged. The plans also required that LRL personnel exposed to potentially harmful lunar material due to a rupture of the vacuum cabinet system needed to be quarantined along with the astronauts.

LRL handled lunar sample release in a similar manner. If all the test protocols produced completely negative results, in that no viable organisms were isolated and no pathogenic effects were noted in the exposed animals and plants, then ICBC would recommend to NASA that the samples be released from quarantine after 30 days. If a replicating organism was detected in the samples but did not exhibit any deleterious effects on the life systems tested, then a series of analyses would be carried out to determine whether the organism was of terrestrial origin, unmodified by lunar exposure and nonpathogenic, or whether it was not readily classifiable and of potential danger.[208] If "definite deleterious effects"[209] were observed in one or more life systems, LRL personnel would carry out experiments to determine whether this was from chemical toxicity, a replicating microbe, or both. Quarantine would be prolonged for at least some of the lunar samples.

Photographic film and data tapes from the spacecraft were to be sent to LRL in the same manner as lunar material and placed in quarantine. The film was to be processed inside the quarantine area but printed through the biological barrier so that prints could be handled

---

206. ICBC, "Quarantine Schemes for Manned Lunar Missions," an attachment to John E. Pickering to Wilmot N. Hess, 16 June 1967, JSC Historical Archives, "June 1967" folder, box 76-15, Lunar Receiving Lab.
207. Ibid., p. 2.
208. Ibid., pp. 3–4.
209. Ibid., p. 5.

as noncontaminated. If, however, tests proved that the film could be sterilized without degradation using ethylene oxide, this would allow immediate release of the film.[210]

The spacecraft was to be quarantined in LRL next to the CRA and handled similarly to the lunar samples. It would also be available for extensive biosampling if required.[211]

Practice-Testing, Staff Training, Readiness Inspections, and Operational Simulations

LRL performed tasks that included sample receiving and distribution; geological, biological, and chemical analysis; and astronaut, sample, and spacecraft quarantining. Before LRL was ready to receive lunar samples and house the Apollo crew, the facility's staff had to receive training in the use of all equipment, then test it to ensure that it functioned correctly. The staff also had to receive thorough training in the safety and quarantining procedures that ICBC and the Baylor study stipulated. The practice-testing and staff-training period was a stressful time for LRL staff, for they were under intense deadline pressure to complete their training and testing before the launch of the first Moon landing mission. If they didn't finish this work on schedule, the concerns of ICBC and the agencies that it represented would not be satisfied and the Moon landing mission would have to be delayed.[212]

Many technical problems surfaced during the installation and testing of LRL equipment. This was not at all surprising, given the complexity of the facility and its operations. By mid-September 1968, however, the only serious problems that remained were with the autoclaves—pressure vessels for sterilizing tools and materials. These had been installed to sterilize items before they were transferred out of the biological containment area. They would present problems to LRL staff for many months.[213]

---

210. Ibid., pp. 8–9.
211. Ibid., p. 9.
212. Mangus and Larsen, *Lunar Receiving Laboratory Project History*, p. 12; Compton, *Where No Man Has Gone Before*, p. 2.
213. Compton, *Where No Man Has Gone Before*, p. 119.

Late in October 1968, Apollo science teams met at MSC to conduct training and simulation exercises on the operation of the sample receiving and processing sections of LRL. These exercises uncovered still more faults with the equipment, some of them major and affecting operation of the vacuum system in which lunar sample containers were to be opened. The vacuum chamber as well as most of the cabinets making up LRL's primary biological barrier were *glove boxes*, enclosures designed so that a range of tools inside the chambers could be manipulated by staff through a pair of impermeable gloves attached to ports in the chamber wall. In LRL, these gloves had to be strong enough to withstand a pressure difference of about 15 pounds per square inch (psi) formed by a high vacuum within the chamber and normal atmospheric pressure without. The glove material thus had to be thick and tough and, as a result, was quite stiff, making it hard for operators to use small hand tools effectively. Another problem that was identified during simulation exercises related to visibility into the chamber. Operators looked through viewing ports to observe chamber operations and these ports did not allow full coverage of all parts of the chamber, but left blind spots in some corners. These and other issues resulted in over 80 major and minor modifications of the vacuum system.

By October 1968, enough of the problems had been ironed out to conduct an operational readiness inspection that would identify any final issues that had to be addressed. MSC Director Robert R. Gilruth appointed a team of 10 people from MSC, Langley Research Center, Fort Detrick (which LRL had been partially modeled after), and NASA Headquarters. This operational readiness team spent a month starting in November 1968 investigating LRL physical facilities, the staff selected, their level of training, and the operational procedures they were following. Each section of the LRL organization had to present detailed descriptions of its equipment and operational procedures to the operational readiness team, who reported back in mid-December 1968, identifying 72 mandatory and 91 desirable modifications to make LRL ready to begin full operations. LRL staff responded quickly to these findings.[214]

---

214. Compton, *Where No Man Has Gone Before*, pp. 119–120; Mangus and Larsen, *Lunar Receiving Laboratory Project History*, p. 42.

During this period, public interest remained high in the quarantine facility, and many people continued to seek tours of the facilities where astronauts would be isolated and Moon rocks examined. As mentioned earlier, managing these visitors was time consuming and drained MSC resources needed elsewhere, especially when visitors such as members of Congress made operational recommendations that they expected to be taken seriously.

When Apollo 8 successfully achieved lunar orbit during the Christmas season of 1968, it underlined the fact that LRL needed to become certified for full operation as soon as possible. On 15 December 1968, quarantine testing trials involving biological agents as well as intensive, continuous personnel training activities were initiated at LRL. From this date until the end of quarantine following Apollo 11, LRL staff planned to operate the sample laboratory and Crew Reception Area as they would be during the mission. (These areas would not be officially certified as mission ready, however, until months later.) Wilmot Hess, MSC Director of Science and Applications, issued a notice limiting tours of the sample laboratory and Crew Reception Area only to "very special VIP visitors"[215] and made sure this policy was followed by locking all entrances to the area except one, which he kept guarded.

As LRL activity ramped up in preparation for the first Moon landing, NASA Headquarters staff began to get more involved with MSC's planetary protection program and needed to visit the receiving facility. NASA Apollo program director Sam Phillips traveled to MSC in early February 1969 for an in-depth review of back contamination prevention measures from splashdown of the Command Module until its release from quarantine. George Mueller, Associate Administrator for Manned Space Flight, whom NASA had just given full responsibility for back contamination control, visited LRL 10 days later along with a task force of advisers from various regulatory agencies. These advisers found equipment problems, a technician shortage, insufficiently trained staff, and inadequate biological testing protocols.[216] Wolf Vishniac, the NAS representative on ICBC, also took issue with

215. Wilmot N. Hess, "Visitor Tours in the Lunar Receiving Laboratory," Manned Spacecraft Center Announcement, No. 68-165, 18 November 1968, folder 012990, "Lunar Receiving Lab," NASA Historical Reference Collection.
216. Compton, *Where No Man Has Gone Before*, p. 124.

the status of the quarantine program during this time. In a letter to Frederick Seitz, President of the National Academy of Sciences, he commented that "Routine apparatus does not seem to work properly, so for instance autoclaves tend to fill with water. This was still true on February 27."[217]

During the early part of 1969, preparations for a critical 30-day simulation of the facility's operations in March and April dominated LRL staff and contractor employee activities. This was the most complex test of the facility to date. Central to the simulation were exercises in the chemistry and geology areas, where LRL's vacuum system sat. Simulated Moon materials went through the vacuum chamber during the simulation, from which they were dispersed throughout LRL. Other parts of the simulation involved the biological part of LRL, which ran back contamination experiments.

LRL performed better during this full-scale simulation than it had in previous exercises, but many problems still surfaced. In a letter that David Sencer, chairman of ICBC, sent to the new NASA Administrator Thomas Paine on 7 April 1969, he listed what he considered to be disappointing results of the simulation, stating that LRL would not receive certification as a containment facility unless "rather drastic changes are made in the priority of activities and operations of the laboratory"[218] In Sencer's mind, the most glaring deficiency was malfunctioning autoclaves for sterilizing items moving in and out of the biological barrier that enclosed the sample laboratories and crew quarters. Sencer also complained that personnel were not sufficiently trained nor was "a scientific discipline apparent in the overall operation directed to the quarantine requirements."[219]

Another important issue concerned the glove box system. Gloves for handling lunar samples kept developing holes, which could result in back contamination of Earth as well as terrestrial contamination of

---

217. Wolf Vishniac to Frederick Seitz, 5 March 1969, in a package beginning with the letter, Frederick Seitz to Thomas O. Paine, 24 March 1969, folder 009901, "Lunar Quarantine and Back Contamination," NASA Historical Reference Collection.

218. David J. Sencer to Thomas O. Paine, regarding a critique of LRL's biological protocols, 7 April 1969, part of a collection of letters and memos that begin with a letter from T. O. Paine to David Sencer, dated 9 May 1969, folder 012990, "Lunar Receiving Lab," NASA Historical Reference Collection.

219. Sencer to Paine, 7 April 1969.

the samples, reducing their scientific value. At the conclusion of the full-scale simulation, many in the Apollo program worried that LRL would not be operationally ready by the time of the first scheduled Moon landing.[220]

Sencer enclosed a list of requirements with his letter that would have to be met before LRL could receive certification as a biologically safe containment system. These included requirements on the following aspects of LRL:

1. *Astronaut release from LRL.* The Crew Reception Area needed to be certified as a containment facility. To achieve this, adequate staffing and infrastructure needed to be implemented for personnel and health needs, including diagnostic laboratory capabilities. Also, an approved protocol for an astronaut release decision needed to be developed.

2. *Vacuum chamber operations for opening lunar samples.* The LRL vacuum system needed to receive certification as a containment facility. This required contingency plans for preserving the integrity of the Moon samples as well as emergency and disaster procedures for breakdowns in the vacuum system.

3. *Sample processing.* All sample laboratories needed to be certified as containment facilities. This required procedures for sample transfer under approved quarantine conditions, an adequate biotesting protocol, a sample to be used for quarantine testing, samples for time-dependent studies, and fall-back procedures for abnormal and catastrophic events.

4. *Sample release from LRL.* LRL needed to develop time-temperature curves of sample sterilization characteristics and other test data for conditional and unconditional sample releases. LRL also had to furnish biotesting results needed for an ICBC decision.[221]

---

220. Carter, "Moon Rocks and Moon Germs," 246; Compton, *Where No Man Has Gone Before*, pp. 119–120, 124; Mangus and Larsen, *Lunar Receiving Laboratory Project History*, pp. 50–51.
221. David J. Sencer, "Requirements," an attachment to the 7 April 1969 letter from Sencer to Paine, part of a collection of letters and memos that begin with a letter from T. O. Paine to David Sencer, dated 9 May 1969, folder 012990, "Lunar Receiving Lab," NASA Historical Reference Collection.

The projected time of Apollo 11's launch and attempted lunar landing was now during summer of 1969, and LRL preparations intensified as summer drew nearer. By April 1969, LRL preparations for receiving materials from the Apollo 11 Moon landing reached a state of urgency. NASA Headquarters staff and other personnel conducted a complete facility review on 17–18 April 1969, again examining all aspects of the facility and its equipment, personnel, and procedures.[222] NASA's John Pickering, executive secretary of ICBC, made up a projected schedule for ICBC of key LRL activities through the end of the Apollo 11 mission. Notable projected dates and actions from Pickering's schedule included the following:

9 May–15 July 1969. Simulations of operations of each of LRL's individual laboratories occur.

12 May–15 July 1969. Parts of LRL go into mission mode, performing actions vital for the safe and successful analysis of lunar materials. Examples include the following:

- Testing Apollo lunar sample return containers.
- Growing seedlings to proper levels of maturity for experiments with lunar samples.

19 May–15 June 1969. LRL closes its biobarrier as the facility goes into mission mode. The Apollo 10 mission flies during this time. While it does not land on the Moon, it includes a rehearsal for parts of the landing missions to follow, using LRL staff and facilities to conduct pre- and postmission crew microbiology studies.

14–26 July 1969. Based on the best guess, Apollo 11 flies during this period.

26–28 July 1969. Lunar samples enter LRL.

28–31 July 1969. Sample protocol begins. Note: if samples cannot be delivered within about a 5-day window, LRL biology staff need to restart the growth of cell lines and other test organisms.

27–29 July 1969. Astronauts enter LRL's Crew Reception Area.

11–18 August 1969. ICBC decides, if possible, on the date for quarantine to end and astronauts to be released.

30 July–1 August 1969. Spacecraft enters LRL.

1 September 1969. The first 30 days of spacecraft analysis end. The decision to release the spacecraft may be made.

---

222. Mangus and Larsen, *Lunar Receiving Laboratory Project History*, p. 51.

29 September–6 October 1969. The of 60 days of biotesting end. The release of lunar samples is possible.[223]

On 29 April 1969, George Mueller, NASA's Associate Administrator for Manned Space Flight, distributed a memo referring to David Sencer's 7 April 1969 letter and underlining the seriousness of the deficiencies that ICBC found during its evaluation of LRL's 30-day operations simulation. The NASA memo gave details of the ICBC evaluation, which cited inferior laboratory procedures, unexplained plant and animal deaths, and what appeared to be casual concern for back contamination prevention. NASA's memo stressed that ICBC was the certifying agency for LRL, as stipulated in Interagency Agreement NMI 1052.90, and that meeting ICBC's requirements for certification would remain "high priority tasks."[224] In addition, NASA Administrator Thomas Paine wrote a letter on 9 May 1969 to David Sencer, who was also the Director of PHS's National Communicable Disease Center in Atlanta. Paine told of LRL's significant progress in meeting ICBC recommendations, including correcting the mechanical deficiencies in the autoclaves. Paine also wrote of the priority that LRL was giving to developing procedural documents and test plans and obtaining test results that would meet certification requirements.[225]

Certification and a Final Simulation

ICBC considered quarantine requirements to be absolutely critical to operating LRL and formulated a detailed certification process for

---

223. John E. Pickering to members of ICBC, 21 April 1969, part of a collection of letters and memos that begin with T. O. Paine to David Sencer, 9 May 1969, folder 012990, "Lunar Receiving Lab," NASA Historical Reference Collection; Lunar and Planetary Institute, "Apollo 10 Mission," copyright 2005, *http://www.lpi.usra.edu/expmoon/Apollo10/Apollo10. html* (accessed 16 August 2005).

224. NASA Headquarters, Associate Administrator for Manned Space Flight, "LRL Readiness," Administrator's office briefing memo, 29 April 1969, part of a collection of letters and memos that begin with T. O. Paine to David Sencer, 9 May 1969, folder 012990, "Lunar Receiving Lab," NASA Historical Reference Collection.

225. T. O. Paine to David J. Sencer, regarding LRL's meeting ICBC certification recommendations, 9 May 1969, part of a collection of letters and memos that begin with T. O. Paine to David Sencer, 9 May 1969, folder 012990, "Lunar Receiving Lab," NASA Historical Reference Collection.

the facility to be deemed safe and ready to operate. One of the recurring problems that ICBC chairman David Sencer had with LRL staff, however, was that they seemed to view quarantining requirements as "an imposed operation to be done the easiest way possible while hoping that it would go away."[226] Sencer felt that LRL staff needed an enthusiastic commitment to meeting these requirements if they were to be reliably followed.

ICBC repeatedly sent representatives to the partially completed LRL to monitor progress on it and ensure that planetary protection requirements were strictly followed. ICBC's scrutiny was not particularly welcome to MSC staff, who worried that the committee would delay completion of LRL and the whole Apollo program. As LRL neared completion, basic differences between ICBC's values and those of LRL staff became increasingly apparent. ICBC "tended to define the facility solely in terms of back contamination concerns and focused on developing quarantine requirements and biological testing protocols,"[227] which clashed strongly with the research interests of the Apollo mission's scientists, especially earth scientists such as geologists and geophysicists.

While LRL needed to pass the ICBC certification process in order to achieve readiness mode for receiving Apollo 11 samples, other organizations also wanted to be sure that the facility was prepared for full operation. PHS was one of these organizations. PHS management had researched in depth the legal issues governing quarantining actions and had concluded that the United States Surgeon General retained lawful authority over quarantine matters. In particular, G. Briggs Phillips, PHS's liaison officer at MSC since mid-1965, determined that "the Surgeon General of the U.S. Public Health Service is empowered by law to make and enforce such regulations as may be needed to prevent the introduction and spread of communicable disease into the United States, its Territories or possessions,"[228] and by right of this power, "returned lunar material, until free of possible pathogenic or

226. Mangus and Larsen, *Lunar Receiving Laboratory Project History*, p. 47.
227. Mangus and Larsen, *Lunar Receiving Laboratory Project History*, p. 12; Compton, *Where No Man Has Gone Before*, p. 35.
228. Mangus and Larsen, *Lunar Receiving Laboratory Project History*, p. 12; Compton, *Where No Man Has Gone Before*, pp. 32–33.

infectious materials and otherwise not harmful to man's biosphere, should be controlled, quarantined, and tested in accordance with procedures approved by the Surgeon General or his representative."[229] In short, if PHS was not convinced that LRL and its quarantining actions were doing their job, PHS had a mandate to stop the Apollo program from completing its agenda.

Phillips also identified another organization, the Department of Agriculture, that had authority regarding quarantining. Parallel to PHS's interest in potential dangers to human health and lives, the Department of Agriculture had a mandate to prevent harmful contamination risks, including lunar contamination risks, to plant and animal life.

Both PHS and the Department of Agriculture were represented on ICBC. The ICBC chairman, in fact, was from PHS. These and other regulatory agencies had initially insisted on "exercising final approval authority over all phases of the program that might affect biocontainment." It would have been quite a difficult situation for the U.S. space agenda, however, if one regulatory agency could have stopped work on the program because it didn't approve the Apollo program's planetary protection procedures.

NASA, not wanting to compromise its mission objectives, sought to limit this power of individual regulatory agencies to halt its activities. After more than 18 months of negotiations between NASA and the members of ICBC, they reached a compromise agreement, signed on 24 August 1967, in which the regulatory agencies agreed "not to take any actions that might have an effect on the lunar program (such as refusing to let astronauts back in the country) without the 'unanimous recommendation of the agencies represented on the [Interagency] Committee [on Back Contamination].'" This was a hugely important agreement. Since NASA was itself a member of ICBC, *no actions could be taken without its approval.*

On the other hand, NASA had to agree that it would not take any actions of its own that might compromise biocontainment of lunar materials without first consulting ICBC, unless such actions were mandatory for the safe conduct of the mission. But NASA could, if pressed, have claimed that virtually any action it wanted to perform was necessary in

---

229. Mangus and Larsen, *Lunar Receiving Laboratory Project History*, p. 12; Compton, *Where No Man Has Gone Before*, p. 33.

order to protect a crew or spacecraft, and so it was far less restricted in its actions than were the regulatory agencies on ICBC.[230]

Throughout spring 1969, LRL staff continued to work on bringing their facility, and especially its quarantine and back contamination prevention capabilities, up to full operational readiness. During this period, ICBC formulated a plan for certifying LRL. A team headed by G. Briggs Phillips, PHS's liaison officer at MSC, would inspect LRL and prepare a report detailing any modifications to the facility that were still needed. NASA would make the recommended changes, then submit a certification to ICBC that it had done so, and ICBC would sign the certification. ICBC would not perform any further inspections to verify that NASA had actually made the modifications requested.[231]

The Phillips team inspected LRL on 26–28 May 1969 and prepared a report of its findings that it gave to ICBC on 5 June 1969. LRL management assigned area test directors the responsibility for implementing the Phillips team's recommendations. After completion of the allocated tasks, each director signed a certificate of test-area readiness asserting that the area was "in a state of readiness to conduct the Apollo 11 mission."[232] NASA then forwarded the certificates to ICBC for concurrence.

By June and early July 1969, LRL launch preparation activities reached their peaks of intensity, with "sixty to seventy-hour work weeks . . . the rule rather than the exception."[233] LRL conducted another mission simulation in order to refine sample handling procedures and smooth out the last remaining technical issues. In spite of the pressure and long hours imposed on personnel, however, reports indicated that they remained enthusiastic about the work and excited about the impending Moon landing. Scientists from academia who were part of the mission started to arrive during the weeks before the landing so that they would be fully prepared for the returning samples. LRL was scheduled to attain full readiness mode on 14 July 1969, just in time for the Apollo 11 launch, which was to be on 16 July. The last of the certificates of test-area readiness, however, was not officially

---

230. Carter, "Moon Rocks and Moon Germs," 242–243.
231. Ibid., 247.
232. Ibid.
233. Mangus and Larsen, *Lunar Receiving Laboratory Project History*, p. 51.

signed until less than 24 hours before the Apollo 11 Command Module splashed down in the Pacific Ocean.[234]

Planetary Protection Strategies from Moon Walk Until Arrival at LRL

The success of the Apollo program's planetary protection efforts depended on controlling and carefully handling all extraterrestrial materials being brought back to Earth and isolating lunar samples and all potentially contaminated materials, astronauts, and parts of the spacecraft until they could be thoroughly analyzed and safely released. Critical planetary protection measures were executed starting on the Moon walk itself and continued on the return journey from the Moon to Earth, during the sensitive period after splashdown, and on the trip from the recovery ship to LRL.

Plans for Preventing Back Contamination from Moon Walks, Docking, and the Return to Earth

When the Apollo astronauts finished their Moon walks and prepared to reenter their ship, they took care to brush off as much loose lunar material from their spacesuits as they could and leave their outer shoe coverings and backpacks behind on the Moon, lowering the probability of transporting lunar contaminants back to Earth's environment. Before docking their Lunar Module (LM) with the Command Module (CM) orbiting the Moon, the two astronauts who had visited the surface would clean the LM's insides with a vacuum brush in order to minimize the quantity of material that they might carry into the CM.[235]

---

234. Compton, *Where No Man Has Gone Before*, pp. 125–126; Carter, "Moon Rocks and Moon Germs," 247.

235. Dora Jane Hamblin, "After the Moonflight, A Wall Against the Unknown—A Close Watch on the Men and Their Prizes," *Life* (4 July 1969): 50–58, folder 012990, "Lunar Receiving Lab," NASA Historical Reference Collection; Carter, "Moon Rocks and Moon Germs," 246; "Minutes—Interagency Committee on Back Contamination," 3 October

Extensive cleaning of the LM would also be carried out while it was docked with the CM. This would be a shirt-sleeve procedure and would include LM cabin gas being continuously circulated through lithium hydroxide (LiOH) filter beds intended to remove virtually all particulate matter. A minimum of 5 hours was envisioned as available for LiOH filtration. NASA estimated that this time would be sufficient, assuming a uniform distribution of dust particles in the LM atmosphere, to reduce the original airborne particulate count to 10^{-15} percent. The spacecraft would be under weightless conditions during this period. Past experience in 19 human spaceflights indicated that in weightless environments, dust, dirt particles, and small scraps of various materials tended to float around the cabin rather than cling to surfaces; this was expected to increase the likelihood that such particles would be carried into the LiOH filter and removed from the cabin atmosphere.[236]

To minimize airborne particulates floating from LM to CM during the time they were docked and crew and sample containers were being transferred, atmospheric pressure in the CM would be kept higher than in the LM so that gas would flow *out of* rather than into the CM and exit the spacecraft through the LM's cabin relief valve.[237]

During the trip back to Earth, the astronauts would repeatedly vacuum and wipe the CM interior. Their vacuums were designed to pick up any Moon particles that had gotten into the module and trap them in chemical filters. In addition, the CM atmosphere would be constantly circulated through a series of LiOH filters. Tests demonstrated that these filters trapped particles as small as the size of bacteria. In discussions from 1966 and 1967, ICBC had come to the conclusion that if CM air was filtered through lithium hydroxide canisters prior to splashdown, sufficient lunar material would be removed from the vessel's atmosphere such that no further air filtration would be required

---

1966, in the package of letters and memos beginning with Col. J. E. Pickering to David J. Sencer, 28 October 1966, folder 4714, "MSC-Lunar Receiving Laboratory," NASA Historical Reference Collection.

236. Richard S. Johnston, Special Assistant to the Director of the MSC, *Report on the Status of the Apollo Back Contamination Program*, 28 April 1969, pp. 4–6, JSC Historical Archives, "April 1969" folder, box 76-24.

237. Johnston, *Report on the Status of the Apollo Back Contamination Program*, pp. 4–6.

to prevent back contamination.[238] ICBC based this conclusion on calculations of spacecraft system characteristics—in particular, the LiOH filter efficiency reported to ICBC, 43–50 percent, and the number of air filtrations a day, 180, performed by the CM air circulation system.[239] MSC modeling predicted that after 63 hours of operation at the time of splashdown, 10^{-90} percent, or virtually none, of the original particulates would remain in the CM atmosphere.

Questions arose as to what happened when carbon dioxide in the cabin atmosphere contacted the LiOH filters. Tests indicated that the reaction that followed would produce water and a sticky surface that would enhance the collection of dust particles. Other testing indicated that the environment inside the filters killed at least some varieties of microbes, which would tend to increase the filters' capability to prevent back contamination.[240]

After Splashdown: Plans To Safely Transport the Apollo Astronauts, Command Module, and Samples to the Recovery Ship

One of the most difficult periods in which to prevent back-contamination was from the time the Apollo Command Module splashed down in the Pacific Ocean until the delivery of the module, its crew, and its lunar samples to LRL. If a significant quantity of contaminants escaped to Earth's environment during Apollo CM recovery operations, the basic objective of the planetary protection program would be subverted and the value of LRL greatly reduced. Weighing the risk of such

---

238. ICBC, "Summary," and "Minutes—Interagency Committee on Back Contamination," 16 December 1966, p. 4. Both documents are in a package beginning with John E. Pickering to Wilmot N. Hess, 16 February 1967, JSC Historical Archives, "February 1967" folder, box 76-14.

239. ICBC, "Activities Summary," 31 May 1967, p. 8, JSC Historical Archives, "May 1967" folder, box 76-15.

240. Dora Jane Hamblin, "After the Moonflight, A Wall Against the Unknown—A Close Watch on the Men and Their Prizes," Life (4 July 1969): 50–58, folder 012990, "Lunar Receiving Lab," NASA Historical Reference Collection; Carter, "Moon Rocks and Moon Germs," 246; "Minutes—Interagency Committee on Back Contamination," 3 October 1966, in the package of letters and memos beginning with Col. J. E. Pickering to David J. Sencer, 28 October 1966, folder 4714, "MSC-Lunar Receiving Laboratory," NASA Historical Reference Collection; Johnston, Report on the Status of the Apollo Back Contamination Program, pp. 6–9.

an occurrence against astronaut safety became an issue of debate in 1966. In its 13 April 1966 meeting, ICBC expressed "its concern with the possibility of uncontrolled out-venting of CM atmosphere following splashdown"[241] and recommended prototype testing of the module's return procedures along with the contamination control equipment to be used.

ICBC wanted the CM, containing the crew and lunar samples, to be "hoisted aboard the recovery aircraft carrier unopened and placed inside a large van which would then be hermetically sealed. The sealed van would be transported back to LRL and attached to an airlock in the wall of the Crew Reception Area."[242] Deke Slayton, however, one of the original seven Mercury astronauts and now MSC's Director of Flight Crew Operations, refused to support any quarantine protocol that put the safety of astronauts at risk. One of the major risks to astronauts was the thermal environment inside an unopened, unvented module. Wayne Koons, who was a supervisor in MSC's Landing and Recovery Division and had responsibility for the design and operation of postsplashdown retrieval equipment, emphasized that after splashdown, the CM was "not really designed to be sealed up, because the first thing you have to do is ventilate in order to provide air and cooling for the crew. The basic plan is to be operating in equatorial waters. It's going to get really hot really soon in there."[243] NASA calculations, in fact, indicated that astronauts could not survive in a poorly ventilated module during the postlanding phase of the mission unless supplemental cooling capabilities were added to the spacecraft.[244]

Another risk to the astronauts was associated with plucking an occupied CM from the ocean. During the Gemini program, NASA had

241. John E. Pickering, "Interagency Committee on Back Contamination: Views and Recommendations," in a package beginning with George M. Low for Robert R. Gilruth, to George E. Mueller, 8 June 1966, JSC Historical Archives, "June 1966" folder, box 76-12, LRL Chron. Files.
242. Kent C. Carter, *The Prevention of Back Contamination from the First Manned Lunar Landing: A Study in Organization*, M.A. thesis, University of Cincinnati (1972): 98–99.
243. Wayne Koons, former supervisor in MSC's Landing and Recovery Division with responsibility for design and operation of postsplashdown retrieval equipment, telephone interview by author, 27 January 2006.
244. "Current System Performance Incorporating Bacterial Filter," in the package, "Interagency Committee on Back Contamination—Dr. Berry," JSC Historical Archives, "May 1966" folder, box 76-12, LRL Chron. Files.

tested the procedure of hoisting a module with its crew aboard up from the sea and had concluded that it was too dangerous.[245] NASA doubted whether any aircraft carrier in the U.S. fleet could safely lift an occupied Apollo CM out of the Pacific and onto the deck with the astronauts aboard, especially in rough seas.

According to John Stonesifer, JSC's branch chief for recovery systems during the Apollo mission, this very issue "was roundly discussed over and over and over. I was the one in charge of giving the briefing to the managers—the pros and cons of lifting a spacecraft with astronauts in it versus bringing them back to the ship in the helicopter. We went through our evolution of how difficult it is to bring that spacecraft over along by the ship and hoist it aboard. Wave action, sea action—we practiced many recoveries out here in the Gulf of Mexico,"[246] where conditions were far gentler than in the open ocean. And NASA conducted these exercises with a ship that was far smaller and more easily controlled than the aircraft carrier that would be used for the actual recovery. Even under these less dangerous conditions, it proved difficult to position the ship so that the CM could be safely hoisted aboard.

In Stonesifer's words, an aircraft carrier "is like a sail. I've seen carriers drift down on the spacecraft. I've seen destroyers—which are more maneuverable than the carriers—I've seen them run into spacecraft. There's just no good way, sure way of bringing a carrier up to the spacecraft and retrieve it in the open sea."[247]

Another danger pointed out by Bob Fish, Apollo curator of the USS Hornet Museum (the *Hornet* was the recovery ship for Apollo 11 and 12), was that if there was a high sea state, the onboard crane would need to lift the CM just at the moment when it was on top of a swell. If it was sliding down a swell when lifted, then the CM would swing back and forth like a pendulum and could crash into

---

245. Carter, *The Prevention of Back Contamination*, p. 99, and "Moon Rocks and Moon Germs," 245; Committee on Planetary and Lunar Exploration, "A History of the Lunar Receiving Laboratory," 2002.

246. John Stonesifer, MSC branch chief for recovery systems during the Apollo mission, telephone interview by author, 13 January 2006.

247. Stonesifer interview, 13 January 2006.

the side of the recovery ship. This had happened on another recovery maneuver.[248]

NASA had an additional concern regarding the actual retrieval equipment the ship would use to haul the CM aboard. The shipboard cranes were primarily designed for dockside or sheltered-water procedures. The estimated dynamic loads that they might encounter lifting the CM in open-sea conditions with 8-foot swells were as high as 44,000 pounds. But in order to provide a 50-percent safety factor, the cranes were prohibited from lifting loads more than 32,000 pounds. In addition, the Navy did not consider these cranes "manrated."[249]

Because of the above dangers, NASA concluded that it would be much safer to bring the astronauts aboard an aircraft carrier recovery ship by means of a helicopter. By doing so, however, the Agency faced the different problem of limiting the release of lunar contaminants into the water and air while the CM was still floating in the ocean, after it had been opened and the astronauts had exited from it.[250]

Richard S. Johnston, special assistant to MSC Director Gilruth, believed that lunar contaminants could be reliably contained even if the CM was opened while in the ocean. He asserted that the planned cleaning activities described in the previous section, which involved extensive vacuuming and wiping of CM surfaces as well as filtration of the module's atmosphere, would ensure that by splashdown, the spacecraft was sufficiently free of contaminants that its hatch could be ajar without endangering the terrestrial environment.[251]

Others did not share Johnston's confidence. ICBC studied the back contamination potential of an ocean-exit recovery strategy and recommended that NASA install biological filters in the CM's air vents. NASA studied such a plan in some detail, examining whether a filter could be put into place without significant modifications to the module's venting fan or power consumption. The analysis showed that this would not be the case, unfortunately. Installing a filter with 99.9

---

248. Bob Fish, trustee of the Aircraft Carrier Hornet Foundation and Apollo curator for the ship, interview by author, Alameda, CA, 17 January 2007.

249. Johnston, *Report on the Status of the Apollo Back Contamination Program*, pp. 16–17.

250. Committee on Planetary and Lunar Exploration, "A History of the Lunar Receiving Laboratory"; Carter, "Moon Rocks and Moon Germs," 245.

251. Hamblin, "After the Moonflight"; Carter, "Moon Rocks and Moon Germs," 246; "Minutes—Interagency Committee on Back Contamination," 3 October 1966.

percent bacterial removal efficiency would dramatically reduce the atmospheric circulation rate within the module by about 83 percent. This would endanger astronaut survivability unless a supplemental cooling capability was added, such as through use of water-cooled garments. But the NASA study estimated that utilizing the current postlanding fan design and adding hardware to water-cool garments for the astronauts would result in a weight penalty for the Command Module of about 80 pounds, provide only marginally adequate thermal control for the astronauts, and result in an uncomfortable cabin humidity of 98 percent. Alternatively, the module's postlanding fan system could be redesigned. To achieve the same thermal cooling capacity that the module had without a bacterial filter, a large, 325-watt postlanding fan would be required (the current power requirement was only 15 watts), which would impose a weight penalty of approximately 470 pounds.[252] NASA rejected both of these plans as impractical.

At its 28 October 1966 meeting, ICBC accepted NASA's decision and decided that if the CM's atmosphere was continuously recirculated "without partition"[253] through the craft's LiOH canisters, no filter would be needed after splashdown for protecting Earth's biosphere. ICBC reaffirmed this position at its 12 January 1967 meeting and again on 4 October 1967. But this approach was not the committee's first choice, and the matter was not yet closed. At its 8–9 February 1968 meeting, ICBC returned to its earlier recommendation for a biological filter and fan that would operate postsplashdown to remove any remaining contaminants from the CM, which would have to remain sealed for some period of time.[254] The committee explained that its new action

---

252. George M. Low for Robert R. Gilruth, to George E. Mueller, 8 June 1966, and the following technical documents: "Current System Performance Incorporating Bacteria Filter," "New Postlanding Fan," "System Incorporating Water Cooled Garment Using Current Fan," and "Conclusions." The letter and all documents are contained in a package entitled, "Interagency Committee on Back Contamination—Dr. Berry," JSC Historical Archives, "May 1966" folder, box 76-12, LRL Chron. Files.

253. Johnston, *Report on the Status of the Apollo Back Contamination Program*, 28 April 1969, p. 11.

254. Sam C. Phillips to Manager, Apollo Spacecraft Program Office, "ICBC," 6 April 1969, in a package beginning with John E. Pickering to "Dear Committee Member," 11 April 1969, JSC Historical Archives, "April 1969" folder, box 76-24.

was motivated by the need to "avoid what could be compromised and untoward decisions."[255]

Wolf Vishniac, the NAS representative on ICBC, wrote a letter to Frederick Seitz, President of the NAS, outlining a plan that met ICBC's recommendation, but did not subject astronauts to dangerous temperature extremes. Seitz forwarded Vishniac's idea to NASA Administrator Thomas Paine. This alternate plan envisioned that after splashdown but before opening the exit hatch, swimmers from the recovery crew would "place a biological filter over the [CM's] vent holes from outside and also provide a power pack to drive sufficiently powerful fans"[256] that would draw the module's atmosphere through the filters and decontaminate it.

Vishniac strongly opposed venting the spacecraft in an uncontrolled manner, believing that it was "irresponsible to leave a large breach in the biological barrier in any part of the recovery procedure."[257] Opening and venting the spacecraft to Earth's atmosphere after splashdown would, in his view, make the rest of Apollo's elaborate quarantine program pointless. NASA did not follow Vishniac's alternate plan, however, believing that its current approach would provide sufficient protection against back contamination. In addition, MSC recommended against Vishniac's scheme because of the possible danger to recovery crew swimmers when trying to install the filter/fan package under open-sea conditions.[258]

Vishniac realized that he was fighting a losing battle on altering NASA's recovery strategy,[259] noting that "the Apollo Program is moving at a pace which we [ICBC] can not stop. It is equally clear that this irresistible progress is being used to brush aside the inconvenient restraints which the Interagency Committee has considered to be an essential part of the Quarantine Program."[260]

255. Johnston, *Report on the Status of the Apollo Back Contamination Program*, p. 11.
256. Wolf Vishniac to Frederick Seitz, 5 March 1969, in a package beginning with Frederick Seitz to Thomas O. Paine, 24 March 1969, folder 009901, "Lunar Quarantine and Back Contamination," NASA Historical Reference Collection.
257. Vishniac to Seitz, 5 March 1969.
258. Johnston, *Report on the Status of the Apollo Back Contamination Program*, p. 11.
259. Carter, "Moon Rocks and Moon Germs," 245.
260. Vishniac to Seitz, 5 March 1969.

ICBC Chairman David Sencer understood this as well and felt that he had to add a strong statement for the record on the importance of preventing back contamination. In a letter to NASA Administrator Thomas O. Paine on 7 April 1969 (a different letter than the one he sent on that day regarding the disappointing results of LRL's March simulation exercise), he asserted that NASA's plan to open the CM after splashdown, allowing its crew to egress while the module was still bobbing in the ocean, violated the concept of biological containment. NASA had not responded adequately to ICBC recommendations and did not apparently recognize the necessity of protecting Earth's environment against any possibility of extraterrestrial contamination.

Nevertheless, Sencer agreed to NASA's recovery plans, perhaps because he did not want to be remembered as the person who prevented the United States from meeting President Kennedy's challenge of reaching the Moon during the 1960s. Speaking for ICBC, he said that although it wanted the recovery to be handled differently, the risk of back contamination could be made acceptably low if "strict 'housekeeping' procedures to minimize contaminating the environment . . . are written for and implemented by the crew."[261] This was not a new idea; ICBC had discussed such procedures as early as 1966 in conjunction with the use of "dust-settling sprays"[262] prior to egress of the astronauts.

Although ICBC agreed to recovery procedures allowing the CM to be opened while still floating in the ocean, and the astronauts to egress into a life raft, the committee did not give up on the idea of keeping the CM sealed during recovery. In its June 1969 meeting, in fact, ICBC decided that "there should be no abandonment of the earlier concept of picking up the Spacecraft with the crew on board," and that the "MSC should actively pursue alternate modes for strict containment during recovery."[263] Louis Locke, a pathologist and veterinarian with the Fish and Wildlife Service of the U.S. Department of

---

261. David J. Sencer to Thomas O. Paine, 7 April 1969, JSC Historical Archives, "April 1969" folder, box 76-24.
262. John E. Pickering, "Interagency Committee on Back Contamination Minutes" of 13 April 1966 meeting, p. 3, in a package beginning with ICBC member list, JSC Historical Archives, "May 1966" folder, box 76-12.
263. Both quotes in the sentence are from John E. Pickering, "Minutes—Interagency Committee on Back Contamination" of 5 June 1969 meeting, p. 3, JSC Historical Archives, "June 1–16, 1969" folder, box 76-24.

the Interior, elaborated on the dangers of the present crew recovery procedure, warning that "should an astronaut fall into the water while making the transfer onto the raft, he would effectively nullify some important quarantine procedures as far as exposing of aquatic life is concerned."[264] Transfer of the crew from CM to raft, where they would be splashed with ocean water, was in his mind an extremely serious breakdown in proper bio-isolation.

Astronaut recovery scenarios. The nominal scenario for recovery was for the CM to splash down within helicopter range of the primary recovery ship. The helicopter would then drop swimmers next to the CM, who would install a flotation collar around it that would provide additional protection to it and its crew. The swimmers would attach a large raft, capable of carrying seven people, to the flotation collar. According to a scenario put forward in April 1969, the helicopter would lower a package of biological isolation garments into the raft. The BIGs were to help guard against infection of Earth during the astronauts' brief trip from ocean to recovery ship. One of the swimmers would don a BIG and, after the CM hatch was opened, assist the astronauts in getting out and into the raft. The CM hatch would then be closed and the astronauts would put on the nylon, head-to-toe BIGs.[265]

This scenario was altered before the mission in an important way that increased the level of planetary protection. The new plan called for one of the recovery crew swimmers to don a BIG and, after opening the CM hatch, throw the BIGs inside for the astronauts, then close the hatch.[266] The astronauts would put on their BIGs while still inside.[267] Only after the astronauts had done so was the hatch to be reopened. The swimmer would then assist the astronauts in getting out of the CM and into the raft.

The BIGs had respirators built into them designed to filter out exhaled particulates as small as 0.3 microns with a 99.98 percent efficiency. This, MSC's Richard Johnston believed, would trap any organisms that the astronauts exhaled. The Crew Systems Division at

264. Louis N. Locke to Richard S. Johnston, 2 June 1969, JSC Historical Archives, "June 1–16, 1969" folder, box 76-24.
265. Johnston, *Report on the Status of the Apollo Back Contamination Program*, pp. 17–18.
266. Bob Fish interview.
267. MSC, "Apollo 11 Back Contamination," 14 July 1969, JSC Historical Archives, "July 14–31, 1969" folder, box 76-25.

MSC had developed the garments and subjected them to human testing in a range of test chambers and at sea under recovery conditions. After the astronauts suited up in the BIGs and stepped into the raft, the recovery crew would scrub them with a decontamination agent. The astronauts would then be lifted into the waiting helicopter and transported to the ship.

Once inside the shipboard quarantine unit, they would doff the clumsy suits and replace them with standard NASA flight suits.[268] The recovery crew would scrub down the area around the module's escape hatch, the flotation collar, and the raft (in which the astronauts had briefly ridden) with a strong acid disinfectant, then sink the raft. These recovery procedures were finally approved by ICBC quite late in the mission planning process, in a telephone conference on 19 May 1969, only two months before the launch of Apollo 11.[269]

Lunar dust issues. What the ICBC analysis did not give importance to was the presence of any remaining lunar dust within the CM. Such dust might have been tracked in by the astronauts during their activities on the Moon, where it adhered to various surfaces and was missed during vacuuming and wiping procedures. Thus, the interior of the CM might still be contaminated. This possible occurrence, however, appeared quite solvable using the existing recovery strategy. After the astronauts left the CM and the recovery crew took the sample containers from it, the module would be sealed and its interior kept isolated throughout the period of quarantine. Only if LRL's analyses determined that the returned lunar material was harmless would the CM be allowed out of quarantine.[270]

Astronaut safety vs. planetary protection. Central to NASA's management of the Apollo program was, as discussed above, the importance given to astronaut safety. NASA simply would not put

---

268. USS Hornet Museum, "USS HORNET Museum Acquires Two Historic Apollo Artifacts!" press release, 12 December 2000, *http://www.uss-hornet.org/news_events/press_releases/media_kit_apollo.html.*

269. NASA, "Splashdown Procedures," press release, 9 July 1969, folder 19.4.1.2, "Lunar Receiving Lab 4715," NASA Historical Reference Collection; Hamblin, "After the Moonflight," 55; Carter, "Moon Rocks and Moon Germs," 246–247; Committee on Planetary and Lunar Exploration, "A History of the Lunar Receiving Laboratory"; Johnston, *Report on the Status of the Apollo Back Contamination Program,* p. 18.

270. Committee on Planetary and Lunar Exploration, "A History of the Lunar Receiving Laboratory."

its spacecraft crews in danger in order to meet quarantine requirements. Decisions based on this policy sometimes, as in the case of the Apollo CMs, ran contradictory to optimal planetary protection actions. It is important to note that when confronted with prioritizing either a very low-probability event with potentially global consequences (back contamination of Earth) or a much higher probability event with a limited potential consequence (injury to one or more of the three astronauts), NASA consistently chose to protect against the latter.

Why was this? John Stonesifer thought it was because NASA did not believe that back contamination would actually occur, whether the Agency practiced planetary protection procedures or not.[271] The astronauts, on the other hand, could be placed in real danger if they stayed in the CM while the USS *Hornet* attempted to maneuver close enough to connect its crane to the module and haul it aboard. The astronauts were people whom many NASA staff personally knew. What's more, many people around the world considered them heroes for conducting the voyage to the Moon. Protecting their lives was, not surprisingly, something that NASA staff strongly wanted to do. By comparison, back contamination was a theoretical danger in that it had never, so far as space scientists knew, ever happened to Earth. Back contamination did not have the recognizable face of Neil Armstrong or Buzz Aldrin; it was a concept, not a human being. It was thus not unexpected that NASA chose to defend against the known, proven danger to several of its friends, rather than against an extremely unlikely risk to strangers.

The Mobile Quarantine Facility

ICBC, recognizing the need to reliably isolate all potentially contaminated personnel and material from the time of spacecraft retrieval until delivery to LRL, developed preliminary concepts and recommendations for such a capability. Two concepts that survived initial critiques were for a mobile, self-contained facility or modification of a

---

271. Stonesifer interview, 13 January 2006.

cabin aboard the recovery ship. The tradeoff analysis between these two options considered the following objectives:

- *Ability to maintain a consistent level of biological isolation.* Use of several quarantine modes (shipboard cabin, aircraft cargo hold, transfer vehicles) would have difficulty meeting the objective. Also, each quarantine mode, and possibly the same mode on different vehicles (such as cabins on different ships) would have to be custom fit with quarantine capabilities, and this would be an expensive proposition. In contrast, if one mobile quarantine unit was employed from the time of recovery until delivery at LRL, it would provide consistency of isolation environment and would preclude the need for fabricating multiple isolation environments.

- *Flexibility of operations.* Installing a quarantine unit in a ship would involve extensive downtime in a shipyard, precluding use of the vessel for a period of time and impairing its "readiness state . . . in the active defense forces."[272] It would also constrain use of a vessel for other applications during spacecraft recovery operations. Use of a mobile quarantine unit, on the other hand, would not require vessel downtime during construction of the unit, and it could be readily moved from one vessel to another, allowing rapid redeployment of a vessel if necessary.

- *Minimal dependence upon the United States Navy and Air Force.* A completely self-contained and movable facility would not only reduce dependence on any one vessel, but also on Defense Department facilities and biological and waste management concurrence and control. In addition, such a unit would give NASA the maximum control of the quarantine effort.

By August 1966, after considerable discussion, NASA decided that the best method of isolating crews and material was to build a well-equipped, self-contained mobile quarantine facility to house and transport the Apollo astronauts and their lunar samples from the time they boarded the Navy aircraft carrier recovery ship until they arrived at

---

272. John E. Pickering, "Minutes—Interagency Committee on Back Contamination," 3 October 1966, p. 4, in a package beginning with John E. Pickering to Wilmot N. Hess, 16 February 1967, JSC Historical Archives, "February 1967" folder, box 76-14.

and entered LRL.[273] NASA and ICBC also decided that a savings in design and fabrication costs could be attained by using a "housetrailer" concept for the quarantine facility employing "basic household-type" equipment and appliances, rather than the initial concept of developing custom isolation units and equipment.[274]

NASA contracted with MelPar, a subsidiary of American Standard, to develop the MQF. In keeping with the housetrailer approach, MelPar subcontracted the actual construction to Airstream, Inc., a manufacturer of commercial travel and vacation trailers. Airstream fabricated four modified 18,700-pound trailers, each of which was meant to prevent the escape of infectious agents. Quarantine conditions were maintained on these MQFs through negative internal air pressure, filtration of effluent air, internal capture and storage of wastes in holding tanks, and a transfer lock system that could bring materials into or out of the facility without letting potentially infected air out. MelPar designed the holding tanks to be capable of containing the fecal and urine wastes of six people for at least five days. Each MQF had an internal backup power system, air conditioning, a communications system, and a medical diagnostic capability.[275]

Two externally mounted exhaust fans maintained negative pressure inside the MQF. All effluent air was passed through two filters capable of 99-percent efficiency in removing particles 0.45 microns (0.45 millionths of a meter) or larger. Airstream made the MQFs skid-mounted, double-walled structures with the outer wall designed to serve as the pressure vessel for preserving negative internal air pressure. Airstream installed insulation between the walls and made the interior wall mostly of durable, vinyl-covered aluminum. There were two doors—one in the rear and one in the side of the MQF. The side door was framed by a flange that provided a smooth surface for bonding to the flexible vinyl shipboard transfer tunnel, which was designed for maintaining biological

---

273. "Features of Mobile Quarantine Facility" and accompanying discussion, in a package, beginning with ICBC member list, JSC Historical Archives, "May 1966" folder, box 76-12.
274. ICBC, "Activities Summary," 31 May 1967, JSC Historical Archives, "May 1967" folder, box 76-15, p. 2.
275. MSC, *Apollo Recovery Equipment Final Report*, prepared by MelPar, Falls Church, VA, under Contract NAS9-6874, July 1969, p. 31, *http://24.73.239.154:8081/moonshot/pdf/MQF%20Report%20July%201969.pdf*; USS Hornet Museum; Carter, "Moon Rocks and Moon Germs," 244.

isolation while conveying equipment to the MQF from the Command Module (once it had been lifted aboard the recovery ship).

The procedures for transferring objects such as medical samples, lunar materials, flight tapes, flight film, and crew equipment *out* of the MQF while it was maintaining biological isolation required vacuum-sealing and heat-sealing the objects into Biological Isolation Containers. After this, MQF personnel put the containers into the MQF's double-door, pass-through decontamination/transfer lock, where they were soaked in a 5,000-parts-per-million (ppm) sodium hypochlorite (bleach) solution for 15 minutes. After draining the solution, technicians outside the MQF opened the outer door of the lock and removed the item. Personnel also transferred items such as meals and various supplies into the MQF through the transfer lock, but decontamination procedures were not required since these objects were not contaminated with lunar material.[276]

In accordance with one of NASA's objectives for the MQF—that it exhibit flexibility of operation—MelPar built into it the capability to employ a range of different power sources from the different environments in which it would be placed. On board the recovery ship, MQF engineers configured it to use 440-volt shipboard power to run its negative-pressure, cooling, lighting, and other systems. While it was being hoisted off the ship, an engine generator kept the MQF's systems operating. While being flown in the hold of an Air Force cargo plane, the MQF ran on the craft's 400-cycle, 115–230-volt system. And when the MQF reached LRL, it used the facility's 120/240-volt, 60-cycle power. At no time could the MQF lose power, because its fans had to keep running in order to maintain the negative pressure within it. Loss of this pressure differential could have resulted in contamination of Earth's environment with lunar microorganisms.[277]

NASA tested and confirmed the performance of one of the MQFs in January 1969 aboard the helicopter landing ship USS *Guadalcanal*, after completion of the Apollo 9 Earth orbit mission (which wasn't expected to bring back any contamination since it never landed on an

---

276. MSC, *Recovery Quarantine Equipment Familiarization Manual*, MSC Report Number 00025—Rev. A, 27 October 1969, pp. 1–9, 25; JSC Historical Archives, "October 27–31, 1969" folder, box 76-31; MSC, *Apollo Recovery Equipment Final Report*, p. 37.

277. Koons interview, 27 January 2006.

extraterrestrial body). After the testing, NASA made plans to transport two MQFs, a primary and a backup, to Pearl Harbor, Hawaii, and put them aboard the USS *Hornet*, which was to be the primary recovery ship for the Apollo 11 Moon landing mission.[278]

The plan for the MQF used during Apollo 11 was to fasten it to the hangar deck of the USS *Hornet*, where it would sit from the time of the Command Module splashdown until the ship's arrival at Pearl Harbor. The astronauts would remain inside the MQF while it was being hoisted off the ship, trucked in a motorcade to Hickam Air Force Base, flown in the hold of a C-141 cargo plane to Ellington Air Force Base in Houston, and then trucked to LRL. Once there, the astronauts would walk from the MQF to LRL through a plastic tunnel.[279]

The Need To Quickly Transport Lunar Samples to LRL

NASA scheduled the lunar dust, rocks, and gases taken from the Moon to arrive at LRL as soon as possible and significantly before the MQF and its astronaut passengers. The samples would be quickly taken to an airport and transported by jet to Ellington Air Force Base, then rushed by ground transport to LRL. This was critical in order to minimize any changes to them that began the moment they were taken from the lunar surface. In particular, the intensity of sample radioactivity caused by cosmic ray bombardment, off-gassing from the samples, and magnetism induced by a lunar field were all expected to decrease rapidly. Delays in analyzing the samples for these possibly faint characteristics might mean losing the chance to detect them.[280]

---

278. USS Hornet Museum.
279. USS Hornet Museum; Carter, "Moon Rocks and Moon Germs," 245.
280. Hamblin, "After the Moonflight," 55.

Planetary Protection Actions Taken During the Apollo 11 Moon Landing Mission

On 16 July 1969 at 9:32 in the morning, eastern daylight time (EDT), Apollo 11 lifted off from Kennedy Space Center with the Moon as its destination. It took the spacecraft a little over four days to get there, touching down in the Sea of Tranquility at 4:18 p.m. EDT on 20 July. Shortly after astronaut Neil Armstrong stepped onto the lunar surface, he collected and stowed the first sample of lunar surface material to assure that "if a contingency required an early end to the planned surface activities, samples of lunar surface material would be returned to Earth."[281] Over the next 2.5 hours, besides emplacing their scientific instruments, astronauts Armstrong and Aldrin "acted like field geologists,"[282] collecting many rock samples and taking two core samples. Obtaining the cores was not easy, since the astronauts found it difficult to drive their sampling tube into the lunar surface.

One of the challenges of the mission was to carry out sampling operations while dealing with forward contamination of the lunar surface with terrestrial materials. As the Lunar Module descended onto the Moon, its descent engine spewed out approximately nine tons of propellant exhaust. While most of this was lost into space, nearly a ton of 100 different byproduct chemicals remained spread around the landing area. In addition, the astronauts' suits were not entirely leakproof; significant quantities of gases exuded from spacesuit joints. According to NASA's Neil Nickle, as the astronauts walked across the lunar surface, they were "virtually jets of gas."[283] To obtain samples as uncontaminated as possible, the astronauts took them from sheltered areas (although this was made more difficult since the landing was in a very flat region) and also used their core tubes to sample below the surface.

---

281. "Apollo 11 Mission Summary," from NASA SP-214, *Preliminary Science Report*, Apollo 11 (AS-506) Lunar Landing Mission, *http://www.nasm.si.edu/collections/imagery/apollo/AS11/a11sum.htm* (accessed 29 July 2005).
282. Compton, *Where No Man Has Gone Before*, p. 144.
283. Victor Cohn, "Lunar Contamination: Growing Worry," *Washington Post* (28 May 1969): A1; JSC Historical Archives, "May 13–31, 1969" folder, box 76-24.

Figure 4.1 Flotation collar attached to the *Columbia* Command Module. The three astronauts plus a Navy swimmer sit in the recovery raft, awaiting pickup by helicopter. All four men are wearing BIGs.

Back Contamination Prevention on the Return Trip to Earth

The astronauts' period of quarantine commenced when they sealed the hatch of the Lunar Module in preparation for lifting off.[284] After liftoff and rendezvous with the *Columbia* Command Module, astronaut Michael Collins, sitting in the CM, slightly opened its oxygen supply valve and Aldrin adjusted the LM's venting valve in order to make sure that the cabin atmosphere flowed from the CM into the LM. This was one of the strategies for minimizing back contamination buildup in the CM and possibly in Earth's atmosphere as well after the return trip.

Aldrin and Armstrong found that the amount of loose dust in the LM was less than NASA had expected, but the dust clung tenaciously to any surface that it touched, including the outsides of spacesuits.

---

284. John Rummel, interview by author, San Francisco, CA, 7 December 2005.

The astronauts used a vacuum cleaner, which had been fabricated by attaching a brush to an exhaust hose of the LM, to try and remove the particles from their suits. This vacuum setup turned out not to be very effective, making it more difficult than planned to thoroughly rid their suits of Moon dust.[285]

Recovery and Quarantine Operations After Splashdown

Early on 24 July 1969, eight days after launching, the *Columbia* Command Module splashed down in the Pacific Ocean 812 nautical miles southwest of Hawaii and less than 13 miles from the USS *Hornet* aircraft carrier, the Apollo 11 prime recovery ship. Recovery operations were directed from the Mission Control Center in Building 30 of Houston's Manned Spacecraft Center (MSC), with support from recovery centers in Norfolk, Virginia, and Kunia, Hawaii. Within half an hour, recovery crews reached the CM, attached a flotation collar to it (see figure 4.1), and passed BIGs into the spacecraft. The recovery crew, which included swimmers, then applied Betadine, an iodine solution, around the postlanding vents in order to kill any microorganisms that might have escaped the inside of the module and might have adhered to the vent area (the Betadine would not, of course, have killed any organisms that had escaped into the atmosphere or the ocean).

After the astronauts put on the BIGs, exited the CM, and boarded the recovery raft, swimmers wiped Betadine onto the hatch of the CM that they'd passed through. Astronauts and recovery crew decontaminated each others' protective suits with a sodium hypochlorite (bleach) solution, which had performed better for this task than Betadine in NASA tests.[286] The astronauts were then hoisted aboard a helicopter and taken to the USS *Hornet* recovery vessel. The helicopter crew wore oxygen masks at all times to guard against inhaling any germs

285. Compton, *Where No Man Has Gone Before*, p. 145.
286. Richard S. Johnston to Special Assistant to the Director, "Quarantine Recovery Operation Changes," 11 July 1969, and attachment, CSD Supporting Development Branch, "Test Results—Biological Insulation [*sic*] Garment—Hydrostatic Pressure Testing," 30 June 1969, JSC Historical Archives, "1–13 July 1969" folder, box 76-25.

Figure 4.2 The Apollo 11 crew, wearing BIGs, leave the helicopter that carried them from the Command Module and walk across the deck of the USS *Hornet* toward the MQF.

from the astronauts. The helicopter itself was later decontaminated with formalin.[287]

The cumbersome BIGs made it difficult for the astronauts to shed heat, and they grew quite uncomfortable during the short helicopter ride to the ship. Armstrong and Collins expressed concern that they had about reached their level of tolerance by the time the helicopter landed on the *Hornet*. They had to wait, however, until entering the MQF before they could strip off the BIGs and don clean flight suits. The BIGs, besides being cumbersome and uncomfortable, may also have leaked, since when the Apollo crew took them off, the suits contained saltwater.[288]

As the astronauts walked across the *Hornet*'s deck (see figure 4.2), television crews broadcast their images to millions of viewers around the world. NASA personnel then sprayed the path they had walked between helicopter and MQF with glutaraldehyde, a sterilant typically used in hospitals to disinfect equipment. This was yet another precaution against back contamination. Television commentators explained to the world that the chances of back contamination were miniscule, but that NASA had to take every precaution.

Public concerns about contamination dangers. During this period of the 1960s, Michael Crichton's novel *The Andromeda Strain* appeared on best seller lists. The book told a story about Earth in danger from microorganisms brought back on a space vehicle. The public response to the book resulted in additional pressure on NASA to implement strict planetary protection measures. Thousands of concerned citizens wrote NASA letters, worried that they were at risk from Moon germs. These fears of the public might today seem alarmist and naïve, but epidemics have occurred repeatedly in the past due to exotic organisms being introduced into an unprotected populace.

---

287. NASA, "Press Kit: Apollo 11 Lunar Landing Mission," Release No. 69-83E, 6 July 1969, pp. 64–65, contained on the CD, *Remembering Apollo*, NASA History Division, NC-2004-06-HQ and SP-2004-4601, July 2004; Apollo Explorer, "View of Mission Control Center Celebrating Conclusion of Apollo 11 Mission," *http://www.apolloexplorer.co.uk/photo/html/AS11/10075305.htm* (accessed 26 January 2011); Apollo Explorer, "Apollo 11 crewmen await pickup by helicopter after landing," *http://www.apolloexplorer.co.uk/photo/html/AS11/10075300.htm* (accessed 26 January 2011); Carter, "Moon Rocks and Moon Germs," 247; Compton, *Where No Man Has Gone Before*, p. 146.
288. Carter, "Moon Rocks and Moon Germs," 247–248; Compton, *Where No Man Has Gone Before*, p. 146.

Figure 4.3 President Nixon welcomes the Apollo 11 astronauts aboard the USS *Hornet*.

One particularly devastating sickness occurred in the Pacific region, just a few thousand miles to the south of the Apollo recovery site. Early in 1875, the king of the Fiji Islands—a man named Cakobau—returned from a visit to Australia suffering from a case of measles. His disease, rarely fatal to Westerners, took root in the Fijis and killed 40,000 of his subjects.[289]

289. National Institute for Occupational Safety and Health, "Glutaraldehyde," Centers for Disease Control and Prevention, NIOSH publication no. 2001-115, September 2001, *http://www.cdc.gov/niosh/docs/2001-115/* (accessed 26 January 2011); Carter, "Moon Rocks and Moon Germs" 247; Compton, *Where No Man Has Gone Before*, p. 146; Hamblin, "After the Moonflight," 50.

Figure 4.4 Apollo 11 Command Module and its flotation collar hoisted aboard the USS *Hornet*. Astronauts have already exited from the craft.

Other examples concerned the native populations of the Americas, who lacked immunity to diseases that had propagated through Europe, Africa, and Asia for centuries. The diseases that explorers and settlers brought included smallpox, measles, chicken pox, typhus, typhoid fever, dysentery, scarlet fever, diphtheria, and cholera. These sicknesses ravaged native populations throughout the New World. Smallpox, for instance, arrived soon after Columbus did and wreaked havoc on Native Americans on the islands of Hispaniola and Puerto Rico, possibly wiping out entire tribes. The smallpox soon spread through many indigenous communities on the mainland as well. Cortez's conquest of the Aztecs was made greatly easier because of the thousands from the tribe who quickly died of the disease. One lesson to take away from the Native American experience is that not

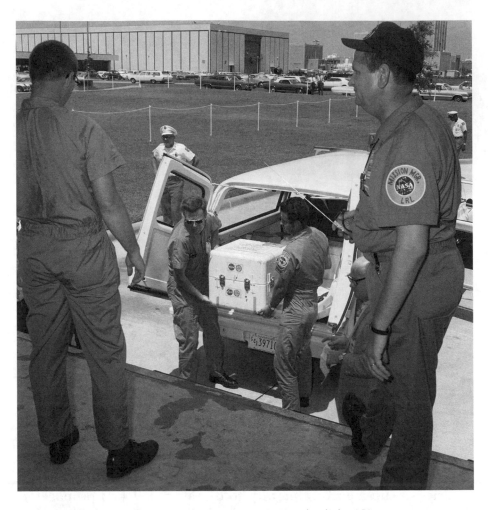

Figure 4.5 The first Apollo 11 sample return container is unloaded at LRL.

only do such epidemics exact an enormous human toll, they also dramatically disrupt existing political and social structures.[290]

President Nixon's welcome to the astronauts. Only after the astronauts were safely sealed in the airtight MQF and the *Hornet's*

---

290. R. S. Bray, *Armies of Pestilence: The Impact of Disease on History* (New York: Barnes & Noble, 2000), pp. 123–128; O. Ned Eddins, "Plains Indian Smallpox," in *Mountain Man Plains Indian Canadian Fur Trade*, *http://www.thefurtrapper.com/indian_smallpox.htm* (last updated 21 October 2007, accessed 29 October 2007); George Childs Kohn, *Encyclopedia of Plague and Pestilence, From Ancient Times to the Present*, 3rd ed. (New York: Facts on File, Inc., 2008), pp. 159–160.

deck disinfected did NASA allow President Richard Nixon, who had flown to the USS *Hornet* to welcome the space travelers home, to approach the large window at the rear of the MQF to give his congratulations (see figure 4.3). During the transfer of astronauts, President Nixon had been kept far away, a helicopter waiting to fly him off the ship should any leaks be detected in the MQF. The *Hornet* also contained a second MQF that could be used to quarantine any of the ship's crew who might have been directly exposed to either astronauts or spacecraft. If a major biocontainment breach had occurred, the *Hornet* itself would have become the isolation unit and would have had to remain at sea for the entire astronaut and lunar sample quarantine period.[291]

Transporting Samples and Equipment from the Splashdown Point to LRL

After the astronauts were helicoptered to the *Hornet*, recovery crews hauled the CM aboard (see figure 4.4), then connected it by a plastic tunnel to the MQF. The film shot on the Moon and the lunar sample containers had to be carried through this tunnel into the MQF, since they, too, were potential sources of back contamination. In order to remove them from the MQF, the recovery crew had to first pass them through a decontamination lock. The recovery crew then packed one of the sample containers as well as the film and tape recorders in shipping containers. These were flown to Johnston Island, 200 nautical miles to the north, loaded onto a C-141 cargo aircraft, taken to Ellington Air Force Base in Texas, delivered to the nearby MSC, and received by LRL. The other sample container was flown to Hickam Air Force Base in Hawaii, and from there to MSC and its LRL.[292]

---

291. Carter, "Moon Rocks and Moon Germs," 247; Compton, *Where No Man Has Gone Before*, p. 146.
292. Compton, *Where No Man Has Gone Before*, p. 147.

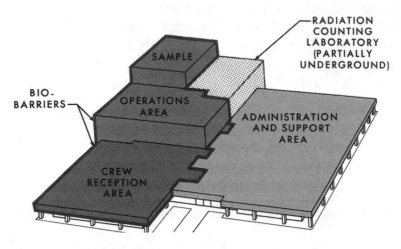

NASA-S-67-696

LRL FUNCTIONAL AREAS

Figure 4.6 LRL functional areas. Biological barriers quarantining certain areas are identified by the heavy black lines.

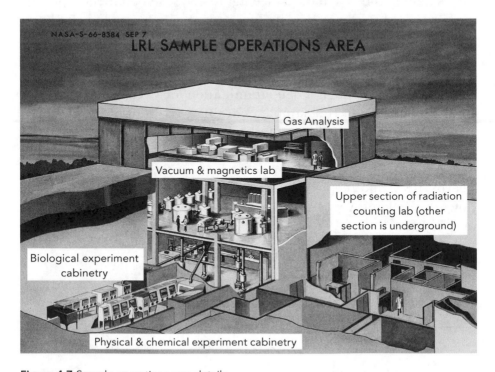

Figure 4.7 Sample operations area details.

Receiving and Handling Lunar Samples and Flight Film at LRL

On 24 July 1969, NASA sent word to LRL that the Apollo 11 spacecraft had been successfully recovered, and the laboratory answered that it was fully operational and prepared to receive the spacecraft and its astronauts and lunar samples. The first Apollo Lunar Sample Return Container (ALSRC), weighing 33 pounds, 6 ounces, arrived at 12:28 p.m. on 25 July 1969 (see figure 4.5). Accompanying it were canisters containing film records from Apollo 11's voyage. The second ALSRC was delivered that night at 9:57 p.m. and weighed 52 pounds.[293] It was because of the "extremely high scientific value and irreplaceable nature of the returned samples"[294] that the each of the two ALSRCs were transported to LRL on separate aircraft. This provided additional assurance that at least one of them would reach the laboratory promptly and undamaged. Minimal transport time was especially important for the low-level gamma radiation counting experiment. Because of the expected rapid decay of sample radiation levels (the radionuclides of interest had half-lives as short as 15 hours), the sooner the lunar material could be delivered, the better the data would be.[295]

LRL (also referred to as MSC's Building 37) included a sample operations area in its northwest corner, a Crew Reception Area in its southwest corner, and its administrative and support area in the eastern half of the building. These functional areas are depicted in figure 4.6; a detailed drawing of the sample operations area is in figure 4.7. The heavy black lines in figure 4.6 around the Crew Reception Area and sample operations area indicate that they were enclosed in their own biologic barriers. Several features made up these barriers, including sealed walls, floors, and ceilings; the control of leaks using differential air pressures between

---

293. MSC, *Apollo 11 Activity*, August 1969, p. 1, JSC Historical Archives, "16–31 August 1969" folder, box 76-25.

294. Joseph V. Piland, Manager, Lunar Receiving Laboratory Program Office, to Director of Flight Operations, "Requirements for Apollo Lunar Sample Return Container (ALSRC) Transport to MSC," 22 August 1966, attachment to Robert O. Piland, Manager, Experiments Program Office, to Director of Flight Operations, "Requirements for Apollo Lunar Sample Return Container (ALSRC)," 13 September 1966, JSC Historical Archives, "09-13-66" folder, box 76-13.

295. Robert O. Piland, Manager, Experiments Program Office, to Director of Flight Operations, "Requirements for Apollo Lunar Sample Return Container (ALSRC)," 13 September 1966, JSC Historical Archives, "09-13-66" folder, box 76-13.

the inside and outside of the isolated areas; special air conditioning features; biologic filters on both the air intake and exhaust systems of the areas; incineration of effluent air; and sterilization of liquid effluents. Staff had to transfer materials into and out of isolated areas using two-door sterilization cabinets, which employed steam and/or ethylene oxide gas.[296]

Film handling. Decontamination procedures for Apollo film canisters began immediately upon their arrival. One of the canisters was delivered covered with dust that was assumed to be of lunar origin. LRL technicians put film canisters from the mission into the two-door sterilization cabinets and exposed them to the ethylene oxide gaseous sterilant for several hours. Only after this treatment did LRL staff send the canisters to the photographic laboratory for processing. The sterilization process decontaminated the film as well as the canisters, which allowed technicians who were not behind the biobarrier to process the film without having to wait several weeks for the quarantine to be lifted.[297]

Handling the first lunar samples. After their delivery to LRL, facility staff brought the ALSRCs into the vacuum laboratory, where technicians gave them preliminary examinations, then passed them through two-stage sterilization consisting of ultraviolet light exposure and contact with peracetic acid, a biocide typically used for sanitizing equipment such as tanks, pipelines, and food contact surfaces.[298] Next, they rinsed the ALSRCs with sterile water, dried them in a nitrogen atmosphere, and passed them through a vacuum lock into the main vacuum chamber glove box.[299] LRL technicians pumped the chamber's atmospheric pressure down to approximate that of the Moon.

296. MSC, *Lunar Receiving Laboratory—MSC Building 37—Facility Description*, MSC, Houston, TX, 10 September 1968, pp. 1, 3-4, folder 012990, "Lunar Receiving Lab," NASA Historical Reference Collection; James C. McLane, Jr., Elbert A. King, Jr., Donald A. Flory, Keith A. Richardson, James P. Dawson, Walter A. Kemmerer, Bennie C. Wooley, "Lunar Receiving Laboratory," *Science* 155 (3 February 1967): 528, folder 4714, "MSC—Lunar Receiving Laboratory," NASA Historical Reference Collection.
297. MSC, *Apollo 11 Activity*, p. 1; MSC, *Lunar Receiving Laboratory*, pp. 18–19; Compton, *Where No Man Has Gone Before*, p. 149.
298. FMC Industrial Chemicals, "MSDS: Peracetic Acid," *http://www.fmcchemicals.com/Industrial/V2/MSDS/0,1881,133,00.html*, 2005 (accessed 19 August 2005).
299. Lunar Receiving Laboratory, "LRL Daily Summary Report No. 3," 26 July 1969, folder 012990, "Lunar Receiving Laboratory," NASA Historical Reference Collection; Lunar Receiving Laboratory, "LRL Daily Summary Report No. 4," 27 July 1969, folder 012990, "Lunar Receiving Laboratory," NASA Historical Reference Collection; Compton, *Where No Man Has Gone Before*, p. 149.

Figure 4.8 The first lunar sample photographed in detail at LRL. This was a granular, fine-grained, mafic (iron- and magnesium-rich) rock, which appeared similar to several igneous rock types found on Earth.

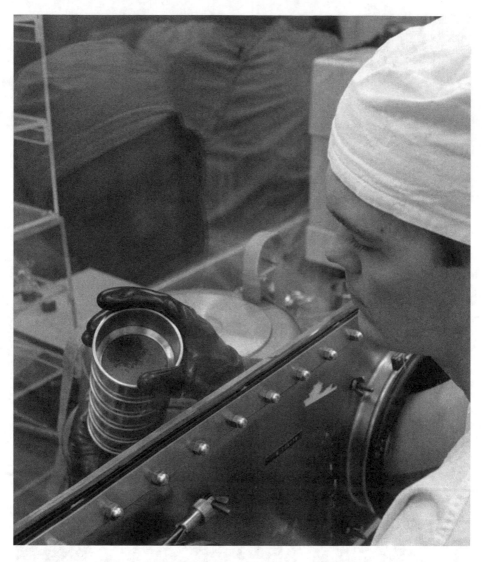

Figure 4.9 Lunar material in a sieve from the bulk sample container being examined in an LRL glove box by a lunar sample preliminary examination team member.

The technicians did not directly touch the samples, but worked from the outside of the vacuum chamber and reached into the interior by putting their hands in glove ports. The gloves and the rest of the vacuum chamber formed one of several lines of defense against contamination of LRL staff or the outside environment with Moon material. The LRL air conditioning system, which had the capability to sterilize all air leaving the facility, formed another contamination barrier. The pressurization system, which maintained a negative pressure[300] within LRL's laboratories, formed yet another barrier.[301]

The staff did not immediately open the sample container because they could not stabilize the pressure in the vacuum chamber. Operators suspected that one of the gloves in the wall of the chamber through which samples were handled was slowly leaking, since they observed several small tears in one of the outer glove's two layers, the Viton non-pressure-retention layer. However by midafternoon of the day after sample delivery, 26 July 1969, the leak had not grown any worse and the staff decided to examine the first ALSRC. Technicians began by puncturing the ALSRC, in order to analyze the pulse of gas that would be released. Later examination showed that the ALSRC had not been punctured sufficiently by the gas analysis probe. As a result, no gas sample was obtained. Procedures were put into place to ensure that a good sample would indeed be obtained from the second ALSRC.[302]

LRL Daily Summary Report records indicate that LRL staff opened the first ALSRC at 3:45 p.m. on 26 July 1969. What they initially saw was not terribly illuminating, since a fine lunar dust obscured the individual characteristics of the rocks. All that was apparent was that they were shaped irregularly with slightly rounded edges (see figure 4.8).[303] Elbert A. King, NASA's first Lunar Sample Curator, said of those first glimpses of Moon rocks, "The moment was truly history, but there was little we could observe or say. We counted the rocks and described the size and shape of each piece, but they looked like lumps of charcoal in the bottom of a backyard barbecue grill."[304] The first lunar rock examined in more detail, however, revealed a fine crystalline structure and an apparently igneous

---

300. Holding the laboratories at negative pressure means that the LRL maintains slightly less air pressure within the laboratories than in the atmosphere outside the LRL. Thus, any leakage of air would occur from outside the laboratories into them, rather than from the laboratories out.

301. Hamblin, "After the Moonflight," 55.

302. MSC, *Apollo 11 Activity*, pp. 1–2.

303. Lunar Receiving Laboratory, "LRL Daily Summary Report No. 3," 26 July 1969, folder 012990, "Lunar Receiving Laboratory," NASA Historical Reference Collection; Lunar Receiving Laboratory, "LRL Daily Summary Report No. 4," 27 July 1969, folder 012990, "Lunar Receiving Laboratory," NASA Historical Reference Collection; Compton, *Where No Man Has Gone Before*, p. 149.

304. Allton, "25 Years of Curating Moon Rocks," 1.

origin. Analysis of the fines showed a substantial number of small glassy spheres 1 millimeter (1/25 inch) in diameter and smaller.[305]

The sample container also held two rock core tubes containing subsurface material and a sheet of aluminum foil that had been laid on the Moon's surface to trap particles of the solar wind. Because the rock cores had received the least exposure to Earth contaminants of any of the samples, they provided the "bioprime"[306] sample—a 100 gram (3.5 ounce) portion of lunar material whose impacts on a range of Earth organisms would be carefully analyzed. MSC staff packed the rock core tubes and solar wind collector in stainless steel cans that they then sterilized and removed from the vacuum laboratory.

Two teams of scientists—the preliminary examination team and the sample analysis planning team—inspected, characterized, photographed, and catalogued the samples. Through August 1969, specimens from the samples were allocated to 142 principal investigators on the project, but they could not immediately receive these specimens. That would have to wait until the quarantine period ended.[307]

Equipment Challenges and Malfunctions

The gloves through which technicians handled materials in the vacuum chambers needed to be quite strong since they had to withstand the very large pressure differential between the outside and inside of the chambers. As a result of this requirement, the gloves were also extremely stiff and unwieldy to use, especially for the delicate parts handling operations that were carried out. Nevertheless, LRL technicians grew amazingly proficient in performing those operations (see figure 4.9). According to Craig Fischer, a NASA pathologist who served as chief of clinical laboratories for the mission, "the technicians got very good at handling the rocks and apportioning the rocks

---

305. MSC, *Apollo 11 Activity*, p. 1. Fines are defined in LRL operations as materials that pass through a 1-centimeter (90.4-inch) sieve. MSC, *Lunar Receiving Laboratory—Sample Flow Directive*, NASA MSC, Houston, TX, 21 May 1969, p. 20, folder 012990, "Lunar Receiving Lab," NASA Historical Reference Collection.
306. Compton, *Where No Man Has Gone Before*, p. 150.
307. Ibid.

and milling the rocks and injecting the mice and exposing the fish and cockroaches and you name it."[308]

On 28 July 1969, a sewage system malfunctioned and appeared in danger of breaking LRL's biological containment, but the problem was corrected without such an occurrence. The torn glove on the vacuum chamber received a temporary fix—to reduce the leaking, technicians slipped another glove over it and taped the two together at the wrists. While this approach temporarily stopped the pressure fluctuations in the vacuum chamber, it didn't last long. Within days, the glove ruptured. The 1 August 1969 LRL Daily Summary Report recorded that on 31 July 1969, "the webbing between the thumb and forefinger of the glove burst with almost no warning."[309] The strong flow of air into the chamber that followed the glove rupture exposed most of the samples in the vacuum system to LRL's atmosphere. This incident also exposed two technicians to the Moon rocks, and these personnel had to go into quarantine. LRL suspended work in the vacuum laboratory until this tendency toward leaks could be solved.

LRL scientists decided to abandon keeping any samples under vacuum except those that absolutely required it, since leakage and subsequent contamination appeared unavoidable with vacuum conditions in the chamber. Instead of keeping the system under vacuum, which would have most closely resembled conditions on the Moon, LRL staff opted to introduce dry nitrogen—a chemically inert and noncontaminating gas—into the chamber. They also put the chamber through a sterilization heating cycle that started on 3 August 1969, when they turned on the chamber's heaters. This sterilization cycle was a necessary part of a glove change operation, which entailed opening up the chamber bulkhead.[310]

---

308. Craig Fischer (MSC's chief of clinical laboratories for all missions from Gemini to Apollo 15, including the three lunar missions that were in quarantine, who also served as chief of the Crew Reception Area during the three quarantine missions), telephone interview by author, 18 January 2006.

309. Lunar Receiving Laboratory, "LRL Daily Summary Report No. 8," 1 August 1969, folder 012990, "Lunar Receiving Laboratory," NASA Historical Reference Collection.

310. Lunar Receiving Laboratory, "LRL Daily Summary Report No. 5," 28 July 1969, folder 012990, "Lunar Receiving Laboratory," NASA Historical Reference Collection; Lunar Receiving Laboratory, "LRL Daily Summary Report No. 6," 29 July 1969, folder 012990, "Lunar Receiving Laboratory," NASA Historical Reference Collection; Lunar Receiving Laboratory, "LRL Daily Summary Report No. 7," 31 July 1969, folder 012990, "Lunar

Quarantining the Astronauts

On the USS *Hornet*'s journey back from the recovery site to
Hawaii, the astronauts shared the MQF with a doctor, William R.
Carpentier, and a mechanical engineer and expert on MQF operation,
John Hirasaki.[311] When the ship reached Pearl Harbor, the MQF, with
the astronauts aboard, was lifted from the deck onto a waiting truck.
After the astronauts received greetings from the Honolulu mayor
and thousands of local citizens, the truck transported the MQF to
Hickam Air Force Base several miles away. There, the MQF was put
into a C-141 cargo plane and flown to Ellington Air Force Base near
Houston, where it landed just after midnight on 27 July 1969. A large
crowd waited, hoping to get a quick look at the returning heroes. The
MQF was transferred onto a special flatbed truck designed for it and
was driven to the Crew Reception Area of LRL. The truck backed up
to the CRA's loading dock, and technicians taped a shroud tightly in
place that effectively mated the MQF to the CRA and formed a sort
of airlock that prevented outside air from being contaminated when
the MQF was opened.

Craig Fischer had this to say about the reception of the first lunar
astronauts:

> We were elated to see them, although it's sort
> of interesting looking back at it. I didn't have the
> sense of history that I do now. It was part of what we
> did. . . . Now . . . it has a totally different meaning than
> it did at the time. All of a sudden you realize that was
> the first lunar crew in the history of mankind, and we
> were privileged enough to work with them. But at the
> time, hey, it's just Neil and Buzz and Mike and we'd

Receiving Laboratory," NASA Historical Reference Collection; Lunar Receiving Laboratory,
"LRL Daily Summary Report No. 9," 2 August, 1969, folder 012990, "Lunar Receiving
Laboratory," NASA Historical Reference Collection; Lunar Receiving Laboratory, "LRL Daily
Summary Report No. 10," 4 August 1969, folder 012990, "Lunar Receiving Laboratory,"
NASA Historical Reference Collection; Lunar Receiving Laboratory, "LRL Daily Summary
Report No. 11," 5 August 1969, folder 012990, "Lunar Receiving Laboratory," NASA
Historical Reference Collection; Compton, *Where No Man Has Gone Before*, pp. 151, 268.
311. John K. Hirasaki, project engineer for the certification and qualification of the MQF,
telephone interview by author, 13 January 2006.

worked with them before on other missions and we were glad to see them. And they were glad to be there. But there wasn't any hoopin' and hollerin' or huggin'. A handshake and in they came.[312]

At 3:30 a.m. on 27 July 1969, the Apollo crew and their companions were "successfully transferred to LRL."[313] Waiting for them in LRL's Crew Reception Area was their support staff—a clinical pathologist (Craig Fischer), five laboratory technicians, a photographer, an MSC public affairs person, a logistics operations officer, and three stewards.[314]

Over the next weeks, medical staff kept a close watch on the astronauts for any sign of infection and illness, conducting frequent examinations and blood tests. The astronauts' quarters had equipment for clinical tests as well as surgical and dental procedures. Doctors and dentists had been identified who could, if needed, enter the CRA and attend to its residents. These professionals would then, of course, have to remain in the CRA until the end of the quarantine.[315]

If a serious medical emergency had occurred that was beyond the capabilities of CRA equipment, NASA would have rushed the afflicted person from LRL to a hospital, regardless of quarantine requirements. Although such a situation did not occur, this was another example of NASA's policy to prioritize the lives of its people above back contamination requirements. (A previously discussed example of NASA's policy was its decision to have the astronauts exit the Command Module, due to concerns for their safety, before the recovery ship's crane hoisted the module aboard the USS Hornet.)[316]

Apollo mission plans, which were in concurrence with the recommendations of ICBC, called for the astronauts to remain in quarantine for at least 21 days after their exposure to lunar material. The air exhaled by them and the other personnel under quarantine in the

---

312. Fischer telephone interview, 18 January 2006.
313. Lunar Receiving Laboratory, "LRL Daily Summary Report No. 4," 27 July 1969, folder 012990, "Lunar Receiving Laboratory," NASA Historical Reference Collection.
314. Compton, *Where No Man Has Gone Before*, pp. 151, 268.
315. Craig Fischer, interview by author, JSC, 25 January 2006.
316. Stonesifer telephone interview, 13 January 2006; Compton, *Where No Man Has Gone Before*, p. 152.

CRA flowed through biological filters and then was heat sterilized in order to ensure that no microbes escaped to the outside world. The body wastes they produced received steam treatment, which heated them to a temperature of 260°F (127°C), while every piece of paper leaving the quarantine facility got exposed to ethylene oxide, a sterilizing gas, for 16 hours. The astronauts' and their staffs' periods of quarantine might have been extended if anyone had exhibited signs of infection by a potentially extraterrestrial organism. The release date from the quarantine would probably not have been altered, however, if the sick person was diagnosed as noninfectious or having a disease of terrestrial origin.[317]

LRL staff conducted extensive medical testing of the astronauts, their support personnel, and the technicians put in quarantine after the breach of the biobarrier in the vacuum laboratory. Staff conducted clinical observations and chemical, biological, and immunological analyses. These did not produce any indications of infections from exotic organisms or other ill effects from the mission. As a result, ICBC agreed to lift the quarantine of the astronauts and other personnel at 1 a.m. on 11 August 1969, one day early. ICBC did recommend, however, that all the personnel be kept under medical surveillance until completion of LRL's biological analyses and release of the Moon samples. NASA chose an early morning hour for ending the quarantine in order to minimize the number of reporters and the scuffle that was expected as the astronauts emerged. After a month of confinement, however, the astronauts sought to get out as soon as they could, and so at 9 p.m. on 10 August 1969, after completing one more medical examination, they walked out of isolation, spoke briefly to the press, and were rushed home to their families.[318]

---

317. ICBC, *Quarantine Schemes for Manned Lunar Missions* (Washington, DC: NASA [GPO 927-741], no date given), p. 1, folder 009901, "Lunar Quarantine and Back Contamination," NASA Historical Reference Collection; Hamblin, "After the Moonflight," 56; Compton, *Where No Man Has Gone Before*, p. 153; MSC, *Lunar Receiving Laboratory*, p. 107.
318. Hamblin, "After the Moonflight," 56; Compton, *Where No Man Has Gone Before*, p. 153.

Decontaminating the Command Module

NASA delivered the Apollo 11 CM to LRL at 6 p.m. on 30 July 1969, where it was placed in the spacecraft storage room.[319] Although the Apollo crew had executed CM cleaning procedures during the mission, NASA still considered the module to be potentially contaminated and thus subjected it to quarantine and decontamination. The aim was to release the CM with the flight crew on the 21st day of quarantine and return the module to the contractor, North American Rockwell Corporation. This would allow Rockwell to conduct postflight testing in ample time to use its data to support the next Apollo mission.

LRL staff first conducted a detailed exterior inspection of the craft, including photographic coverage, and removed access panels to the interior. A recovery engineer entered the CM and removed and bagged all stowed equipment, including clothing, which had to be quarantined along with the astronauts and lunar samples. LRL staff finished removing spacecraft equipment on 2 August 1969. The following day, the recovery engineer completed preparations for decontaminating the craft's water and waste management system by hooking up lines through which a piped-in formalin solution containing 40 percent formaldehyde would flow, remaining in the system for 24 hours. The engineer also opened all of the ship's compartments and wiped them down with disinfectant. LRL staff subsequently heated the interior of the module to 110°F (43°C), evacuated its pressure to 8.5 psi, and filled it with formaldehyde gas for 24 hours. Because the recovery crew that performed the decontamination might have gotten contaminated themselves, they had to be quarantined afterward.[320]

---

319. MSC, *Apollo 11 Activity*, p. 2.
320. MSC, *Spacecraft Quarantine and Release Plan*, MSC 00024, NASA MSC, Houston, TX, 21 May 1969, p. 1-11, folder 012990, "Lunar Receiving Lab," NASA Historical Reference Collection; Lunar Receiving Lab, "LRL Daily Summary Report No. 10," 4 August 1969, folder 012990, "Lunar Receiving Lab," NASA Historical Reference Collection.

NASA-S-67-689

LRL RADIATION COUNTING LABORATORY

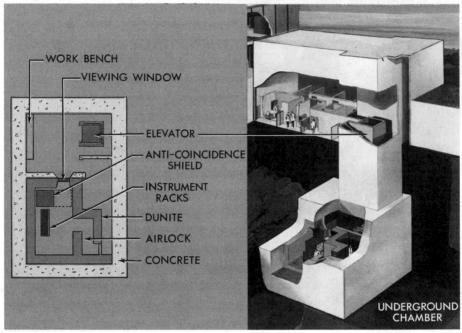

Figure 4.10 Radiation counting laboratory.

Moon Sample Quarantine and Analysis

LRL staff conducted intense sample analysis work in August 1969, aiming at finishing their tests and releasing the samples from quarantine by mid-September. Completing the analyses was estimated to require 50 to 80 days after the samples entered LRL.

Bioprime and biopool samples. LRL staff prepared two batches of lunar material—a bioprime and a biopool batch. They extracted the *bioprime* material—the sample that had least contacted Earth's environment—from the two rock coring tools. They and the principal investigators subjected it to minute analyses that sought to find any evidence of "living organisms or their relics."[321]

---

321. Compton, *Where No Man Has Gone Before*, p. 154.

LRL personnel made up two *biopool* samples from several hundred grams of fines and chips derived from Moon rocks in the bulk sample container. The mission scientists tested the biopool samples in a large range of living systems in order to determine whether Moon material was pathogenic or toxic, and to what extent.[322]

Gas and radiation counting operations. Scientists quickly performed the time-critical effluent-gas analysis and gamma-radiation counting experiments on the lunar material. LRL's gas analysis laboratory staff strove to carefully analyze outgassing from the samples without the data being compromised by terrestrial contaminants. The vacuum-tight sample containers made this possible by providing a barrier against biological and chemical contamination during the trip from the Moon to LRL. Once they received the near-pristine samples, gas laboratory staff performed several closely related analyses of the following:

- Gas from the sample containers. This provided data on the composition as well as quantity of gas released from the samples. It also provided a measure of just how effective a vacuum container's seal was in keeping out terrestrial gases.
- Gas generated during the splitting and preliminary examinations of the lunar rocks. The same gases were expected to be observed as in the above experiment, except perhaps for interstitial and occluded gases[323] released during cleaving of the lunar rocks.
- Gas evolved from heating the samples. These gases could have been adsorbed on rock surfaces or in interstitial spaces.
- Gas released or gaseous reaction products generated during atmospheric-reaction testing in LRL's physical-chemical test laboratory.

Samples collected on the Moon, where no atmosphere or substantial magnetic field is present to shield or deflect cosmic rays and solar wind protons from bombarding surface rocks, were expected to contain induced radioactive nuclides in addition to the isotopes contained in terrestrial rocks. Analysis of the induced gamma radiation

---

322. Compton, *Where No Man Has Gone Before*, pp. 153–154.
323. Interstitial gases refer to the gas in the small, narrow spaces between rock particles. Occluded gas refers to gas absorbed into and retained within the lunar material itself.

in the rocks could yield valuable information on the sample's composition as well as on the nature of the activating radiation that had bombarded it. Derived from this might be clues to the history of the Moon rocks.

Gamma-ray counting experiments had to be performed in LRL as soon after receiving the specimens as feasible, before many of the short-lifetime nuclides possibly decayed to undetectable levels. Scientists suspected this decay would occur because lunar rock materials, they thought, might be similar to the materials in meteorites, and the extremely weak radioactivity levels in those were not able to be adequately analyzed using the capabilities of the time.[324]

Scientists measured the samples' gamma-ray levels in the underground counting laboratory of LRL that had been constructed 50 feet (15 meters) below ground (see figure 4.10). The overburden of earth plus 5-foot (1.5-meter) thick concrete walls eliminated most cosmic ray background from the laboratory. To shield the space from terrestrial background radiation sources that might also obscure faint signals from the rocks, the construction crew lined the laboratory's walls with 36 inches (91 centimeters) of crushed dunite— a material found in Earth's mantle that consists mainly of olivine $[(Mg,Fe)_2SiO_4]$—and held it in place with a 3/8-inch steel liner. A special ventilation system using chilled charcoal beds and other filters provided a radon-free atmosphere in the laboratory. Finally, special detector enclosures that employed a 26-ton lead shield further reduced the radioactive background reaching the detectors by another one to two orders of magnitude.[325]

Biologic examinations. Biological analyses sought to determine whether contamination of Earth environments with lunar material would or would not be harmful. LRL staff sought to determine the presence of any infectious and possibly hazardous lunar microorganisms by observing their ability to replicate in selected animal and

---

324. James C. McLane, Jr., Elbert A. King, Jr., Donald A. Flory, Keith A. Richardson, James P. Dawson, Walter A. Kemmerer, and Bennie C. Wooley, "Lunar Receiving Laboratory," *Science* 155 (3 February 1967): 526–527, folder 4714, "MSC—Lunar Receiving Laboratory," NASA Historical Reference Collection.
325. McLane et al., "Lunar Receiving Laboratory," p. 527; MSC, *Lunar Receiving Laboratory*, p. 48.

plant species.[326] In these experiments, technical staff pulverized tiny pieces of Moon material and fed them to carefully raised colonies of insects, mice, birds, and plants that were bred to be germ-free and extremely sensitive to new forms of microbes.

Some of these colonies proved quite difficult to raise. For instance, developing ordinary mouse specimens was fairly easy, but raising germ-free colonies delivered by Caesarian section under sterile surgical conditions presented many challenges. LRL lost three such colonies during its attempts.[327]

Biologic analyses included the following protocols:

- *Direct sample observations.* Optical microscopes using visible and infrared light sources as well as electron microscopes were used to search for life-forms.
- *Bacteriology.* Attempts were made to grow microbes from lunar samples incubated in air, carbon dioxide, and nitrogen environments, using cultural media that included blood agar, glucose yeast extract, and thiogycolate broth.
- *Virology.* Identification of viruses was attempted by observing lunar material's ability to replicate in cell cultures, embryonated eggs, and other media.

The lunar material was also tested to determine whether it contained life-forms capable of generating a disease or reproducing in specific categories of organisms. These categories, and the particular species used in the tests, included the following:

- *Mammalian vertebrates:* Germ-free mice.
- *Avian vertebrates:* Japanese quail.
- *Invertebrates and fish:* Cockroaches, houseflies, wax moths, brown shrimp, commercial oysters, and fathead minnows.
- *Plants:* 33 varieties with economic importance.[328]

Physical and chemical analyses. Investigators subjected samples to physical and chemical examinations as well as biological tests. The physical-chemical test laboratory contained 36 linear feet of

---

326. MSC, *Lunar Receiving Laboratory—Sample Flow Directive*, NASA MSC, Houston, TX, 21 May 1969, p. 13, folder 012990, "Lunar Receiving Lab," NASA Historical Reference Collection.

327. Hamblin, "After the Moonflight," 58.

328. MSC, *Report on the Status of the Apollo Back-Contamination Program*, Rev. A, June 1969, pp. 35–36, JSC Historical Archives, "June 17–30, 1969" folder, box 76-25.

double-sided, gas-tight, dry nitrogen atmosphere cabinets providing a biologic barrier, with glove ports for 20 different operator positions. Most of the minerologic,[329] petrologic,[330] and geochemical[331] investigations during the quarantine period took place here.

Equipment designed for operating through the biologic barrier included petrographic microscopes, stereoscopic binocular microscopes, balances, a gas comparison pycnometer (an instrument for measuring the density or specific gravity of materials), gas reaction equipment, x-ray diffraction and fluorescence units, photomicrography capabilities, and the hand tools, reagents, and other paraphernalia of a chemistry laboratory. Laboratory staff used microscopy to examine and characterize the different types of lunar material collected. They exposed the rock fragments to common Earth atmospheric elements, including nitrogen, oxygen, and water vapor. None of these elements would elicit a dramatic reaction from an Earth rock, but the lunar material might never have encountered water, oxygen, or nitrogen. Violent reactions could not be ruled out, such as rapid disintegration, sharp temperature rises, or, less likely, explosions. Significant changes in minerologic composition or physical properties were also possible.

When such dramatic reactions to Earth elements did not occur, scientists subjected lunar rock chips to a battery of analyses that gave clues to the Moon's history and origin, its surface materials and processes, the forces that formed the surface, and the period of the solar system's history in which it was formed. Many laboratories throughout the scientific community outside LRL performed more detailed investigations once the quarantine was lifted. These experiments included comprehensive analyses of elements and isotopes, which generated mineral identifications and descriptions.[332]

When investigators could not determine the structures and characteristics of rock fragments using microscopes, they employed analytical methods such as x-ray diffraction or ignition of the sample and

---

329. "Mineralogic" refers to the study of the distribution, identification, and properties of the minerals found in the samples.
330. "Petrologic" refers to studies dealing with the origin, composition, and structure of rocks."
331. "Geochemical" refers to the study of the chemical composition and properties of rocks.
332. MSC, *Lunar Receiving Laboratory*, pp. 60–61; McLane et al., "Lunar Receiving Laboratory," 526; Hamblin, "After the Moonflight," 56.

analysis of the flame's spectrum. This was useful because different elements can be identified by the characteristic spectral "fingerprints" they emit when burned.[333]

Releasing the Lunar Samples from Quarantine

LRL staff conducted intense testing throughout August 1969. The intention was to generate enough data on the lunar samples that they could be released from quarantine in mid-September with confidence that they would do no harm to Earth's environment.

None of the biological testing during August revealed any indication of the existence of lunar microorganisms nor any indication of hazards to Earth organisms. One result, however, was that lunar samples appeared to *stimulate* the growth of plants used in the testing.

ICBC reviewed the extensive data from the testing and concluded that the lunar samples brought back by Apollo 11 presented no biological danger to Earth life. ICBC notified MSC Director Robert Gilruth of its finding and gave permission to release the samples on 12 September 1969. The principal investigators (PIs) picked up their samples in person at LRL and began their in-depth, exhaustive studies of the lunar material.[334]

Apollo 12

The scientific community was satisfied for the most part with the way that LRL staff processed Apollo 11 samples, although many complained about the long delay in releasing samples to PIs, attributing this to inefficient procedures, inexperience, and problems with equipment. NASA wanted to correct these issues before the Apollo 12 launch, which was scheduled for mid-November 1969.

In order to meet this deadline, LRL began to work on the corrections and fixes while they were still completing their Apollo 11 tasks.

---

333. Hamblin, "After the Moonflight," 58.
334. Compton, *Where No Man Has Gone Before*, p. 154.

In early September 1969, NASA's Lunar Sample Analysis Planning Team sent recommended changes to MSC's new Director of Science and Applications, Anthony J. Calio, which covered procedures such as photographing samples and weighing and transferring material during quarantine as well as displaying data on each sample's history, status, and location. Efficient display of current data was necessary because during Apollo 11, problems had emerged regarding where each sample was located and what operations had been performed on it.

Staff who had handled Apollo 11 samples gave recommendations for reducing the number of personnel in LRL and streamlining sample preparation for quarantine testing.[335] U.S. Department of the Interior–Geological Survey personnel in particular developed sample handling procedures that they claimed would "drastically improve the speed and completeness of the preliminary examination."[336] One of these suggestions was to eliminate the biopool sample and use only a single bioprime sample. This would help to correct a serious bottleneck in Apollo 11 sample processing—the time-consuming preparation and handling of both bioprime and biopool samples. The U. S. Geological Survey personnel justified this proposed action by claiming that bio-pool sample testing was "merely a more extensive verification of the tests made on the bioprime sample."[337]

The LRL vacuum system required significant modifications. As discussed above, problems had arisen regarding effective sample handling, and the glove boxes themselves had leaked, potentially contaminating workers and the outside environment. In addition, organic materials from the vacuum pumps had contaminated the vacuum system. Because of these issues, Lunar Sample Analysis Planning Team members thought it necessary to change policy and open at least one of the two lunar sample containers in a dry nitrogen atmosphere, avoiding the need for a vacuum system for that batch of samples.[338]

---

335. Compton, *Where No Man Has Gone Before*, p. 174.
336. E. C. T. Chao and R. L. Smith to Wilmot Hess and/or A. J. Calio, "Recommendations and Suggestions for Preliminary Examination of Apollo 12 Returned Lunar Samples," 8 September 1969, attachment to Richard S. Johnston to E. C. T. Chao and R. L. Smith, 23 September 1969, JSC Historical Archives, "22–24 September 1969" folder, box 71-63.
337. Ibid.
338. Compton, *Where No Man Has Gone Before*, p. 174.

Figure 4.11 USS *Hornet*, prime recovery vessel for Apollo 12, moves toward the Command Module. A helicopter from the recovery ship, which took part in recovery operations, hovers over the scene of the splashdown.

NASA officially decided many of these issues in late September 1969. Richard Johnston reported in a 23 September letter that MSC was attempting, as suggested by USGS personnel, to develop an adequate sampling criterion that would permit the use of only a single sample for Apollo 12 bioprotocols.[339] ICBC concurred with the use of a single pooled lunar material sample, as long as it was representative of the returned rock material from the Moon.[340]

NASA ultimately decided to prepare only one sample. Furthermore, the two lunar sample return containers would go through simultaneous processing, but this processing would be conducted for one container in a sterile nitrogen atmosphere and for the other in a vacuum. NASA refused, however, to fulfill an ICBC request to install a bacterial filter on the Command Module's postlanding ventilation system, which would have given greater protection against back contamination but

---

339. Richard S. Johnston to E. C. T. Chao and R. L. Smith, 23 September 1969, JSC Historical Archives, "22–24 September 1969" folder, box 71-63.
340. ICBC, "Minutes," 30 October 1969, JSC Historical Archives, "27–31 October 1969" folder, box 76-31.

would also have generated the problems discussed above regarding the Apollo 11 CM.[341]

In its 30 October 1969 meeting, ICBC studied whether results from the Apollo 11 mission warranted elimination of the BIGs during recovery operations on future Apollo missions. The Apollo 11 astronauts had found the BIGs oppressively uncomfortable. John Hirasaki, the MQF engineer on Apollo, commented that NASA had "a significant concern about the use of BIGs compromising the health of the crew because of thermal overload."[342] ICBC recommended eliminating use of the BIGs if "the Apollo 12 crew condition is normal at earth landing."[343] NASA concurred with this recommendation and implemented it for Apollo 12. In lieu of BIGs, the recovery crew would provide astronauts with clean flight suits and oral-nasal respiratory masks to put on before leaving the Command Module. These masks were equipped with bacterial filters to prevent back contamination due to exhalations of the astronauts. BIGs would be available, however, "for use as a contingency in case of unexplained crew illness."[344]

At 11:22 a.m. eastern standard time (EST), 14 November 1969, Apollo 12 took off from Kennedy Space Center, lifted by a Saturn 5 launch vehicle. On 20 November, astronauts Pete Conrad and Alan Bean conducted a 4-hour Moon walk. At its conclusion, Bean and Conrad dusted each other off as best they could, but they had not taken any brushes or other equipment to help with the task. From this and the following day's Moon walk, they carried considerably more lunar dust into the *Intrepid* Lunar Module than Armstrong and Aldrin did during Apollo 11. Conrad described himself and Bean as looking like "a couple of bituminous coal miners . . . ."[345]

The astronauts' quarantine period officially began on the lunar surface the moment they closed the Lunar Module's hatch.[346] Once the *Intrepid* Lunar Module lifted off the Moon and attained lunar orbit, the dust carried in by the astronauts began to float throughout

341. Compton, *Where No Man Has Gone Before*, p. 174.
342. Hirasaki interview, 13 January 2006.
343. ICBC, "Update of Apollo 12 Quarantine Procedures," 31 October 1969, LBJ SCHC, folder 076-31, "27–31 October 1969."
344. ICBC, "Update of Apollo 12 Quarantine Procedures."
345. Compton, *Where No Man Has Gone Before*, p. 185.
346. John Rummel, interview by author, San Francisco, CA, 7 December 2005.

the cabin and was thick enough to be visible. After docking with the *Yankee Clipper* Command Module, the astronauts tried to vacuum the dust up, but had little success, so they removed and stowed their suits in an attempt to minimize the amount of contamination that would enter the CM. In spite of their efforts, the CM returned to Earth with significant amounts of dust clinging to every surface and remaining suspended in the cabin's atmosphere. The filters in the ship's environmental control system proved not capable of removing it from the ship's air as thoroughly as NASA had hoped.[347]

The Apollo 12 CM splashed down in the Pacific Ocean on 24 November 1969, 375 miles (600 kilometers) east of Pago Pago, American Samoa, and 2 miles (3.5 kilometers) from the USS *Hornet* recovery vessel, which then approached the CM (see figure 4.11). When the recovery crew swimmers arrived, they tossed respirators and coveralls into the module but, as mentioned above, did not furnish the astronauts with BIGs. Half an hour later, the crew reached the *Hornet* via helicopter and entered the MQF. The recovery crew removed lunar sample containers and film magazines from the CM, which were flown to Pago Pago and then on to Houston and LRL. The astronauts did not have quite so speedy a journey. They remained in their MQF during the 4-day sea journey to Hawaii and the 9-hour cargo plane ride to Houston. Astronauts Conrad, Bean, and Richard Gordon finally reached LRL, and continued their quarantine, on the morning of 29 November 1969.[348]

Once again, problems arose with LRL's glove box isolation system. During a preliminary examination of lunar samples, a small cut was found in one of the gloves. This potentially contaminated the area and sent 11 people into quarantine. Several scientists were among those isolated, and they were quite disappointed that they had been separated from their work. Other than the problem with the glove, however, LRL operations during Apollo 12 went quite smoothly. The staff finished inventorying and conducting preliminary examinations of all samples within a week after they had arrived at LRL.

---

347. "Minutes—Interagency Committee on Back Contamination," 3 October 1966, in the package of letters and memos beginning with Col. J. E. Pickering to David J. Sencer, 28 October 1966, folder 4714, "MSC—Lunar Receiving Laboratory," NASA Historical Reference Collection; Compton, *Where No Man Has Gone Before*, pp. 185–187.
348. Compton, *Where No Man Has Gone Before*, pp. 186–187.

The astronauts and their companions spent a largely uneventful quarantine period. Periodic medical examinations showed no change in the conditions of any of the bio-isolated personnel. On 10 December 1969, slightly before NASA's official end of the quarantine period, they were released but kept under medical surveillance until data from the biological testing confirmed that they presented no threat to Earth.[349]

LRL's intensive biological testing continued into January 1970. No evidence of living or fossilized microorganisms was detected. No indications of any sort of lunar-material-related pathology were observed among Apollo 12 astronauts, the personnel who had contact with them or with lunar material, or with any of the test species of plants and animals.[350]

After the release of the samples, outside scientists did find carbon and carbon compounds in the samples, but they did not find any indication of molecules that could "clearly be identified as derived from living organisms."[351] Investigators as well as the Lunar Sample Analysis Planning Team, however, recognized that only an extremely small fraction of the Moon's surface had been sampled and understood the importance of obtaining materials from a diversity of terrains and locations. In the words of one scientist, "what I looked at was equal to the size of three postage stamps and a thimbleful of material . . . . Our results are based on the wanton destruction of two grams of lunar material."[352] NASA aimed in future missions to partially correct this problem by exploring a greater range of sites and terrains. This would be critical for drawing any conclusions about the Moon and its potential for back contamination of Earth.

---

349. Charles A. Berry to Maj. Gen. J. W. Humphreys, Jr., 8 December 1969, JSC Historical Archives, "December 5–18, 1969" folder, box 76-31; Compton, *Where No Man Has Gone Before*, pp. 188–189.
350. SSB, "Report of Meeting on Review of Lunar Quarantine Program," 17 February 1970, JSC Historical Archives, "February 1970" folder, box 76-32.
351. Compton, *Where No Man Has Gone Before*, p. 190.
352. Ibid., pp. 190–191.

Apollo 13 and 14

In early 1970, ICBC used data from Apollo 11 and 12 to reexamine the need for quarantine on future missions. The environment that had been found on the Moon and that probably had existed for billions of years suggested to ICBC that quarantine was perhaps not needed. ICBC decided, after an extended dialogue, "to recommend to the NASA Administrator that crew quarantine be discontinued"[353] and to advise the SSB of this recommendation. But in February 1970, the SSB examined this issue and did not agree that crew quarantine for Apollo 13, scheduled to launch in two months, should be discontinued. Apollo 13 was to land on a lunar highland site believed to be significantly different from the mare[354] sites that Apollo 11 and 12 sampled. Nonetheless, the SSB agreed with ICBC that crews on future missions need not be quarantined unless the anticipated differences in landing sites required it.

Apollo 13 launched on 11 April 1970, aiming to land at a spot on the Moon just north of the Fra Mauro crater. This would allow the astronauts to sample the Fra Mauro Formation, which scientists believed was made up of ejected material from the Imbrium basin caused by a large impact in the Moon's past. Thus, the astronauts would be collecting samples from deep inside the Moon that had been forcibly excavated during the Imbrium event. Apollo 13 never landed on the Moon, however. Hours after launch, an oxygen tank in the Service Module exploded, resulting in damage to other systems and loss of most of the spacecraft's electrical power and oxygen. NASA had to order the mission aborted; only a brilliant, real-time replanning of procedures by the mission operations team saved the lives of the crew.[355]

---

353. Compton, "The End of Quarantine," in *Where No Man Has Gone Before*, Chap. 12, pp. 242–243.
354. A mare is a large, dark, basaltic plain on the Moon formed from ancient volcanic eruptions.
355. Lunar and Planetary Institute, "Apollo 13 Mission," *http://www.lpi.usra.edu/expmoon/Apollo13/Apollo13.html* (accessed 8 August 2005); NASA, "Apollo 13 Command and Service Module (CSM)," *http://nssdc.gsfc.nasa.gov/database/MasterCatalog?sc=1970-029A,NSSDCMasterCatalog: Spacecraft,NSSDC ID 1970-029A* (accessed 8 August 2005); Compton, *Where No Man Has Gone Before*, pp. 198–199.

After NASA had to abort Apollo 13, the Apollo site selection board almost unanimously agreed to send Apollo 14 to the previous mission's target destination, Fra Mauro. Crew quarantine remained in place for Apollo 14 as well. The rationale given was the same as for Apollo 13: the mission was tasked to excavate a deep lunar core sample, whose material might differ from the surface samples of the mare sites taken by the previous lander missions.[356]

Apollo 14 took off on 31 January 1971 and, after mission commander Alan B. Shepard, Jr., and Lunar Module pilot Edgar D. Mitchell conducted two Moon walks, returned to Earth and splashed down on 9 February 1971. Recovery operations began with swimmers deployed to the CM to install the flotation collar, pass flight suits and respirators to the astronauts still inside the CM, then assist them from the module into the life raft. After this, the recovery helicopter flew the astronauts to the USS *New Orleans* helicopter carrier recovery vessel, where they entered an MQF. During the Apollo 14 mission, however, they did not remain in this initial MQF until it reached LRL, but left it while it was still on the recovery ship, and they were helicoptered to Pago Pago, American Samoa, where they transferred to a second MQF aboard a C-141 aircraft. This aircraft flew them to Ellington Air Force Base near MSC.[357]

The new approach considerably shortened crew return time. During Apollo 12, the CM splashed down on 24 November 1969, but the astronauts did not reach LRL until 29 November. On Apollo 14, splashdown occurred 9 February 1971; the astronauts reached LRL about 60 hours later.[358]

NASA carried out its new return procedures over the strenuous objections of Charles Berry, who pointed out that within the MQF on the recovery ship, astronauts would remove their masks, allowing their exhalations to mix freely with the atmosphere inside the mobile facility. The entire recovery vessel and its crew would become exposed to lunar material when the outer MQF door was opened. Berry held that

---

356. Compton, "The End of Quarantine," in *Where No Man Has Gone Before*, Chap. 12, pp. 242–243.
357. MSC, *Recovery Requirements—Apollo 14*, MSC-03666, 7 December 1970, JSC Historical Archives, "7 December 1970" folder, box 79-44/45.
358. MSC, "Mission Support Performance," in *Apollo 14 Mission Report*, MSC-04112, May 1971, Chap. 11, *http://www.hq.nasa.gov/office/pao/History/alsj/a14/a14mr11.htm* (accessed 26 January 2011).

the procedures "clearly violate any pretext of maintaining an effective biological barrier."[359]

●●●

The End of Quarantine

Life-detection experiments on Apollo 14's return samples produced negative results, as had the samples from Apollo 11 and 12. Furthermore, no evidence emerged from any test system of any type of hazard associated with Apollo 14 samples.[360] Thus after Apollo 14, NASA revisited whether quarantine procedures were needed on future lunar landing missions.

Quarantining operations were very expensive. They were also troublesome and inconvenient to technical staff because they impeded postflight briefings and delayed the release of eagerly sought lunar samples to investigative teams. Considerable evidence suggested that quarantines simply were not required for the protection of Earth and that life had either never existed on the Moon, or at least appeared to have left no trace at the sites examined. On 26 April 1971, based heavily on the recommendation of ICBC,[361] Acting NASA Administrator George M. Low discontinued the quarantine for future Apollo flights to the Moon, stating that

> The analysis of quarantine information from Apollo 14 has now been completed. On the basis of this analysis, as well as the results from the Apollo 11 and Apollo 12

---

359. Charles A. Berry to Director of Flight Operations, "Quarantine Plans for Apollo 14," 30 October 1970, JSC Historical Archives, "1–31 October 1970" folder, box 75-44.
360. MSC, "Biological Sciences Operations," final report for Apollo 14 activities in the LRL, no date given, JSC Historical Archives, "April–July 1971" folder, box 76-36.
361. Ivan D. Ertel and Roland W. Newkirk, with Courtney G. Brooks, "April 26," in Part 3 (H)—"Man Circles the Moon, the *Eagle* Lands, and Manned Lunar Exploration—1971," in *The Apollo Spacecraft—A Chronology*, Volume IV (Washington, DC: NASA SP-4009, 1978), *http://www.hq.nasa.gov/office/pao/History/SP-4009/contents.htm* (accessed 26 January 2011).

flights, it has been concluded that there is no hazard to man, animal or plants in the lunar material.[362]

Although quarantine procedures were terminated, biomedical analyses of returned lunar samples would continue. In addition, the careful procedures for protecting lunar samples from Earth contamination on the return journey from the Moon as well as during transport to LRL would not significantly change.[363]

The Apollo mission was an extremely complex endeavor with many technical and political challenges. A concise summary of major events from this mission is included in the timeline in Appendix E. Specific planetary protection approaches employed by Apollo as well as other missions are summarized in Appendix F.

---

362. George M. Low, "Decision to Terminate Quarantine Under NMI 1052.90 (Attachment A, Change 1, 2)," 26 April 1971, attachment to Dale D. Myers to Manned Spacecraft Center, 10 May 1971, JSC Historical Archives, "April–July 1971" folder, box 76-36; NASA, "Apollo Quarantine Discontinued," NASA news release 71-78, 28 April 1971, folder 4714, "MSC—Lunar Receiving Laboratory," NASA Historical Reference Collection.

363. SSB, *The Quarantine and Certification of Martian Samples* (Washington, DC: National Academies Press, 2002), p. 80, *http://books.nap.edu/openbook/0309075718/html/80.html#p200053958960080001* (accessed 26 January 2011); NASA, "Apollo 14 Lunar Module/ALSEP," *http://nssdc.gsfc.nasa.gov/database/MasterCatalog?sc=1971-008C,NSSDCMasterCatalog:Spacecraft,NSSDCID1971-008C* (accessed 8 August 2005); Compton, *Where No Man Has Gone Before*, p. 269.

PLANETARY PROTECTION FOR MARS

5

The Viking Experience

Landing on Mars was something nobody had ever done before. Looking for life on Mars is a pretty exciting venture. . . . It probably didn't hurt that the science requirements and the planetary quarantine requirements somewhat overlapped.

—*Tom Young, former Viking Mission Director*[1]

The first United States effort to send soft landing spacecraft to another planet began with two Viking launches to Mars in 1975.[2] These were our country's most ambitious robotic space ventures up until that time. Furthermore, the Viking mission employed "the most stringent planetary protection requirements imposed to date on any U.S. flight project."[3] NASA estimated the probability of growth of an Earth microorganism in Mars's environment to be relatively large. This necessitated rigorous sterilization procedures for the Viking Lander Capsules (VLCs), since they would directly contact the Martian surface.[4]

---

1. Tom Young, former Viking Mission Director, telephone interview by author, 20 February 2006.
2. G. R. Hintz, D. L. Farless, and M. J. Adams, "Orbit Trim Maneuver Design and Implementation for the 1975 Mars Viking Mission" (paper no. AIAA-1978-1394, American Institute of Aeronautics and Astronautics and American Astronautical Society, Astrodynamics Conference, Palo Alto, CA, 7–9 August 1978).
3. J. Barengoltz and P. D. Stabekis, "U.S. Planetary Protection Program: Implementation Highlights," *Advances in Space Research* 3(8) (1983): 7.
4. Exotech Systems, Inc., *Planetary Quarantine Parameter Specification Book*, Gaithersburg, MD, 1973, and later revisions as reported in J. Barengoltz and P. D. Stabekis, "U.S. Planetary Protection Program: Implementation Highlights," *Advances in Space Research* 3(8) (1983): 7.

Bionetics Corporation ("Bionetics") was of particular help in attaining effective planetary protection. The company had been founded in 1969 specifically to "perform planetary quarantine support for NASA Langley Research Center's Viking Project"[5] and, in 1972, Bionetics was awarded a $1 million contract to manage Viking's planetary quarantine and microbiological assay facility. Bionetics aided NASA in minimizing spacecraft contamination that could have led to false positive indications of life on Mars. Bionetics also assisted NASA and Martin Marietta, the manufacturer of the spacecraft, in developing cleaning approaches and verification procedures.[6]

Each Viking spacecraft consisted of an orbiter, or "VO," and a lander capsule, or "VLC" (see figure 5.1). The VLC, in turn, included the actual **lander** (a detail of which is given in figure 5.2) that touched down on Mars, surrounded by the other components depicted in figure 5.1: a **descent capsule** made up of an aeroshell–heat shield–parachute system that enclosed, protected, and slowed down the lander during its descent through the Martian atmosphere, and a **bioshield** cap and base assembly around the descent capsule that protected it from terrestrial contamination before and during launch.

The Viking spacecraft journeyed 815 million kilometers (505 million miles) in approximately one year to land on Mars in the summer of 1976. The entire mission (i.e., the two craft's journeys) cost approximately $1 billion.[7] JPL engineers used the knowledge they gained developing the Mariner spacecraft in the 1960s and 1970s to design and fabricate the VO, and they completed this task with relatively few technical difficulties. The VLC team, on the other hand, "was tackling a new field . . . [and] . . . breaking much new technological ground."[8] The prime contractor for developing the VLC, Martin Marietta, had to work through a long list of engineering challenges before it produced a reliable capsule (see figure 5.3 depicting assembly of the lander).

---

5. "Joseph A. Stern," *Virginian-Pilot* (3 February 1996): B4, *http://scholar.lib.vt.edu/VA-news/VA-Pilot/issues/1996/vp960203/02030273.htm*.

6. "Bionetics Wins $1-M NASA Pact," *Times-Herald* (no city or state given) (28 January 1972), LaRC Archives, Viking News Clippings—1971–1974.

7. Steven J. Dick and James E. Strick, *The Living Universe: NASA and the Development of Astrobiology* (Piscataway, NJ: Rutgers University Press, 2004), p. 80.

8. Edward Clinton Ezell and Linda Neuman Ezell, "Viking Lander: Creating the Science Teams," in *On Mars: Exploration of the Red Planet—1958–1978* (Washington, DC: NASA SP-4212, 1984), Chap. 7, available online at *http://history.nasa.gov/SP-4212/ch7.html* (accessed 21 March 2011).

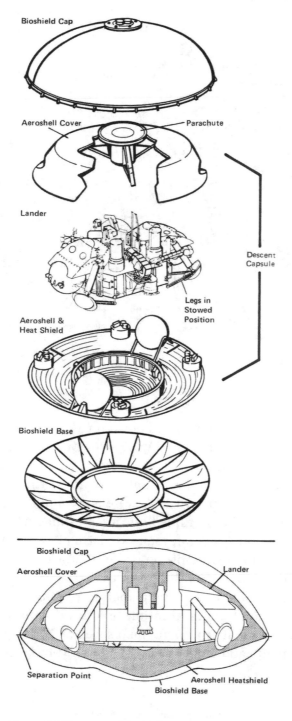

Figure 5.1 Elements of the VLC.

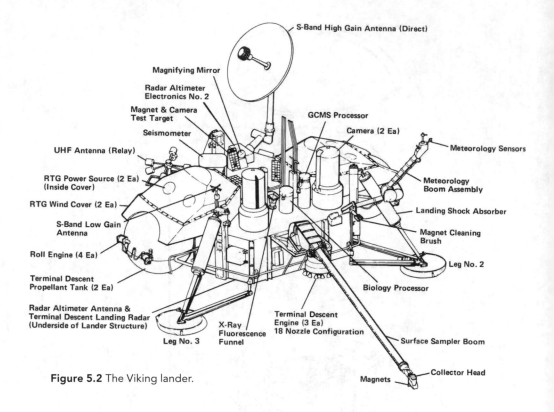

Figure 5.2 The Viking lander.

The Viking lander that touched down on Mars contained far more sophisticated equipment than NASA's previous robotic lander, the lunar Surveyor, and weighed twice as much. The Viking lander had to carry out a soft landing on a celestial body many times farther away than our Moon, and its mission goals were more ambitious. The Viking lander performed studies in three basic areas—imaging, organic analysis, and life detection. The craft took extensive pictures of the Martian surface and contained state-of-the-art laboratories and instruments that conducted detailed scientific analyses.[9]

The Viking mission had a vital planetary protection goal as well—to prevent biological contamination of Mars during the mission's search for life. To satisfy international as well as NASA planetary protection agreements and guidelines, "all launch vehicle and spacecraft

---

9. Nicholas Panagakos and Maurice Parker, "Viking Mars Launch Set for August 11," in *Viking Press Kit,* NASA news release no. 75-183, 1975, p. 1.

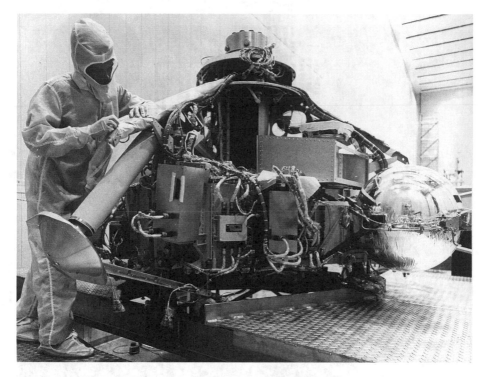

Figure 5.3 Viking spacecraft under assembly at Martin Marietta Aerospace near Denver, Colorado.

hardware injected into a Mars flight trajectory [were] considered to be possible sources of contamination for Mars."[10] This was NASA's first opportunity to protect the environment of a planet that many scientists thought capable of supporting life.

Astronomers believed that Mars was the planet most like Earth. Data taken by Mariner 9 in 1971 and 1972 revealed possible evidence of flowing, liquid water on the Martian surface during some period in its past. The Mariner evidence included photographs of braided channels that were reminiscent of dry riverbeds, which suggested that water had flowed there and increased the chances that Mars might harbor, or might have once harbored, life.[11] Carl Sagan and Stanford geneticist

---

10. Bionetics, *Lessons Learned from the Viking Planetary Quarantine and Contamination Control Experience,* 1990, p. 2.2, folder 006697, "Quarantine/Sterilization," NASA Historical Reference Collection.
11. Nicholas Panagakos and Maurice Parker, *Viking Press Kit,* NASA news release no. 75-183, 1975, pp. 2, 11–12.

Figure 5.4 Joshua Lederberg with a model of the Viking spacecraft and launch vehicle. This picture was taken in an exobiology exhibit at the 1975 pre-Viking "Earth and Mars: The Role of Life" symposium held at the Kennedy Space Center.

and Nobel Laureate Joshua Lederberg (figure 5.4) noted that organisms "which extract their water requirements from hydrated minerals [minerals whose molecules are chemically combined with those of water] or from ice are considered possible on Mars."[12]

During the Viking mission, scientists needed to walk a tightrope between preventing contamination of Mars with Earth organisms and conducting a rigorous scientific investigation that would generate important data on the Red Planet. This was not an easy line to walk. Microorganisms exist in large numbers almost everywhere on Earth, and thus NASA had to take extreme protective measures to

---

12. Carl Sagan and Joshua Lederberg, "The Prospects for Life on Mars: A Pre-Viking Assessment," *Icarus* 28 (1976): 291.

minimize their transport to Mars.[13] NASA scientists believed it inevitable that some microorganisms would survive on or inside the spacecraft and took special precautions to avoid introducing these entities to the Martian surface. If such organisms contaminated Martian soil and then multiplied and spread from the original point of entry across the planet's surface, a later mission might detect and mistake them for indigenous forms of life. Such terran life-forms might also impact and even destroy any existing Martian organisms. A 1966 statement by the American biologist K. C. Atwood expressed what a tragedy such an occurrence would be:

> It is possible that Mars may provide our only chance to study life-forms of an entirely separate descent. The importance of this opportunity outranks every prospect; the effect on world thought of the discoveries to be made on Mars may compare with that of the works of Copernicus or Darwin. A careless blunder would destroy the opportunity. The reasons for the sterilization of spacecraft are clear: we do not know enough about Mars to predict with confidence the outcome of microbial contamination of its surface. Regrettable and irreversible results can be imagined. The means of avoiding these results are known; hence we must employ such means.[14]

To protect against the scientific tragedy that K. C. Atwood warned of, spacecraft to Mars had to be thoroughly sterilized. But that was not the only reason to remove microbial contamination from the vehicles. Martin Favero, a microbiologist who headed CDC's spacecraft sterilization and planetary quarantine activities from 1964 to 1972,[15] commented that "with Viking, there were two reasons to sterilize the spacecraft. One was to make sure that the life detection

---

13. Bionetics, *Lessons Learned*, p. 2.1.
14. Samuel Glasstone, "Avoidance of Contamination of Mars," in *The Book of Mars* (Washington, DC: NASA SP-179, 1968), Chap. XII, pp. 247–248.
15. Faculty of 1000 Medicine, "Biography of Martin Favero," Faculty of 1000 Medicine Web site, *http://www.f1000medicine.com/about/biography/9376667068581463* (accessed 18 March 2006).

systems didn't detect self-contamination, and secondly to prevent contamination of the planet."[16] On the Viking project, life-detection experiments and planetary protection activities had overlapping needs, in that they both required extreme microbial contamination control.

●●●

Probabilistic Contamination Control Guidelines

Although directives written in 1959 by Abe Silverstein, NASA's Director of Space Flight Programs at the time, called for absolute sterilization of spacecraft, U.S. space scientists soon proposed more realistic guidelines using probabilities of contamination as the basis for designing planetary protection procedures (see Chapter 2).[17] Instead of the "abstract concept of sterilization,"[18] NASA required microbe burden reductions estimated as effective for planetary protection, but not so severe that attaining them would damage spacecraft reliability.

During the first decades of the space program, NASA and COSPAR issued forward contamination control guidelines for Mars and other missions that followed this approach. For instance, in 1964, COSPAR issued Resolution No. 26, which called for a probability of less than 1×10^{-4} that a spacecraft intended for planetary landing or atmospheric penetration would contain even one viable organism. In May 1966, COSPAR issued a more attainable recommendation that the contamination probability of a planet of biological interest be no more than 1×10^{-3}. COSPAR also recommended that this standard be adopted by all countries conducting space exploration. The United States adopted the standard on 6 September 1967.

---

16. Martin Favero, CDC microbiologist and planetary protection specialist on both the Apollo and Viking missions, interview by author, 31 December 2005.

17. Charles W. Craven, "Part I: Planetary Quarantine Analysis," *Astronautics & Aeronautics* (August 1968): 20. Also available in folder 006695, "Sterilization/Decontamination," NASA Historical Reference Collection.

18. Barengoltz and Stabekis, "U.S. Planetary Protection Program: Implementation Highlights," 5.

COSPAR apportioned the contamination risk among different nations and gave the United States a maximum allowable risk of 4.4×10^{-4}. In other words, NASA had to ensure that during a planet's period of biological interest (originally 20 years but eventually lengthened to 50 years), the total contamination risk from all U.S. spacecraft did not exceed 4.4×10^{-4}.[19] It was NASA's responsibility how to suballocate this risk among its various missions. See Chapter 3 for a more detailed discussion of these probabilistic approaches to planetary protection.[20]

To satisfy COSPAR and NASA planetary protection policies, all launch vehicles and all spacecraft hardware that were sent on Mars flight trajectories had to be viewed as potential contamination sources for the planet. For the Viking mission, vehicles of concern included not only the spacecraft (consisting of the VO and VLC), but also the Centaur upper rocket stage that carried the spacecraft onto a trajectory toward Mars. After the spacecraft separated from it, Centaur's flight path needed to be such that it would not impact and possibly contaminate Mars.[21]

Apportioning Contamination Probability Limits for a Mars Mission

NASA suballocated maximum contamination risks to each U.S. mission to Mars in a manner to conform with the COSPAR guideline that the total contamination probability by all U.S. spacecraft not be greater than 4.4×10^{-4}. NASA's suballocations varied depending on the class of the mission. Landers, for instance, were assigned a larger maximum contamination risk than orbiters and flybys.[22] For Viking launches, NASA further divided this risk probability into three

---

19. Ibid.
20. Homer E. Newell, "Political Context," in *Beyond the Atmosphere: Early Years of Space Science* (Washington, DC: NASA SP-4211, 1980), Chap. 18, http://www.hq.nasa.gov/office/pao/History/SP-4211/cover.htm (accessed 26 January 2011); Committee on Preventing the Forward Contamination of Mars, SSB, National Research Council, *Preventing the Forward Contamination of Mars* (unedited prepublication copy) (Washington, DC: National Academies Press, 2005), Chap. 2.
21. NASA, "Viking Pre-Launch Test Flight," Great Images in NASA (GRIN), 20 January 1974, http://grin.hq.nasa.gov/ABSTRACTS/GPN-2003-00047.html (accessed 26 January 2011).
22. Perry Stabekis, telephone interview by author, 7 September 2007.

subprobabilities, each of which applied to a different mechanism of potential planetary contamination:

- *Large impactibles: Parts of the spacecraft that might, through errors in trajectory determination or malfunctions, hit and contaminate Mars.* NASA designed its procedures for deflecting and making midcourse corrections and trimming maneuvers to the spacecraft's trajectory[23] to minimize the risk of accidental planetary impacts.

- *Ejecta-flux: Organisms that may get dislodged from the sterile surfaces of the spacecraft and directly impact the planet.* Such dislodgings could result from collisions with micrometeorites or other events.

- *Lander sources of contamination.* Organisms on the VLC that survive terminal sterilization procedures on Earth or organisms such as from other parts of the spacecraft that recontaminate the VLC after it is sterilized.[24]

In order for terrestrial organisms to actually contaminate Mars, they had to survive 1) space vacuum and temperatures, 2) space ultraviolet radiation flux, 3) Mars atmospheric entry, and 4) landing and subsequent release from the spacecraft. After this, the organisms had to actually grow and proliferate on Mars. NASA calculated the probability of contamination of Mars that took into account its estimates of the probabilities of these occurrences.[25]

The probability that viable terrestrial organisms escaping from the Viking lander would actually proliferate and spread, forming infected microenvironments on the Martian surface, depended not only on the characteristics of the organisms but also on the characteristics of the planet's surface and microenvironments that they encountered. First, a microbe ejected from the lander needed to survive the following conditions:

---

23. G. R. Hintz, D. L. Farless, and M. J. Adams, "Orbit Trim Maneuver Design and Implementation for the 1975 Mars Viking Mission" (paper no. AIAA-1978-1394, American Institute of Aeronautics and Astronautics and American Astronautical Society, Astrodynamics Conference, Palo Alto, CA, 7–9 August 1978), p. 1.

24. Bionetics, *Lessons Learned*, pp. 2.1–2.2.

25. Donald. L. DeVincenzi, M. Race, and H. P. Klein, "Planetary Protection, Sample Return Missions and Mars Exploration: History, Status, and Future Needs," *Journal of Geophysical Research* 103(E12) (25 November 1998): 28577–28585.

- Exposure to UV radiation.
- Exposure to extreme diurnal cycles of temperature.

Second, it required some way of being transported into a microenvironment capable of supporting it. Third, that microenvironment, according to the opinions of a 1970 SSB study, needed to have the following factors occurring nearly simultaneously to support the propagation of a hardy terrestrial microorganism:

- Water activity.
- Temperature above 0°C for at least 30 minutes during a Martian day.
- Nutrients including small amounts of water-soluble nitrogen, phosphorus, sulfur, and carbon.
- Fairly neutral pH values (between 5 and 8).
- Attenuation of UV flux by at least a factor of 1000.
- Absence of antimetabolites.[26]

The probability of growth was a key factor in a long chain of factors making up the probability that a planetary surface would be contaminated.[27] The SSB study estimated the probability at 3×10^{-9} that growth and spreading of terrestrial organisms released on the Martian surface would occur. For the Viking project, NASA in consultation with SSB adopted a more conservative value of 10^{-6}.[28]

Early Concepts for Sterilizing Robot Mars Vehicles

NASA performed exhaustive studies in order to choose the best sterilization approaches. The Agency weighed costs, engineering issues,

---

26. Space Science Board, "Review of Sterilization Parameter—Probability of Growth (P_g)" (draft), 16–17 July 1970, attached to Dean R. Kastel to John E. Naugle, 13 November 1970, Record 5483—LPMB/SSB, NASA Historical Reference Collection.
27. Perry Stabekis, telephone interview by author, 7 September 2007.
28. Task Group on Planetary Protection, SSB, National Research Council, *Biological Contamination of Mars: Issues and Recommendations* (Washington, DC: National Academies Press, 1992), Appendix D, *http://www7.nationalacademies.org/ssb/bcmarssummary. html*, and *http://www7.nationalacademies.org/ssb/bcmarsch1.html* through *http://www7. nationalacademies.org/ssb/bcmarsch6.html*. Also available from Steven Dick Unprocessed Collection, NASA Historical Reference Collection.

available technology, and the "biological assurance"[29] of reliable planetary protection that could be attained with different technologies. When NASA and its contractors began designing Viking, they considered chemical sterilization, and, in particular, ethylene oxide sterilization, as a possible approach. Ethylene oxide had serious dangers associated with it, however, in that it was quite a toxic and corrosive material[30] as well as flammable;[31] its vapors could readily form explosive mixtures in air.[32]

NASA also examined heat sterilization approaches, whose impacts were more controllable but required heat-resistant, robust spacecraft parts. In the words of Israel Taback, a Bionetics Corporation engineer and consultant to NASA, heat sterilization was "a clean and relatively painless solution that produced no overpowering problems beyond . . . building components capable of surviving the qualification testing and the sterilization temperatures."[33]

NASA eventually decided that heat was the most predictable, reliable, and overall best sterilization environment to use.[34] Part of the reason was that sterilization temperatures could be closely managed and kept within the acceptable range of all the components installed in the spacecraft. Other methods such as gas sterilization involved "unknown degrees of risk for those components and might require new technologies that carry their own burden of compromise."[35]

In the mid-1960s, NASA worked with other space science-related organizations to develop a multistep approach applicable to preventing

29. Bionetics, *Lessons Learned*, p. 2.7.
30. Joseph A. Stern, President and CEO of Bionetics Corporation and manager of planetary quarantine and organic contamination control support for the Viking Project Office at LaRC, interview in Bionetics, *Lessons Learned*, p. 3.21.
31. Richard W. Davies and Marcus G. Comuntzis, "The Sterilization of Space Vehicles to Prevent Extraterrestrial Biological Contamination" (JPL external publication no. 698, 31 August 1959, presented at the 10th International Astronautics Congress, London, 31 August–5 September 1959), pp. 8–9, *http://ntrs.nasa.gov/archive/nasa/casi.ntrs.nasa.gov/19630042956_1963042956.pdf* (accessed 26 January 2011).
32. U.S. Department of Labor, Occupational Safety & Health Administration, "Substance Safety Data Sheet for Ethylene Oxide (non-mandatory)," Regulations (Standards - 29 CFR), Standard Number: 1910.1047 App A, 8 January 1998, *http://www.osha.gov/pls/oshaweb/owadisp.show_document?p_table=STANDARDS&p_id=10071* (accessed 26 January 2011).
33. Israel Taback, Bionetics Corporation Technical Consultant and Chief Engineer, interview in Bionetics, *Lessons Learned*, p. 3.14.
34. John D. Goodlette, Vice President and Chief Engineer for Space Systems at Martin Marietta's Space Systems Group, Denver, CO, interview in Bionetics, *Lessons Learned*, p. 3.17.
35. Taback, interview in Bionetics, *Lessons Learned*, p. 3.15.

forward contamination carried by robot landers to Mars. The approach involved four steps:

- Selecting or developing materials for spacecraft components that were as heat-resistant as possible in order to resist damage from thermal sterilization techniques.
- Engineering manufacturing processes to minimize biological loads accruing on VLCs.
- Applying sterilizing heat to a range of components and assemblies.
- Conducting terminal sterilization of the entire VLC, then hermetically sealing it in order to maintain sterility until it had launched and left Earth's atmosphere.[36]

NASA eventually included all of these steps in its sequential approach to preparing VLCs for their mission to Mars. All personnel levels of the Viking project, including NASA upper management, space and research center staffs, contractors, and vendors, engaged in efforts to protect the Martian environment from contamination by terrestrial microbes. Their efforts were applied to individual components as well as to entire VLCs. Details of the NASA process are described in the next section.

The Incremental Sterilization Procedure for Viking Lander Capsules

NASA's sterilization plan[37] for reducing the bioburden on the complex Viking Lander Capsules were formulated to be consistent with the *Viking '75 Program Quarantine Plan* (M75-149-0) and responsive to NASA regulation NHB 8020.12, *Planetary Quarantine Provisions for Unmanned Planetary Missions*. NASA's approach for the VLC employed an incremental procedure to reduce bioburden (microbial)

---

36. Lawrence B. Hall, "Sterilizing Space Probes," *International Science and Technology* (April 1966): 50–56, 61, folder 006695, "Sterilization/Decontamination," NASA Historical Reference Collection.

37. NASA—Viking Project Office, *Viking '75 Program Lander Capsule Sterilization Plan*, M75-147-0, 27 February 1974, Washington Federal Records Center 255-84-0634, box 14.

levels. The procedure began with parts selection and development as well as clean manufacturing efforts. It also used a series of sterilization steps in order to minimize the severity of any one sterilization increment, thus lowering the chances of damaging Viking equipment.

NASA and its contractors applied some planetary protection measures to the unsterilized Viking Orbiter (VO) such as using clean assembly procedures, conducting careful bioassays of VO surfaces and factoring the results into the total bioburden estimates for the spacecraft, and designing the VO trajectory to avoid impact with the target planet. But the lion's share of NASA's forward contamination prevention effort went into the VLC design and included procedures for the actual landers that touched down on Mars as well as their descent capsules' aeroshell–heat shield–parachute systems and their bioshield cap and base assembly (see figure 5.1).

Each Viking lander was itself a complex piece of equipment, containing approximately 61,000 parts, none of which could be allowed to carry terrestrial microbes to Mars.[38] One of the first steps in NASA's approach that it carried out years before the first Viking craft launched was to select or develop the most robust parts for the VLC that were able to withstand sterilization environments without incurring damage.

Selecting and Developing Heat-Resistant Materials and Parts

NASA, and in particular JPL, began a testing and development program in the mid-1960s to determine which off-the-shelf materials and components withstood sterilization heating environments and continued to function properly. JPL oversaw the examination of more than 40,000 parts of 262 different types. In addition, several aerospace contractors, including Boeing, conducted their own research efforts in this area.[39]

When available parts and materials failed the testing, development efforts to produce better alternatives began. Serious product development efforts were an extensive part of the Viking project, for "in the engineering of Mars-mission payloads, sterility and reliability are

---

38. Bionetics, *Lessons Learned*, pp. 2.2–2.4.
39. C. S. Bartholomew and D. C. Porter, "Reliability and Sterilization," *J. Spacecraft* 3(12) (December 1966): 1762.

related so inseparably that the problems of reliability control appear in the management of sterility assurance and planetary quarantine programs as well."[40]

NASA and its contractors sought to procure general-usage parts or develop new parts demonstrating high reliability and insensitivity to dry-heat sterilization temperatures of up to 135°C. Nevertheless, Viking engineers had concerns about post-sterilization performances of certain materials and parts. Among the most difficult parts to make more heat-resistant were batteries and tape recorders. It was also problematic to sterilize the insides of transistors, capacitors, vacuum tubes, switches, propellants, and igniters without damaging them.[41]

It became clear that sterilizing a spacecraft containing extensive electronics and a spectrum of different materials and subassemblies constituted "a significantly different task from the problem of sterilizing a surgeon's instruments or canned foods."[42] Examples of sterilization-related issues encountered for particular materials and parts are discussed below.

Solder. At the temperatures employed for sterilization, the tensile strength of the eutectic[43] solder used on Viking components was low enough to be of concern.[44] Sterilization procedures could cause some of these solders to become brittle and crack. It was thus critical to identify types of solder that could withstand the heating environment.[45]

40. John B. Opfell and Temple W. Neumann, "A System Engineering Approach to Spacecraft Sterilization Requirements," *J. Spacecraft* 3(11) (November 1966): 1603.
41. A. M. Nowitzky, "The Influence of Sterilization on the Reliability of Interplanetary Spacecraft Systems" (AIAA Paper No. 67-776, AIAA 4th Annual Meeting and Technical Display, Anaheim, CA, 23–27 October 1967); Hall, "Sterilizing Space Probes," 53, 56.
42. "NASA Announces Spacecraft Decontamination Procedures," NASA news release no. 63-200, 13 September 1963, folder 006695, "Sterilization/Decontamination," NASA Historical Reference Collection.
43. *Eutectic* refers to the lowest temperature at which a mix of two materials will melt. Often the melting temperature is significantly lower than the individual melting temperature of either material in the mixture. Lead-tin solder is an example. Lead melts at 327°C, tin at 231°C. The lowest melting combination of the two metals, however—67% lead and 33% tin—has a melting temperature of only 180°C. This information was taken from "Glossary," *http://www.digitalfire.com/education/glossary/*, DigitalFireCorporation (accessed 9 October 2005).
44. Bionetics, *Lessons Learned*, p. 2.6.
45. Charlie King, materials and processes, and Ansel Butterfield, electronic parts qualification, Viking Project Office, LaRC, interview in Bionetics, *Lessons Learned*, p. 3.16.

Conformal coatings. The problem with conformal coatings[46] was that they were frequently applied to components in layers that were too thick. When subjected to heat, the coating then expanded and deformed the components, potentially causing them to fail. To avoid this problem, NASA demanded standardized thicknesses of conformal coatings.[47]

Electronic parts. Materials used in electronic assemblies such as capacitors could explode or become damaged during sterilization procedures; qualification tests needed to ensure that the parts selected were not prone to this problem.[48]

Fasteners. Attachment fasteners such as bolts sometimes had different compositions and rates of thermal expansion than the nuts or holes into which they were threaded. Such bolts could "creep," or in other words, loosen and back out. Thus, project engineers needed to fully understand "the amount of heat-induced elongation"[49] of different alloys so that precision estimates of sterilization impacts on parts made of different materials could be predicted.

Gyroscopes. Project engineers worried about temperature-induced outgassing and decay of epoxy resins within gyroscopes, which could seriously degrade reliability and performance. Also of concern was the mechanical integrity of entire gyroscope assemblies due to internal pressure buildup. The fluid used in the gyroscopes presented issues as well. It had to be heat-tolerant, able to flow readily through gyroscope orifices at a range of different temperatures, and dense enough to float the gimbals of the device.[50]

Recorder. Viking managers decided to design a new recorder that would meet heat compatibility requirements rather than modify existing equipment. The development program tested various materials, assembly processes, and components. Testing activities included evaluations of electronic parts, seals, lubricants, ball bearings, magnetic tape, and heads. Organic materials were used only when unavoidable, and

---

46. *Conformal coatings* are insulating, protective coverings that "conform" to the shapes of the objects coated (typically printed circuit boards or other electronic components). These coatings provide mechanical protection and barriers against environmental conditions. The coatings may be composed of plastic or inorganic materials.
47. King and Butterfield, interview in Bionetics, *Lessons Learned,* p. 3.16.
48. Goodlette, interview in Bionetics, *Lessons Learned,* p. 3.18.
49. Goodlette, interview in Bionetics, *Lessons Learned,* p. 3.18.
50. Bionetics, *Lessons Learned,* p. 3.16.

then only materials that had low outgassing properties. NASA designed the flight recorder's transport assembly so that it was virtually all metal. Motor housing and capstan were fabricated from titanium in order to eliminate differences in coefficients of expansion that could result in damage to the recorder during heat sterilization. Engineers chose metallic recording tape in order to achieve heat tolerance. Dry lubricants were used on the tape, bearings, and gears, which tended to be more resistant to degradation under high temperatures than oil-based lubricants.[51]

Computer. The memory medium of the computer included a densely packed matrix of plated wire with individual strands only two-thousandths of an inch in diameter. This design made the memory easily impacted by heat. Engineers built a capability into the computer to recognize temperature-dependent behavior and employ temperature-compensation devices.[52]

Batteries. Project engineers initially considered two battery types for the spacecraft—nickel-cadmium (NiCd) and silver-zinc (AgZn). Whichever batteries they chose had to have several features, including the following:

- Capacity to be sterilized.
- Long life (about 22 months).
- Capability to survive an 80-percent discharge during terminal descent and repeated 50-percent discharges per day during operations on the Martian surface.

Table 5.1 summarizes the positive and negative features of both battery types. AgZn batteries had continuing problems with leakage after being subjected to heat sterilization environments. NiCd batteries, although of heavier weight, had long lives and the ability to be charged quickly and cycled thousands of times, and they were relatively tough. As a result, NASA selected NiCd batteries for use on missions.[53]

---

51. Bionetics, *Lessons Learned*, pp. 2.7–2.9, 3.16, 3.18–3.19.
52. Bionetics, *Lessons Learned*, pp. 3.19–3.20.
53. Institute of Biomedical Engineering Technology, "Rechargeable Batteries," *http://ibet.asttbc.org/batterys.htm* (last updated 30 March 2002, accessed 12 September 2006); Bionetics, *Lessons Learned*, pp. 2.9–2.10.

Table 5.1 A comparison of NiCd and AgZn batteries.

	NiCd	AgZn
Advantages	Very long life, due to insoluble electrodes	Light weight for the amount of energy stored (high energy density)
	Able to be cycled thousands of times	Low self-discharge rate
	Capable of rapid charging	
	Robust, able to survive rough handling	
	Easily sealable	
Disadvantages	Relatively heavy for the amount of energy stored (low energy density)	Difficult to design these batteries to achieve a long life
	Rapid self-discharge rate	Typically doesn't even get a few hundred cycles
	Requires a minimum rate of change	Difficult to reliably seal; subject to chemical leaks
Problems Observed	Nylon parts disintegrated (successfully replaced with polypropylene)	Cellophane parts disintegrated (successfully replaced with polypropylene)
		Plastic cases cracked
		Epoxy seals leaked when exposed to sterilization environments

Bioshield. Integral to the VLC was a contamination control barrier impervious to microbes that provided a reliable impediment to recontamination after the sterilization process ended. This barrier, termed a "bioshield" (see figure 5.5), remained sealed during and after terminal sterilization of the VLC and was not breached until postlaunch. Otherwise, resterilization would have been required to kill any biological contamination deposited on lander or aeroshell surfaces as well as those of the bioshield's interior.

NASA spent considerable effort choosing the most appropriate types of materials for the bioshield. The materials needed to withstand sterilization regimes and reliably protect the interior of the VLC from recontamination, which could lead to forward contamination of Mars. Both rigid and flexible materials were considered as well as combinations of the two.

NASA seriously examined flexible film materials for the bioshield. Thin-film bags were currently used to isolate animals in a germ-free state for prolonged periods of time, and NASA thought they might be applicable to containing a lander as well. Film materials such as 0.01-inch thicknesses of cellulose acetate or fluorinated ethylene propylene (FEP) fluorocarbon, for instance, were capable of withstanding 400°F sterilization temperatures or higher—far more than needed for steriliz-ing a Mars landing capsule. Transparent films had the advantage of per-mitting interior spaces to be viewed, and the films' flexibilities allowed objects within to be manipulated. Furthermore, the weight of such films was very low—a distinct advantage for spaceflight applications.

These films were, however, susceptible to penetration and tearing, which could lead to recontamination of the lander. In addition, the remotely controlled separation and jettisoning of a thin-film bag from a complex-shaped object was a difficult operation to design. The prob-lems associated with handling such films, as well as the possibility of compromising an entire space mission through damage to a thin-film container, argued strongly against using such a containment material as a bioshield.

The container in which the Ranger spacecraft had been sterilized was completely rigid (the Ranger lunar exploration program is dis-cussed in Chapter 2). Although sterilization was accomplished on the Ranger through the use of gaseous ethylene oxide, a similarly designed container was also possible for VLC terminal sterilization. One of the challenges of designing any bioshield container for the VLC, however, was that the internal pressure that built up during heating could rup-ture it and expose the lander to microbes. A dependable venting sys-tem thus had to be included in the design. Additionally, after cooling, the pressure within the bioshield would drop, leading to its possible collapse unless sterile gas was pumped in.[54]

Another approach that NASA considered was the combined use of rigid and flexible materials, which offered certain potential advantages. Where access to the interior space was required, flexible work stations could be incorporated in the wall of a rigid canister, rendering it a sort of flying glove box.[55]

54. J. B. Tenney, Jr., and R. G. Crawford, "Design Requirements for the Sterilization Containers of Planetary Landers" (AIAA paper no. 65-387, AIAA Second Annual Meeting, 26–29 July 1965).
55. Tenney and Crawford, "Design Requirements for the Sterilization Containers of Planetary Landers."

NASA eventually chose a material for the bioshield that was some-what pliable, but also very tough and heat resistant (see figure 5.5). The Viking team constructed the bioshield cap and base out of coated, woven fiberglass only 0.005 inches (0.13 millimeters) thick, which cut down on weight. It would not crumple like a thin-film bag, however, because it was supported by stays of aluminum tubes.[56]

To provide added protection against recontamination of the lander, a positive air pressure was maintained within the bioshield during and following terminal sterilization. During the actual launch ascent, when outside pressure rapidly dropped, excess pressure within the VLC was vented through a biofilter that did not allow microbes to reenter.

●●●

The above parts and materials were critical for dependable VLC operations during the mission. Viking personnel were also concerned about proper operation of the mission's science experiments after sterilization. A particularly vital piece of equipment for analyzing Martian substances was the gas chromatograph mass spectrometer.

Gas chromatograph mass spectrometer (GCMS). Viking's organic compound analysis capability depended on the proper functioning of its GCMS. The mission's GCMS experiment cost $41 million to construct, about 4 percent of the entire mission budget.[57] When examination of the GCMS indicated that it would have difficulty meeting planetary protection heat sterilization requirements, the Viking team grew quite concerned.

The team had designed the spacecraft's GCMS to analyze the composition of the Martian atmosphere and to identify chemical substances emitted by Martian soil when heated.[58] These analyses were

---

56. William R. Corliss, *The Viking Mission to Mars* (Washington, DC: NASA SP-334, 1974), p. 30, record 5503, NASA Historical Reference Collection; Holmberg et al., *Viking '75 Spacecraft Design*, p. 21.

57. Dick and Strick, *The Living Universe: NASA and the Development of Astrobiology*, p. 80.

58. D. R. Rushneck, A. V. Diaz, D. W. Howarth, J. Rampacek, K. W. Olson, W. D. Dencker, P. Smith, L. McDavid, A. Tomassian, M. Harris, K. Bulota, K. Biemann, A. L. LaFleur, J. E. Biller, and T. Owen, "Viking Gas Chromatograph–Mass Spectrometer," *Review of Scientific Instruments* 49(6) (June 1978): 817–834.

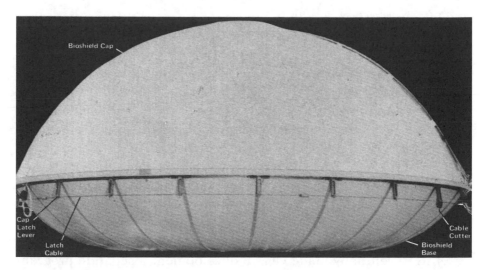

Figure 5.5 The Viking bioshield.

key to Viking's search for life. The instrument drew vaporized chemicals into its gas chromatograph (GC) section, which separated them into individual compounds. These were then drawn into the mass spectrometer (MS) section of the instrument, where they were ionized, the constituent compounds identified, and their individual quantities determined. The GCMS generated profiles for each compound and sent these data to Earth.

Results of the organic chemistry analysis gave scientists insights into chemicals that may have been produced by forms of Martian life as well as by purely chemical means. The major aim of the organic chemistry investigation was to help determine if any Martian life had existed or currently existed, or "if the right organic compounds were present for the evolution of life in the future."[59] The organic chemistry experiment constituted a critical cross-check on the biological life-detection experiments.

The main thrust of GCMS work was to support organic chemical evaluations, but it had other uses as well. Its analyses also yielded data on concentrations of volatile inorganic compounds such as ammonia, carbon dioxide, carbon monoxide, nitrogen dioxide, nitric oxide, sulfur

---

59. Ezell and Ezell, *On Mars: Exploration of the Red Planet—1958–1978*, Chap. 7.

dioxide, and hydrogen sulfide. It was thus extremely important that the instrument operate correctly during the mission.

The GCMS used on Viking was not an off-the-shelf item, and its development ran into some serious snags. Langley Research Center (LaRC) initially assigned the development of a GCMS prototype suitable for Viking's needs to JPL in August 1968.[60] But defects in the Viking project's management structure led to a serious planetary-protection-related problem with the instrument. JPL independently developed the GCMS for the mission without paying sufficient regard to prime contractor Martin Marietta's specifications. An unfortunate result of this was that JPL constructed the GCMS prototype using sensitive electronics whose tolerances to Viking's sterilization environment were questionable.

NASA eventually made GCMS development a separate program that it could better control, but qualifying the GCMS for the planetary protection heat sterilization regime remained an ongoing issue.[61] Recurring troubles included seals and gaskets that became unstable when heat was applied. The most significant development issue according to Klause Biemann, the GCMS/Molecular Analysis Team Leader, concerned the instrument's potting compounds.[62] These repeatedly broke down under heat sterilization conditions. This resulted in parts pulling away from the potting compounds, creating a danger of arcing (high-voltage electrical discharge) that could have severely damaged the GCMS. The GCMS team eventually dealt with this problem by obtaining more stable potting materials from Hughes Aircraft.[63] A schematic of the GCMS is depicted in figure 5.6.

NASA's heat sterilization protocol was a likely cause of problems in both Viking GCMSs. These problems were not corrected before the

---

60. Ibid.
61. King and Butterfield, interview in Bionetics, *Lessons Learned*, p. 3.16.
62. A potting compound is an electrically nonconductive material employed to encapsulate electronic parts, conductors, or assemblies or to fill space between them. Potting compounds, which are typically liquid resins that are allowed to harden, provide strain relief for parts and circuits and help protect them from moisture damage, abrasion, and impacts.
63. Klause Biemann, GCMS/Molecular Analysis Team Leader and a faculty member of the Massachusetts Institute of Technology, interview in Bionetics, *Lessons Learned*, p. 3.36; Dale Rushneck, GCMS Principal Technical Manager, interview in Bionetics, *Lessons Learned*, pp. 3.39–3.41.

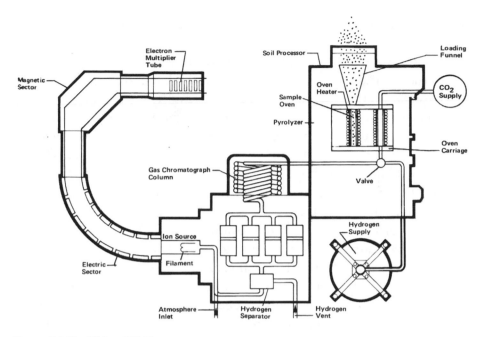

Figure 5.6 The Viking GCMS.

two Viking spacecraft launched and ended up impacting, although fortunately in a minimal way, the search-for-life investigation. The GCMS aboard Viking's Lander Capsule 1 (VLC-1) had gone through the heat qualification procedure for individual components and did not appear at that time to have a problem. But later, during terminal sterilization of the entire spacecraft, the VLC-1 GCMS developed a leak in its vacuum system. NASA had to make a difficult decision—leave the GCMS in place and live with the leak, or open up VLC-1's bioshield and change out the instrument, which would have required that the entire VLC go through terminal sterilization another time. This might have resulted in missing the launch window.[64]

Dale Rushneck, the GCMS Principal Technical Manager, calculated precisely where the leak was in the GCMS, based on his knowledge of the flow dynamics through the instrument. He determined where the pressure had to be in order to produce the observed leak and narrowed down the problem to a particular fitting. If that fitting could

---

64. Rushneck, interview in Bionetics, *Lessons Learned,* pp. 3.39–3.41.

have been tightened, Rushneck was confident that the leak would disappear. But this couldn't be done without removing the GCMS from the spacecraft and disassembling it. Instead, Rushneck calculated the magnitude of the leak, its contribution to the background noise in the instrument, and its estimated effect on the atmospheric analysis that the GCMS was going to do on Mars. He determined that the leak was small enough that it would not compromise any but the most sensitive of measurements. Furthermore, NASA had installed another GCMS on Viking Lander Capsule 2 (VLC-2), which could obtain some of the data that the first lander could not. NASA concluded that it was best to leave the GCMS in place in VLC-1.[65]

This decision turned out to be sound. Although the VLC-1 GCMS was not able to make some of the most sensitive of its planned measurements, the GCMS in VLC-2 successfully carried out similar measurements. One lesson learned from this experience, in Rushneck's view, was that having a second spacecraft with redundant instrumentation land on the planet (even in a very different location) can greatly lessen the scientific impact of an instrument failure in either craft.

The GCMS in VLC-2 did not, however, perform flawlessly. Because VLC-2 landed much closer to the Martian northern polar region than VLC-1, it was subjected to wider temperature variations, and the GCMS on board eventually failed. The GCMS team traced the cause of the failure to a high-voltage arc discharge in the power supply of a pump that kept the instrument clean. Critical connectors on this component had been potted during their fabrication back on Earth, and Rushneck postulated that a failure somewhere in this potting material led to the high-voltage arc discharge. He also believed that this probably would not have occurred if the material had not been subjected to, and weakened by, NASA's heat sterilization regime.[66] Thus, NASA's planetary protection procedures may have hurt GCMS performance in *both* Viking landers. It is important to emphasize, however, that this was never proven and that both GCMSs performed up to a high enough standard that NASA staff considered them key tools for organic compound analysis and, in particular, Viking's search for life.[67]

---

65. Ibid., pp. 3.39–3.40.
66. Rushneck, interview in Bionetics, *Lessons Learned,* p. 3.40.
67. Perry Stabekis, telephone interview by author, 7 September 2007.

Reducing Biological Loads Through Improved Manufacturing Processes

An important step in the planetary protection process was to minimize the biological loads that accumulated on spacecraft components during manufacturing activities. The more successful these efforts were, the less heat sterilization of those components was required, which reduced the amount of thermal stress that they were subjected to. These potentially useful steps for improving manufacturing processes included

- Maximizing the cleanliness of the basic raw materials and parts used in the manufacturing process.
- Controlling contamination within the manufacturing environment—for instance, reducing dust and contamination from personnel.
- Using cleanrooms for manufacturing whenever feasible.
- Physically cleaning subassemblies on a regular basis—wiping down, vacuuming, and blowing off all accessible surfaces.[68]
- Applying standard spacecraft cleaning procedures to all surfaces prior to mating.
- Implementing component packaging requirements that preserved their level of cleanliness.[69]

VLC planetary protection and heat sterilization activities were carried out at multiple locations, including the following:

- Subcontractor facilities, which supplied parts as basic as individual fasteners and as complex as the biology instrument.
- Martin Marietta Corporation's VLC assembly site in Denver.
- NASA's Kennedy Space Center, which eventually launched the vehicles.

Details of the sterilization efforts and responsibilities of these different facilities are discussed below and depicted in figure 5.7.

---

68. Hall, "Sterilizing Space Probes," 53, 56.
69. Martin Marietta Corporation, *Viking Lander System and Project Integration—Technical Proposal*, April 1969, Washington Federal Records Center 255-84-0634, box 6.

Sterilization Activity Sequence

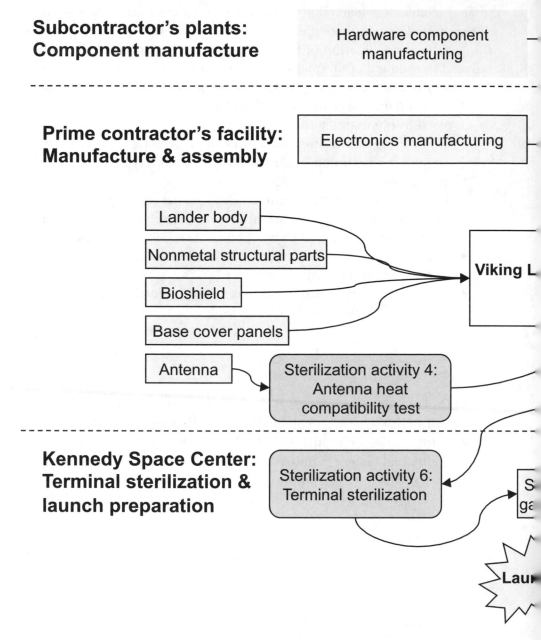

Subcontractor's plants: Component manufacture

Hardware component manufacturing

Prime contractor's facility: Manufacture & assembly

Electronics manufacturing

Lander body

Nonmetal structural parts

Bioshield

Base cover panels

Antenna

Viking L

Sterilization activity 4: Antenna heat compatibility test

Kennedy Space Center: Terminal sterilization & launch preparation

Sterilization activity 6: Terminal sterilization

S ga

Lau

Figure 5.7 Sterilization activity sequence.

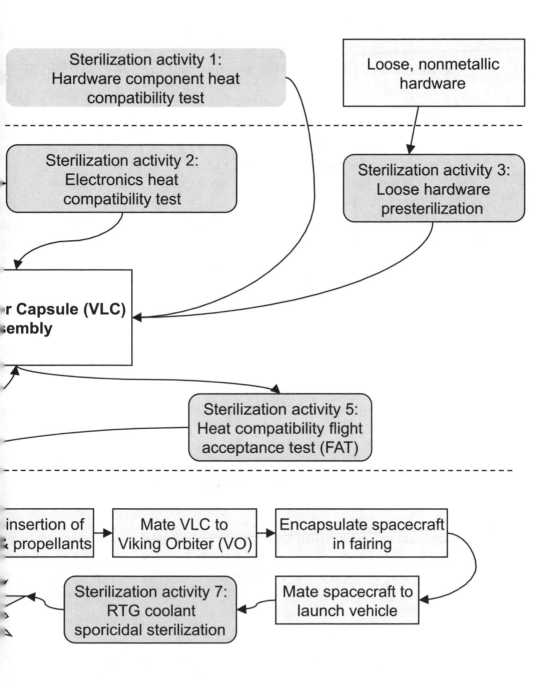

Subcontractor Responsibilities

NASA required manufacturers of many types of flight hardware, including replacement parts, to subject their components to a flight acceptance heat cycle. This had dual benefits:

- Verification of parts' abilities to withstand heat sterilization environments.
- Bioburden reduction, especially of the following:
    - Encapsulated microbes—those located in the interior of parts and completely surrounded by a nonmetallic, solid material impervious to water vapor transmission.
    - Mated-surface microbes—those hidden within interfaces between two surfaces.

Vendor flight acceptance heat cycling became a key step for incrementally achieving the required bioburden reduction on the VLC, thereby helping to minimize the necessary exposure time, and thermal stresses, of the terminal sterilization activity. But component suppliers that performed the heat cycling needed the equipment and the staff to do so. This added to their products' costs and reduced the control that the prime contractor, Martin Marietta, had over the manufacturing process. Martin Marietta thus needed to consider all the pros and cons of heat cycling at component vendor facilities. In its VLC integration proposal, Martin Marietta specified that nonmetallic loose hardware would not undergo the flight acceptance tests at vendor sites but would instead be heat-screened at its Denver facility. As will be discussed in the next section, Martin Marietta also performed many VLC component sterilization procedures in Denver, and this eliminated the need for bioassays by subcontractors and vendors.[70]

Prime Contractor Responsibilities

The major mission of prime contractor Martin Marietta Corporation was to assemble and thoroughly test the VLC, preparing it for shipment

---

70. Martin Marietta Corporation, *Viking Lander System and Project Integration—Technical Proposal,* April 1969, Washington Federal Records Center 255-84-0634, box 6; NASA—Viking Project Office, *Viking '75 Program Lander Capsule Sterilization Plan,* pp. 1–3, 10, 23–24.

to KSC. To render the VLC ready for shipment, Martin Marietta needed to include several key planetary protection activities, including heat sterilization of many of the components.

While Martin Marietta mainly assembled different parts into a finished VLC, it also had some module manufacturing responsibilities, in particular building electronics assemblies, after which it subjected them to a heat sterilization environment followed by a post-heating checkout to verify that the modules were compatible with the sterilization. Martin Marietta also subjected loose, nonmetallic hardware received from subcontractors, as well as the VLC antenna, to heat sterilization.[71]

Further planetary protection measures involved the design of the manufacturing environment. Periodic surface cleaning of spacecraft assemblies were scheduled in order to aid in bioburden reduction. All cleaning, assembly, and test operations of VLC systems as well as of the capsule itself were conducted in cleanrooms with personnel working under strict handling constraints and procedures. This minimized the bioburden accumulation on the VLC prior to terminal sterilization. In addition, NASA's sterilization plan called for a heat compatibility flight acceptance test (FAT) on the entire VLC before sending it to KSC, in order to verify that it operated properly after heat processing. The FAT was primarily an engineering test rather than one designed for planetary protection purposes, but it had the side benefit of further reducing the VLC bioburden.[72]

Martin Marietta also needed to closely monitor the activities of Viking subcontractors who were manufacturing key parts of the spacecraft. Martin Marietta was "responsible for really making sure that the subcontractors all produced and delivered essentially clean items like the instruments and the spacecraft components."[73] To help accomplish this, Martin Marietta developed an approved parts list that subcontractors had to follow and also assigned a cadre of personnel "that lived with the [sub]contractors when they needed it from time to time and overlooked all of their activities."[74]

---

71. Martin Marietta Corporation, *Viking Lander System and Project Integration*; NASA—Viking Project Office, *Viking '75 Program Lander Capsule Sterilization Plan*.
72. Martin Marietta Corporation, *Viking Lander System and Project Integration*; NASA—Viking Project Office, *Viking '75 Program Lander Capsule Sterilization Plan*, p. 3.
73. Israel Taback, Viking Chief Engineer, LaRC, telephone interview by author, 10 February 2006.
74. Ibid.

Langley and Bionetics Roles: Management and Verification

Langley Research Center's Viking Project Office was officially responsible for everything that was done on the mission but actually acted chiefly in a management function. Langley oversaw Martin Marietta work on the VLC, JPL's design and fabrication activities on the VO, and the design work of Glenn Research Center in Cleveland (known then as Lewis Research Center) on the Titan-Centaur launch vehicles. But very little technical work was actually performed by Langley—that was the responsibility of other NASA Centers, Martin Marietta, and the subcontractors. Langley primarily tracked and oversaw other groups' efforts.[75]

Langley was greatly aided in its monitoring efforts by Bionetics Corporation, a company that was formed in 1969 for the purpose of supporting the Viking mission and, in particular, its planetary quarantine function. Langley was the company's first customer. Bionetics acted as an independent verifier, largely to ensure that Martin Marietta put together the spacecraft that NASA wanted. Bionetics provided much-needed expertise in spacecraft sterilization, contamination control, and microbial analysis.[76]

NASA and its contractors conducted continuing microbiological assays and monitoring throughout the VLC assembly and sterilization process (see the section later in the chapter on bioassaying). Project staff employed data from these assays to predict the interim and endpoint burdens on the VLC, using a bioburden model. The data also predicted potential trouble spots in the assembly process, where microbial samples especially needed to be taken. The bioburden-model calculation prior to terminal sterilization and VLC models of temperature distribution during terminal sterilization were used to help determine the required time that the VLC needed to undergo the terminal sterilization process in order for its microbial burden to be reduced to NASA limits.[77]

---

75. NASA, "Viking: Trailblazer for All Mars Research," NASA Fact Sheet, 22 June 2006, http://www.nasa.gov/mission_pages/viking/viking30_fs.html (accessed 26 January 2011); Taback telephone interview; NASA, "Viking Mission to Mars," NASA Facts, 1988, http://www.jpl.nasa.gov/news/fact_sheets/viking.pdf (accessed 11 July 2006); NASA, "About Glenn," http://www.nasa.gov/centers/glenn/about/index.html (last updated 14 November 2007, accessed 25 March 2008).
76. Tom Martin, former Viking Mission Director, telephone interview by author, 20 February 2006; Bionetics Web site, http://www.bionetics.com/AboutUs/heritage.html (accessed 11 July 2006).
77. Martin Marietta Corporation, Viking Lander System and Project Integration.

Kennedy Space Center's Terminal Sterilization of the Viking Lander Capsule

KSC received the first Viking flight components during November and December of 1974, including the Titan 3E liquid-fueled core launch vehicle, the solid-fueled booster rockets, and the Centaur upper stage. Shortly thereafter on 4 January 1975, the first of the Viking Lander Capsules (VLC-1) arrived. It and its sister VLC were subjected to a battery of verification tests.[78] KSC staff disassembled, inspected, and functionally checked each of the two VLCs to confirm that they were ready for their mission. During reassembly, KSC staff performed detailed bioassays in order to estimate the microbe load on each of the VLCs. These assays were key in determining "the duration of the terminal sterilization cycle necessary to satisfy both PP [planetary protection] and scientific investigation requirements."[79]

The KSC terminal sterilization procedure consisted of a dry-heat process using an inert gas atmosphere of 97 percent nitrogen, 2.5 percent oxygen, and 0.5 percent other gases. Thermal analysis of the VLC had indicated that its interior would heat very slowly during terminal sterilization if heat was only applied to its exterior. This presented a serious problem to the integrity of its systems. As the VLC's interior slowly rose to sterilization temperatures, components near its surface would heat much faster and be subjected to extremely long heating times, endangering their reliability.

The Viking team found a way to bring interior temperatures up more quickly by injecting the hot nitrogen gas mixture directly into the VLC, including into its heating and coolant lines. Viking engineers helped in this effort by designing effective gas flow paths within the VLC.[80] Internal circulation of the hot gases was very effective at lowering the total length of the heating cycle and reduced prolonged temperature gradients within the VLC that could generate thermal stresses and damage electronics and structural materials. Nitrogen was excel-

---

78. Edward Clinton Ezell and Linda Neuman Ezell, "Viking Lander: Building a Complex Spacecraft," in *On Mars: Exploration of the Red Planet—1958–1978* (Washington, DC: NASA SP-4212, 1984), Chap. 8, p. 271.
79. Bionetics, *Lessons Learned,* p. 2.12.
80. Erwin Fried and Richard J. Kepple, "Spacecraft Sterilization—Thermal Considerations" (AIAA paper no. 65-427, AIAA Second Annual Meeting, San Francisco, CA, 26–29 July 1965).

lent for internally conducting the heat because it is an inert gas that was unlikely to react with the materials of the VLC.[81] Heat was also supplied to the VLC interior by the radioisotope thermoelectric generators meant to provide electric power during the mission, as well as through use of their cooling loops within the bioshield. The end result was to heat the entire VLC much more uniformly and quickly than if heat had only been applied from the exterior, avoiding stress to certain of its components or inadequate sterilization of others.

KSC staff performed the terminal sterilization procedure in a thermal test chamber within the Center's Spacecraft Assembly and Encapsulation Building, with the VLC bioshield sealed and pressurized. The VLC contained no propellants or tanks filled with pressurized gases at this point, although gas fill lines as well as the line filters were placed in the thermal chamber so that they could be sterilized for future use.

The parameters of a particular VLC sterilization cycle were in part derived from bioload results of the last assay done on the lander prior to encapsulation into the bioshield.[82] Selecting the best temperature profile for sterilization meant walking a tightrope between temperatures high enough to be lethal to microorganisms aboard the VLC, but low enough so as not to damage equipment. Based on NASA-funded research, lethality could be achieved for temperatures of at least 104°C. NASA established sterilization specifications for the temperature range of 104° to 125°C. Viking project management (with the concurrence of NASA's Planetary Quarantine Office) selected a value between these temperatures, specifying that each VLC be exposed to 111.7°C for 30 hours *after* the coldest contaminated point reached that temperature. Some parts of a VLC heated sooner than others and experienced slightly higher temperatures.[83]

KSC staff placed VLC-2 in the thermal chamber on 15 June 1975 and, including ramp-up and ramp-down, subjected it to terminal sterilization for over 43 hours. Temperatures within VLC-2 reached as high as 116°C (241°F). The sterilization procedure for VLC-2 ended on 18 June, and the Viking staff completed its software and hardware operational readiness

81. J. Barengoltz and P. D. Stabekis, "U.S. Planetary Protection Program: Implementation Highlights," *Advances in Space Research* 3(8) (1983): 7.

82. Perry Stabekis, telephone interview by author, 7 September 2007.

83. Perry Stabekis e-mail to author, 8 October 2007.

testing on 19 June, after which capsule verification testing began. NASA staff moved VLC-1 into the thermal chamber by 20 June, but terminal sterilization did not begin until the Viking project team confirmed that VLC-2 had survived the procedure and continued to function properly.[84] VLC-1's terminal sterilization spanned a longer time period than VLC-2, almost 50 hours, over a period from 20 to 22 June.

Mitigating recontamination. While the insides of individual components of the VLCs should have been adequately sterilized during heating procedures at sub- and prime contractor facilities, NASA expected that some of the surfaces of these components would get recontaminated during handling and transport. The terminal sterilization environment, however, took care of such situations by reducing "the population of microorganisms accruing on the exposed surfaces during assembly and test of the VLC."[85]

Preventing further recontamination of the VLC after final assembly. The lander within the VLC needed to get electrically and mechanically connected to the VO in order for the entire spacecraft to operate correctly during the mission. These connections had to be made using procedures that would not recontaminate any part of the VLC's interior, including the lander. To prevent such recontamination, the lander was designed to never directly interface with the VO. Connections from the lander interfaced instead with the bioshield that surrounded it, and the bioshield then interfaced with the VO.

A component called the bioshield power assembly, or BPA, assisted in isolating the interior of the VLC. The BPA, which was mounted on the outside of the bioshield, provided the power interface between lander and VO. The BPA remained attached to the VO after separation from the VLC.[86]

---

84. G. D. Sands, ed., "Viking Science Activities, No. 87," LaRC, 20 June 1975, record 5503, NASA Historical Reference Collection; NASA—Viking Project Office, *Viking '75 Program Lander Capsule Sterilization Plan*, pp. 1–3, 16–21; Martin Marietta Corporation, *Viking Lander System*; Ezell and Ezell, *On Mars: Exploration of the Red Planet—1958–1978*, p. 274.

85. Bionetics, *Lessons Learned*, pp. 2.7, 2.12–2.14.

86. Lawrence B. Hall, James R. Miles, Carl W. Bruch, and Paul Tarver, "The Objectives and Technology of Spacecraft Sterilization," NASA news release, 9 February 1965, folder 006695, "Sterilization/Decontamination," NASA Historical Reference Collection; Hall, "Sterilizing Space Probes," 56; Neil A. Holmberg, Robert P. Faust, and H. Milton Holt, *Viking '75 Spacecraft Design and Test Summary, Volume I—Lander Design*, NASA Reference Publication 1027, November 1980, p. 111.

To provide added protection against recontamination of the interior of the VLC, positive air pressure was maintained within the bioshield during and following terminal sterilization that continued until after the launch. The details of this procedure will be discussed later in the chapter, in the section on planetary protection procedures during flight.

Gas and propellant insertion. After the functional checkout of the VLC, NASA staff transferred propellants and pressurized gases into the capsule. These materials, which were used in VLC descent and landing engines as well for attitude control, needed to be loaded through fill lines that pierced the bioshield. Thus, NASA had to develop procedures that did not carry contamination into the spacecraft lander capsule, either in the fuel and pressurized gases themselves or in the equipment used for fueling.[87]

Pressurant gas fill lines for the VLC as well as the line filters had been sterilized at the same time as the VLC, within the thermal chamber. The pressurant gases themselves were sterilized by being passed through the filters. The hydrazine propellant was self-sterilizing, in that it was toxic to terrestrial microbes and thus did not need additional sterilization.

RTG coolant treatment. Finally, after the VLC was mated to the VO and it to the launch vehicle, NASA introduced a sporicide into the coolant loop for the radioisotope thermal generators, a source of shipboard power. The sporicide was made up of an alcohol-formaldehyde mixture whose sterilization effectiveness had been previously identified in tests.[88]

Factors in Selecting Sterilization Temperature Regimes

When Viking staff was setting sterilization temperature regimes, it had to weigh various considerations to arrive at the best values. Planetary protection requirements and existing test data indicated that component sterilization temperatures had to be at least 100°C, the minimum temperature at which microorganism lethality could be

---

87. King and Butterfield, interview in Bionetics, *Lessons Learned*, p. 3.17.
88. NASA—Viking Project Office, *Viking '75 Program Lander Capsule Sterilization Plan*, pp. 2–4, 9, 17–18.

achieved. Mission staff who researched specialized time/temperature regimes for qualifying Viking parts, however, proposed temperatures as high as 145°C. "We began with a temperature that anyone would consider safe," commented G. Calvin Broome, the Project Lander Science Instruments Manager, in order to be "pretty sure there won't be any microbes left alive in there."[89]

Many of the mission's engineers were opposed to such high temperatures. Electronics engineers in particular recognized the difficulty of finding or designing electronic components and circuit boards able to withstand such a sterilization environment. Materials engineers also worried whether some substances would stand up to the intense heat but could not always identify adequate alternatives. Materials of concern included nonmetallics such as coatings, adhesives, and rubber-like parts used in O-rings, gaskets, and other types of seals.[90] Such materials might develop weaknesses and flaws during the rigorous heat qualification tests and end up failing prematurely, possibly compromising the mission.

Some nonmetallics were also prone to outgassing (emitting volatile compounds). These gases would condense on cooler surfaces and could contaminate and corrupt the results of science experiments.[91] Furthermore, different metals and other materials had different expansion rates when heated. High temperatures could lead to mechanical strains on the equipment and eventual damage.

Lower temperatures can adequately sterilize parts and materials if longer exposure times are employed. Lower temperatures also make the technical challenges of meeting planetary protection requirements easier—in particular, finding reliable parts and materials while keeping the probability of planetary contamination acceptably low. Extensive analyses of the impacts of time/temperature trades on die-offs of microbial populations identified the particular trades that could be made without compromising contamination control[92] and led NASA

---

89. G. Calvin Broome (both quotes in the sentence), JPL, Project Lander Science Instruments Manager on the Molecular Analysis Team, interview in Bionetics, *Lessons Learned*, p. 3.37.
90. John D. Goodlette, interview in Bionetics, *Lessons Learned*, p. 3.17.
91. Bionetics, *Lessons Learned*, p. 3.7.
92. Taback, interview in Bionetics, *Lessons Learned*, p. 3.15.

to an effective approach for killing microorganisms that also allowed low enough temperatures—104° to 125°C—to not damage hardware.

The exact temperature selected was a matter of choice for the mission staff, but different temperature choices necessitated different heating times. Sterilizing at 104°C required an *order of magnitude more time* to attain a certain bacterial reduction than sterilizing at 125°C. For instance, if at 125°C, an acceptable bacterial reduction could be achieved in 0.5 hours, it would take 5 hours at 104°C to attain the same reduction.

Sterilizing the spacecraft at lower temperatures definitely had its advantages, but this practice also had a risk associated with it. The biologists on the project insisted that "we must be sure we don't contaminate the biological instrument,"[93] which had been built to detect the presence of life. Proper functioning of this instrument was central to the success of the mission, and concerns regarding its possible contamination (including by other parts of the VLC itself that had not received as rigorous a sterilization protocol as the biology instrument) led the biology instrument team to incorporate additional sterilization treatment of the biology package prior to the landers' terminal sterilization.[94]

In the end, NASA decided on a temperature of at least 111.7°C for the Viking landers' terminal sterilization.[95] For the qualification of individual parts and components, however, the project used higher temperatures in the range of 125° to 145°C, depending on the heat sensitivity of the particular object.[96]

NASA engineers believed that their strategy would ensure a high probability of adequate sterilization and extremely dependable mission hardware performance. One aspect of this strategy was to implement long temperature *ramp times*—the periods employed to heat components from ambient to sterilization temperatures and to cool them

93. Walter O. Lowrie, aerospace consultant and retired president of the Missiles and Electronics Group, Martin Marietta Corporation, Orlando, FL, interview in Bionetics, *Lessons Learned*, p. 3.10.
94. Perry Stabekis, e-mail to author, 30 October 2007.
95. Perry Stabekis, telephone interview by author, 7 September 2007; Broome, interview in Bionetics, *Lessons Learned*, p. 3.37.
96. Sands, "Viking Science Activities"; NASA—Viking Project Office, *Viking '75 Program Lander Capsule Sterilization Plan*, pp. 1–3, 16–21; Martin Marietta Corporation, *Viking Lander System*; Ezell and Ezell, *On Mars: Exploration of the Red Planet—1958–1978*, p. 274; Perry Stabekis, e-mail to author, 30 October 2007.

back down to ambient. The longer the ramp times, the less *thermal shock* was visited on the electronic components. The term "thermal shock" refers to the uneven expansion or contraction of a component resulting from rapid or uneven heating or cooling. Cracking and other damage can occur when one part of a component has been raised or cooled to a different temperature than another part and therefore has expanded or contracted by a different amount.[97]

Another factor on which adequate sterilization depended had to do with a part's or a spacecraft's *thermal inertia,* a measure of its response time to external temperature changes. The higher the thermal inertia, the slower the response to external temperature changes. Thermal inertia depends on factors such as density, thermal conductivity (ability to transfer heat), and heat capacity (amount of heat required to raise the object's temperature one degree). The importance of these factors for planetary protection was that if the details of an object's thermal inertia were not understood, a sterilization protocol could not be chosen that would dependably elevate temperatures in regions buried deep within the object to levels that would kill any microorganisms present.[98]

Engineering Planetary Protection into Viking Component Designs

Viking components needed to do more than perform well and survive sterilization environments; they also needed to perform without doing damage to the target planet. The workings of every aspect of the spacecraft had to be analyzed to verify that this was so. While microbiological contamination was the chief danger to be prevented, the spacecraft was also capable of visiting chemical and mechanical damage on the Martian surface, and these potential problems too had to be mitigated. Different areas of the spacecraft that required forward contamination prevention actions are discussed below.

Propulsive fuel. Planetary protection requirements limited the lander's acceptable fuel options. A type of fuel was needed that was unlikely to harbor or, even worse, nurture terrestrial microorganisms.

---

97. Broome, Bionetics, interview in *Lessons Learned*, p. 3.37.
98. Ibid.

NASA eventually chose hydrazine as the propulsive fuel, partly because of its toxicity to such microorganisms.

In its analyses of the lander's exhaust plume chemistry resulting from hydrazine combustion, NASA initially found problematic substances that could potentially contaminate the Martian surface in the areas where samples were going to be taken.[99] Bob Polutchko, a Martin Marietta engineer who worked on the Viking project from 1968 to 1978 and led many of its task groups commented that the team detected hydrogen cyanide in the exhaust,[100] which is often involved in amino acid reactions. Since Viking was going to Mars to look for life, mission engineers wanted the lander exhaust to be more pure. Examination of the hydrazine propellant manufacturing process revealed that small quantities of aniline had been employed to get the final amounts of water out of the fuel. When aniline was burned in the presence of carbon dioxide, which was the case in the lander's combustion chamber, hydrogen cyanide was produced. To correct this situation, NASA separated out the aniline through a repeated freezing and thawing process that produced hydrazine ice cubes.

Lander engine nozzles. Martin Marietta staff tested their lander engine concept at NASA's New Mexico White Sands Test Facility. They envisioned using three rocket engines to control the Viking lander as it descended its final 5,200 feet to the Martian surface.[101] Multiple engines were needed in order to control the craft's pitch and yaw. But during their testing of just one of the engines, they discovered a very disturbing occurrence—one that did not contaminate the Martian surface with bacteria but did chemically contaminate and severely disturb the surface. The engine's exhaust during the test impacted soil that NASA believed was similar to what would be found on Mars. As the engine fired, it dug a 4-foot crater—a totally unacceptable situation. The lander was supposed to analyze an undisturbed, pristine Martian site for the possible presence of life. A jumbled, cratered, chemically contaminated

---

99. King and Butterfield, interview in Bionetics, *Lessons Learned*, p. 3.17.
100. Bob Polutchko, telephone interview by author, 21 February 2006.
101. T. D. Reisert, "Development Testing of Terminal Engine Nozzle Configurations for Viking Mars Lander" (draft of paper to be presented at the 38th meeting of the Supersonic Tunnel Association, 11–12 September 1972, General Dynamics Aerospace Division, San Diego, CA); Attachment to Hayden H. Parrott to Viking Project Office, 10 August 1972, record 5494, Viking Papers, NASA Historical Reference Collection.

site that had been made so by lander engine exhaust was hardly a pristine location in which to search for life. Furthermore, the lander was not capable of reaching out far enough beyond the cratering to examine an undisturbed section of soil. This problem "became a program-stopper for the whole Viking"[102] mission. It simply had to be solved.

When analyzing the fluid dynamics of the engine exhaust, Bob Polutchko and his team determined that "in a vacuum like in space the plume of that engine expands broadly—almost to 130 degrees from the axis of the engine. So it's not a focused kind of plume. But you fire such an engine with an atmosphere around it that has pressure—now the atmosphere focuses that plume and, like a launch vehicle taking off, you see . . . the flame coming out in a very directional way, [similar to] a Bunsen burner or a propane torch."[103] Even the thin atmosphere on Mars had enough pressure to focus the engine's exhaust plume sufficiently to dramatically disturb the planet's virgin surface.

Martin Marietta personnel conducted an intense redesign effort aimed at minimizing the "impingement effects"[104] of the engine's exhaust plume on the Martian surface. They evaluated 10 different engine nozzle configurations by measuring impingement pressures on a flat blast plate fitted with pressure sensors and calorimeters.

Martin Marietta designed seven of these nozzle configurations. Another of those examined was a conventional bell nozzle, and the others were supplied by Rocket Research Corporation and TRW. The best design turned out to be one that Martin Marietta had come up with—to change the exhaust nozzle from each of lander's three engines into 18 small nozzles and point each one slightly outward. This approach cost the lander only 3 percent in engine performance but reduced disruption of the Martian surface by a factor of 20 or more. The team's final design "disturbed less than a millimeter [of depth] of the surface. The plume hardly touched the native surface. . . . There was no churning action of that plume . . . the combustion products of the engine kind of moved horizontally along the

---

102. Polutchko telephone interview.
103. Ibid.
104. Reisert, "Development Testing of Terminal Engine Nozzle Configurations for Viking Mars Lander."

ground without disturbing the surface material. A little bit of dust was about all it ever kicked up."[105]

Planetary protection issues and limitations for the biology package. The Viking instruments that conducted searches for life needed special scrutiny. Even a small number of terrestrial organisms that managed to stow away within these instruments could have caused false positive results for the presence of life on the planet. They could also have presented a higher contamination risk to Mars than terrestrial organisms within other parts of the lander that did not contact or operate in such close proximity to the surface.

Not surprisingly, the Viking biology team chose to implement more stringent requirements than the planetary protection guidelines required for the biology package. The team decided that the probability of contamination of the biology experiments had to be kept below one chance in one million (1×10^{-6}).[106] This implied that if the standard was met, there would be no more than a 1×10^{-6} chance of the biology package depositing a living organism on Mars. Such a stringent standard required that the biology package be subjected to additional cleaning and sterilization that was not needed for the rest of the VLC.

TRW Systems of Redondo Beach, California, began development of the Viking biology package in 1970 under contract to Martin Marietta Corporation, the VLC prime contractor. Design, fabrication, and testing of the biology package took five years; it was delivered in the spring of 1975. The TRW team built the biology package under cleanroom conditions that possibly exceeded the cleanliness of Martin Marietta's facilities when the VLC was being assembled.[107]

---

105. Polutchko telephone interview.
106. Lawrence B. Hall to Harry Eagle, 30 October 1972, Washington Federal Records Center, accession no. 255-84-0634, box 26.
107. J. Barengoltz and P. D. Stabekis, "U.S. Planetary Protection Program: Implementation Highlights," *Advances in Space Research* 3(8) (1983): 8; Gilbert Levin, a principal investigator on the Biology Science Team and president of Biospherics, Inc., interview in Bionetics, *Lessons Learned*, p. 3.30; Harold P. Klein, Viking Biology Team Leader and member of Department of Biology, Santa Clara University, Santa Clara, CA, interview in Bionetics, *Lessons Learned*, pp. 3.32–3.35; Harrison Wroton, Resident Manager for the Biology Instrument at TRW, interview in Bionetics, *Lessons Learned*, p. 3.35; F. S. Brown, H. E. Adelson, M. C. Chapman, O. W. Clausen, A. J. Cole, J. T. Cragin, R. J. Day, C. H. Debenham, R. E. Fortney, R. I. Gilje, D. W. Harvey, J. L. Kropp, S. J. Loer, J. L. Logan, Jr., W. D. Potter, and G. T. Rosiak, "The Biology Instrument for the Viking Mars Mission," *Review of Scientific Instruments* 49(2) (February 1978): 143; Nicholas Panagakos and Maurice Parker,

Thorough removal of organic materials from all parts was exceedingly important, and it was accomplished largely by using plasma cleaning methods, in which the object was immersed in an ionized gas. Bombardment with energetic ions is quite effective for surface contaminant removal, especially for very thin films of organic materials that remain after conventional cleaning.[108]

After its construction, TRW sterilized the biology package in a 120°C (248°F) dry-nitrogen-atmosphere environment for 54 hours. This was a higher temperature environment than the biology package and the rest of the VLC would experience during terminal sterilization (which is discussed below). While the biology package was not the only component to be heat-qualified by its subcontractor, the biology package and its accompanying sample-path hardware were the only components to be sealed in a biological barrier during their sterilization and transport procedures in order to lower the risk of recontamination.

All gases admitted to chambers in which TRW tested the instrument poststerilization were filtered through 0.45-micrometer membranes. Personnel handling the instrument wore sterilized caps, gowns, and gloves, and the instrument was shipped within a sterilized nylon bag to protect it from contamination sources such as airborne particles. Preventing recontamination required a very sensitive operation when the mission team had to take the package out of its biological enclosure and install it in the VLC. It was the time at which the instrument was most at risk of recontamination.[109]

Viking's biology package cost $59 million, or 6 percent of the entire mission cost.[110] The instrument included three automated laboratories, a computer, and sophisticated electronics, all contained in a

---

"Mars Lander Successfully Completes Sterilization," NASA news release no. 75-181, 20 June 1975, record 5503, NASA Historical Reference Collection; Walter O. Lowrie, aerospace consultant and retired president of the Missiles and Electronics Group, Martin Marietta Corporation, Orlando, FL, interview in Bionetics, *Lessons Learned*, p. 3.10.

108. Pamela P. Ward, "Plasma Cleaning Techniques and Future Applications in Environmentally Conscious Manufacturing," Sandia National Laboratories, no date given, *http://www.sc.doe. gov/epic/docs/555.pdf* (accessed 8 October 2007); Sebastian Deiries, Armin Silber, Olaf Iwert, Evi Hummel, and Jean Louis Lizon, "Plasma Cleaning," Optical Detector Team, European Southern Observatory, Garching, Germany, 2005, *http://www.eso.org/projects/odt/ODTnew/ documentations/docs/paper_sd_2005.pdf.*

109. Brown et al., "The Biology Instrument," pp. 140, 143.

110. Dick and Strick, *The Living Universe: NASA and the Development of Astrobiology*, p. 80.

box roughly 1 cubic foot in volume. The team considered the biology package so complex that they sometimes referred to it as a miniature spacecraft. There was a price to pay, however, for its strict heat sterilization protocol. One impact was that it prevented glucose, a simple energy source, from being included in the experiment, for the high-temperature sterilization environment would have caused the glucose to decompose. Glucose's sensitivity to high temperatures was unfortunate because most known organisms—even algae—can use glucose as a nutrient. Biologists believed that glucose would have been likely to nurture the growth of any existing Martian life and would have proved very useful in Viking's life detection experiments.[111]

Enzymes are another type of material that is not typically heat stable, and the biology package's sterilization environment generally precluded experiments that required them as well. In particular, the "J-band experiment,"[112] which sought to detect a reaction between an organic dye and Martian organic macromolecules such as proteins or nucleic acids, had been under consideration for the biology package but was dropped because it required the use of an enzyme.[113]

Assessing Microbial Burdens at Each Stage of the Sterilization Process

NASA and its contractors conducted a continuing microbiological assay and monitoring program from the start of VLC assembly up until terminal sterilization. Data from these assays were essential for two reasons relating to prevention of the forward contamination of Mars:
- The data were necessary in order to calculate the required parameters of the terminal sterilization process.

111. Levin, interview in Bionetics, *Lessons Learned*, p. 3.30; Klein, interview in Bionetics, *Lessons Learned*, pp. 3.32–3.35; Wroton, interview in Bionetics, *Lessons Learned*, p. 3.35.
112. Richard S. Young, former NASA Planetary Quarantine Officer and Exobiology Manager as well as Viking Program Scientist, interview in Bionetics, *Lessons Learned*, p. 3.43.
113. SSB, *Signs of Life: A Report Based on the April 2000 Workshop on Life Detection Techniques* (Washington, DC: National Academies Press, 2002), p. 56, *http://www.nap.edu/openbook/0309083060/html/56.html* (accessed 26 January 2011).

- The data also were needed to confirm that the probabilities of Martian contamination arising from the Viking missions were within their allocations.

The Viking team needed to control, and monitor where appropriate, three types of microbial burdens on VLC hardware: encapsulated, mated, and surface burdens. *Encapsulated* bioburdens—those contained within a component (such as a resistor or capacitor) would have been extremely difficult to monitor. However, each VLC flight component, as well as spare replacement components, was subjected to flight acceptance heat compatibility tests. That procedure eliminated the need for bioassays at the component level,[114] many of which would have had to be performed at vendor facilities, and saved a considerable amount of personnel effort and expense.

Mated bioburdens are those VLC areas where contaminated surfaces are pressed together so that subsequent handling cannot increase the microbial level. The Viking team assayed VLCs by taking a minimum of 25 different samples from these surfaces just before they were mated. All flight VLCs were assayed as well as the proof test capsule (PTC), a complete VLC assembly fabricated from flight-style parts and used for system-level qualification tests.[115]

Surface burdens on the VLC constituted the major type of microbial contamination used for terminal sterilization planning and calculations. A total of 250 surface burden samples were required to be taken for each flight VLC or PTC each time that it was assayed. The Viking team was required to conduct bioassays of an assembled capsule at each of the following milestones:

Martin Marietta Operations
- Just before environmental testing.
- Just before preparations to ship the capsule to KSC.

KSC Operations
- During disassembly after dry-mating the capsule to the VO.
- During reassembly.
- During integration of the VLC and VO.

---

114. LaRC, *Viking '75 Program—Microbiological Assay and Monitoring Plan*, M75-148-0, 27 February 1974, p. 5, Washington Federal Records Center 255-84-0634, box 14.
115. Ezell and Ezell, *On Mars*, Chap. 8.

- Just before the VLC-VO assembly was encapsulated in the spacecraft fairing (nose cone).[116]

Sampling procedure. In order to bioassay VLC hardware, the Viking team used the swab technique specified in the *Viking Uniform Microbiological Assay and Monitoring Procedures,* PL-3701042. Whenever possible, 4 square inches of VLC surface was sampled by each swab, which was typically a small piece of absorbent material such as cotton attached to the end of a stick. On areas less than 4 square inches, the actual area sampled was estimated from component drawings or measurements on similar hardware.[117]

Viking Orbiter Bioassessment

NASA analyzed the forward contamination risks presented by the VO and decided that it did not need sterilization. The Agency estimated that the VO's mated and encapsulated bioburden would not contribute significantly to Mars contamination risks, although ejecta from VO surface bioburden due to meteoroid flux and particle dispersion could conceivably cause planetary contamination. Because of this possibility, NASA decided that monitoring was necessary and designated the VO's large, exposed areas as critical zones within which bioassays were to be concentrated. The critical zones included the following:

- Structural areas, such as the scan platform.
- Solar panels and devices.
- Thermal shields and blankets.
- High gain antenna.

These zones constituted the majority of the VO surface area. NASA required that at least 250 samples be taken from these zones using a swab technique.[118]

Because the VO was not subjected to terminal sterilization, microbes possibly survived within the craft, and in the unlikely event that it crashed onto Mars, could have been released. An SSB study

---

116. LaRC, *Viking '75 Program—Microbiological Assay and Monitoring Plan,* pp. 5–9.
117. Ibid., pp. 9–11.
118. Ibid., pp. 15–17.

addressed possible bioburden within the VO's interior,[119] where most microbes were subjected to temperatures estimated at 10° to 38°C. Although some cell dehydration would have occurred due to high vacuum, the SSB study believed that a fraction of the microbes—perhaps 1 to 10 percent—still survived.[120]

Sterilization of Bioassay Equipment and Materials

All equipment and materials used for bioassays had to themselves be thoroughly sterilized, or else they could recontaminate the parts of the spacecraft being analyzed. Viking staff sterilized equipment such as test tubes and their racks by autoclaving it at temperatures of at least 121°C (250°F) for 20 minutes or more. The nutrient media used within these test tubes also had to be certified as sterile. This was done by letting it incubate at ambient temperature as well as at a warmer temperature of 32°C (90°F) for several days, then inspecting the media for any signs of microbe growth. Viking staff sterilized the isopropyl alcohol used in the assays by passing it through a certified sterile 0.25-micrometer filter into a presterilized Teflon or polypropylene squeeze bottle.[121]

Side Benefits of Planetary Protection Procedures

While planetary protection heat sterilization procedures were necessary for the integrity of scientific investigations, they also

---

119. Space Science Board, Committee on Planetary Biology and Chemical Evolution, *Recommendations on Quarantine Policy for Mars, Jupiter, Saturn, Uranus, Neptune, and Titan* (Washington, DC: National Academies Press, 1978), pp. 11–12, folder 006697, "Quarantine/Sterilization," NASA Historical Reference Collection.

120. References for this conclusion mentioned in Space Science Board, *Recommendations on Quarantine Policy,* include P. Mazur, "Survival of Fungi After Freezing and Desiccation," in *The Fungi,* ed. C. Ainsworth and A. S. Sussman, vol. 3 (New York: Academic Press, 1968), pp. 325–394; R. M. Fry, "Freezing and Drying of Bacteria," in *Cryobiology,* ed. H. T. Meryman (London: Academic Press, 1966), pp. 665–696; R. E. Strange and C. S. Cox, "Survival of Dried and Airborne Bacteria," in Symposium No. 26, Society for General Microbiology, ed. T. R. G. Gray and J. R. Postgate (Cambridge: Cambridge University Press, 1976), pp. 111–154.

121. LaRC, *Viking '75 Program—Microbiological Assay and Monitoring Plan,* pp. A5-1 through A5-5.

produced valuable side benefits. The severe requirements on VLC components that rigorous sterilization treatments imposed led to use of tougher materials with enhanced dependabilities. The end result was a robust, "uniformly and highly reliable spacecraft"[122] with a longer useful lifetime. This was borne out by the strong technical and scientific performances of both Viking landers, as well as their longevities. The lessons learned in accomplishing these improvements helped to advance electronics and materials manufacturing techniques.[123]

●●●

Planetary Protection Measures Implemented During the Journey to Mars

During the period from launch until touchdown on Mars, NASA introduced several procedures, described below, that proved useful for preventing forward contamination.

The Launch

As mentioned earlier, positive air pressure was maintained within the VLC's bioshield during and following terminal sterilization in order to prevent recontamination. During the actual launch ascent, when outside pressure rapidly dropped, excess pressure within the VLC was vented through a biofilter that did not allow microbes to reenter. This venting was necessary in order to avoid the danger of the bioshield rupturing due to the pressure differential between its interior and exterior.

The bioshields were kept in place during the launch ascent and while the craft was still in Earth's atmosphere in order to prevent any possibility of recontamination by atmospheric microbes. The space science community proposed various altitudes at which the bioshield could be opened without fear of contaminating the lander with

---

122. Young, interview in Bionetics, *Lessons Learned*, p. 3.43.
123. Hall, "Sterilizing Space Probes," 53, 56; Bartholomew and Porter, "Reliability and Sterilization," 1762.

microbes. Tenney and Crawford[124] suggested that a conservative value was at least 300,000 feet (roughly 60 miles or 90 kilometers).

The Cruise Phase

A Titan/Centaur launch vehicle lifted the spacecraft from KSC into a 115-mile-high (185-kilometer) parking orbit. After the craft coasted in this orbit for 30 minutes, the Centaur rocket reignited to inject the spacecraft onto a trajectory toward Mars. The spacecraft then separated from the Centaur, which was not a sterilized piece of flight hardware. Afterward, the Centaur was deflected off of its flight path in order to prevent an eventual impact with, and possible contamination of, Mars.

Subsequent to the spacecraft beginning its cruise phase to Mars, NASA sent the command for explosive bolts on the craft to eject the cap of its bioshield (see figure 5.1). This took place well above the minimum altitude for this action that Tenney and Crawford had recommended. The explosive bolts as well as other separation explosive devices had to be carefully designed to avoid generating shock waves through the spacecraft or creating flying fragments and shrapnel, all of which could damage sensitive equipment.[125]

An additional risk associated with opening the bioshield was whether the VO, which was unsterilized, would recontaminate the VLC. Particles, chemicals, and possibly biotic material from the VO, including from the propellants of its rocket engines and from the gases emitted by its coatings, greases, and paints, were all potentially capable of contaminating the VLC.[126] Extensive calculations and experimental study indicated that this was not likely to occur, but mission staff planned an extra prevention measure just to make sure. Before the spacecraft reached Mars, mission control turned it so that the VLC

---

124. Tenney and Crawford, "Design Requirements for the Sterilization Containers of Planetary Landers."
125. NASA, "Viking Mission to Mars," NASA Facts NF-76/6-75, 1975, pp. 3–5, LaRC Archives—Viking booklets; Bionetics, *Lessons Learned*, pp. 2.2–2.3; Hall, "Sterilizing Space Probes," 56, 61; Tenney and Crawford, "Design Requirements for the Sterilization Containers of Planetary Landers."
126. Hall, "Sterilizing Space Probes," 56, 61; Tenney and Crawford, "Design Requirements for the Sterilization Containers of Planetary Landers."

faced toward the Sun, allowing ultraviolet solar radiation to provide a sterilant function on the capsule's exposed surfaces.[127]

The Descent to Mars

When the Viking spacecraft reached the vicinity of Mars, the lander, nested in its descent capsule, separated from its bioshield base and the VO and headed toward the planet's surface. Following this action, the VO ejected the bioshield base into space.[128] To ensure that the unsterilized VO did not eventually intercept, impact, and contaminate Mars, its trajectory had to be carefully calculated and controlled.[129]

As the lander-descent capsule assembly approached the Martian surface, it passed through an entry heating phase caused by friction with the atmosphere. During this heating phase, the descent capsule's aeroshell protected the lander's sensitive equipment from overheating. After the heating phase ended, at an altitude of about 20,000 feet (6 kilometers), the descent capsule's parachute opened and the aeroshell was ejected. The parachute itself was jettisoned at about 1 mile (1.6 kilometers) above the surface. The radar-controlled terminal propulsion system's three engines then fired for 30 seconds to slow the lander for a gentle touchdown on Mars.[130]

The lander's biology package included instruments sensitive enough to detect a single microbe. After the lander's touchdown but before it took samples, these instruments detected no trace whatsoever of contaminants within the biology package itself. NASA believed this to be an indication that its sterilization procedure, at least for the biology package, was entirely effective.[131]

---

127. Barengoltz and Stabekis, "U.S. Planetary Protection Program: Implementation Highlights," 8.
128. NASA, "Viking Mission to Mars," NASA Facts NF-76/6-75, 1975, pp. 3–5, LaRC Archives—Viking booklets; Bionetics, Lessons Learned, pp. 2.2–2.3.
129. Hall, "Sterilizing Space Probes," 56, 61; Tenney and Crawford, "Design Requirements for the Sterilization Containers of Planetary Landers."
130. NASA, "Viking Mission to Mars," p. 5; Bionetics, Lessons Learned, pp. 2.2–2.3.
131. Levin, interview in Bionetics, Lessons Learned, p. 3.30; Klein, interview in Bionetics, Lessons Learned, pp. 3.32–3.35; Wroton, interview in Bionetics, Lessons Learned, p. 3.35.

Viking Findings Pertinent to Planetary Protection

Estimating the likelihood of terrestrial organisms propagating on Mars required a comparison between known physical and chemical limits to terrestrial growth and known and inferred conditions present on or just below the Martian surface.[132] The two Viking landers sampled only several square meters of the Martian surface, and these were located at subpolar (nonpolar) sites. The landers conducted searches for life on soil sampled during the Martian summer and early fall, but only as deep as 6 centimeters (about 2.4 inches) below the surface. In other words, the Viking sites were hardly typical of all Martian climates and surface and subsurface characteristics. Nevertheless, NASA staff made certain conclusions regarding quarantine issues that it extrapolated to other regions of the planet, greater depths, and other seasons of the Martian year. These are summarized below for the specified regions of the planet.[133]

Subpolar Regions Within 6 Centimeters of the Surface

Viking scientists concluded that no terrestrial organisms could grow within a few centimeters of the surface in regions lying between Mars's two residual polar caps. They based this judgment on the following:
- The presence of strong oxidating chemicals in samples taken by both Vikings.
- The absence of detectable organic compounds, which the oxidants were assumed to have decomposed.[134]

---

132. SSB Task Group on Planetary Protection, *Biological Contamination of Mars: Issues and Recommendations*, Appendix D.
133. Space Science Board, Committee on Planetary Biology and Chemical Evolution, *Recommendations on Quarantine Policy for Mars, Jupiter, Saturn, Uranus, Neptune, and Titan* (Washington, DC: National Academies Press, 1978), p. 4, folder 006697, "Quarantine/ Sterilization," NASA Historical Reference Collection.
134. Noel W. Hinners to Alastair G. W. Cameron, 4 February 1977, in Space Science Board, *Recommendations on Quarantine Policy*, Appendix D.

- The belief that conditions found by both Vikings probably prevailed throughout Mars's subpolar regions, based on
  - The similarity in findings at both widely separated Viking landing sites.
  - The strong probability that the oxidants found were products of atmospheric or atmosphere-regolith[135] reactions, which would have been very pervasive throughout other regions of Mars as well.
  - The result of extensive infrared mapping of the Martian surface, which did not reveal thermal heterogeneities indicative of any regions whose temperatures were more favorable to growth of terrestrial organisms.[136]

Viking scientists did recognize the possibility that small "oases" of warmer climates existed with environments more friendly to life and that these had been missed by the thermal mapping. Such locations would also be accessible, however, to the oxidating chemicals that decomposed organic substances. While Mars's subpolar regions do contain large channels that space scientists believe were carved by flowing liquid water, this probably occurred more than a billion years ago. Thus the channels would have little impact on the present suitability of the planet for supporting Earth organisms.[137]

Subpolar Regions Deeper Than 6 Centimeters

The supposedly lethal surface conditions observed by Viking might not extend to deeper layers. The necessary depth to escape lethal conditions on the surface was not known to Viking scientists, mainly because the relationship between oxidating chemical quantities and their depth below the surface was unknown. What was believed, however, was that maximum temperatures fell rapidly with depth. In the northern hemisphere of Mars, at depths of 4 centimeters, Viking scientists estimated that maximum temperatures were 20°C below the

---

135. The regolith is the layer of loose material forming much of the Martian surface.
136. Space Science Board, *Recommendations on Quarantine Policy*, pp. 5–6, 19.
137. Ibid., p. 22.

"minimum confirmed terrestrial growth temperature"[138] of –15°C for terrestrial organisms.[139] And at a depth of 24 centimeters, the maximum temperature was estimated at –50°C, which was 35°C below the minimum confirmed terrestrial growth temperature. At increased depths, it became more likely that an ice layer would be encountered, but due to the temperatures, scientists did not believe that it or any liquid water present would support the growth of any known terrestrial organisms.

Thus, due to temperature alone, the prevailing opinion of the Viking team was that terrestrial organisms would not propagate at depths deeper than 6 centimeters, although scientists did admit the possibility that some anomalous subsurface regions could be warmer and that there was a remote chance of some unknown microbe growing at temperatures below –15°C.[140]

Residual Polar Caps

As with the subpolar regions, the low temperatures mapped in the residual polar caps by infrared surveys convinced most scientists that known terrestrial microbes would not grow. There were, however, some anomalous areas that, while still far below freezing, raised the remote possibility that other, yet undiscovered thermal heterogeneities might be warm enough to support life. But if warmer regions did exist, they would probably be drier regions as well because water would rapidly distill or sublime unless special conditions were present to allow water to liquefy, such as a freezing point depression due to electrolytes in the water. Such a water reservoir, however, would be too cold in the opinion of the SSB committee to allow growth of terrestrial organisms.[141]

---

138. Ibid., p. 7.
139. H. D. Michener and R. P. Elliott, "Minimum Growth Temperatures for Food-Poisoning, Fecal-Indicator, and Psychrophilic Microorganisms," *Advanced Food Research* 13 (1964): 349–396. This paper was reported on page 22 of the Space Science Board's *Recommendations on Quarantine Policy* report as the most thorough review of terrestrial organisms' minimum growth temperatures.
140. Space Science Board, *Recommendations on Quarantine Policy,* pp. 7–8, 23–25.
141. Ibid., pp. 8–9.

Note: The above conclusions were based on Viking results and theories of the 1970s on what habitats would support life. See Chapter 7 on later missions to Mars to see whether these conclusions changed.

Recommended Quarantine Strategy for Future Mars Missions

According to the SSB study, post-Viking flights to Mars should analyze sites shielded from ultraviolet radiation and the strong oxidant chemicals found on the surface. Thorough analysis would require subsurface sampling, and thus future landers would need to have equipment to penetrate the Martian surface to greater depths than Viking. Of particular interest are sites at the interface between the regolith and ice strata.

Regarding forward contamination prevention actions, the SSB study concluded that the probabilities of terrestrial organism growth on Mars were so low that "for the mission that conducts the first exploratory phase"[142] of a subpolar region, *terminal heat sterilization of the lander is not necessary.* Ben Clark, Lockheed Martin's Chief Scientist of Space Exploration Systems, explained that this did not refer to missions whose primary aims would include the search for life. Such missions would indeed "trigger the Viking protocol," including the need for terminal sterilization.[143]

The SSB study was not opposed to the sterilization of exploratory-phase spacecraft, provided that it did not increase mission cost and had no impact on the lander's scientific payload. But an otherwise beneficial experiment for an exploratory-phase mission should not be eliminated or its performance degraded because of "the imposition of unessential sterilization requirements."[144] The SSB recommendation was important in designing planetary protection protocols for Mars missions of the 1990s and 21st century, as is discussed in Chapter 7.

The VLCs were the first planetary landers designed to be fully sterilized. To accomplish this, the Viking program had to make great strides in advancing space vehicle technology to be able to withstand

---

142. Ibid., pp. 12–13.
143. Ben Clark, Lockheed Martin's Chief Scientist of Space Exploration Systems, interview by author, Denver, CO, 15 August 2006.
144. Space Science Board, *Recommendations on Quarantine Policy,* pp. 12–13.

the heat sterilization environment. The Viking mission was, in many respects, "the trail-blazer and pathfinder for NASA's important future [lander] missions."[145] It generated invaluable data and guidelines for the field of planetary protection. In the words of Jim Martin, Viking Project Manager, Viking "needs to be out in front as both a standard and a filter for systems-level implementation."[146]

For a timeline of major events from the Viking as well as other missions, see Appendix E. For a summary of planetary protection methodologies used on the Viking as well as other missions, see Appendix F.

---

145. Bionetics, *Lessons Learned*, p. 3.52.
146. Ibid., p. 3.53.

SMALL BODIES OF THE SOLAR SYSTEM

6

Deep within volatile-rich bodies, certain environments . . . might provide protection of dormant organisms . . .

—*"Survival of Life on Asteroids, Comets and Other Small Bodies"*[1]

An ejected rock could cocoon viable microbes for millions of years, delivering them safely to another planet.

—*Review of* Looking for Life, Searching the Solar System[2]

The solar system's small bodies include planetary satellites, asteroids, meteoroids, and comets as well as particles from sources such as interplanetary and interstellar dust and the solar wind.[3] These bodies typically have little or no atmosphere[4] and thus cannot support a surface ecosystem likely to resemble that of our own planet. In fact, we have no direct evidence that a living organism evolved or currently exists on any small solar system body. Nevertheless, certain conditions, such as the presence of water on or under the surface of the

---

1. B. C. Clark, A. L. Baker, A. F. Cheng, S. J. Clemett, D. McKay, H. Y. McSween, Jr., C. M. Pieters, P. Thomas, and M. Zolensky, "Survival of Life on Asteroids, Comets and Other Small Bodies," *Origins of Life and Evolution of the Biosphere* 29(5) (October 1999): 521–545.
2. Paul Davies, "The Greatest Mysteries of All," *New Scientist* (25 June 2005). This article is a review of Paul Clancy, Andre Brack, and Gerda Horneck, *Looking for Life, Searching the Solar System* (Cambridge: Cambridge University Press, 2005).
3. Some references to small solar system bodies distinguish them from planetary satellites, while various others do not. In this book, small bodies of the solar system will include planetary satellites.
4. Saturn's massive moon Titan is an exception to this.

body, might make it possible for life to exist. Some of the solar system's small objects also contain abundant organic material and hydrothermal processes. Speculation that such bodies could serve as reservoirs of life has become of increasing interest to space scientists because of the identification of material transfer mechanisms between them and planets.[5] Thus, the need for planetary protection measures to prevent both forward contamination to and back contamination from small bodies must at least be considered when missions are planned.

The Task Group on Sample Return from Small Solar System Bodies

Advances in planetary science and technology in recent decades enable increasingly active space exploration, including the capacity to collect and return samples to Earth from all over the solar system. Recognizing this, NASA asked the Space Studies Board (SSB), formerly called the Space Science Board, of the National Research Council (NRC) to "assess the potential for a living entity to be contained in or on samples returned from planetary satellites and other small solar system bodies such as asteroids and comets."[6] In response, SSB established the Task Group on Sample Return from Small Solar System Bodies, which extensively reviewed existing technical reports, interviewed NASA representatives and space scientists, and held a workshop to understand the wide range of perspectives. The Task Group attempted to identify the following:

- Detectable differences among small solar system bodies that would affect potential-for-life assessments.
- Scientific investigations that needed to be conducted to reduce the uncertainty in such assessments.

The Task Group asked the key planetary protection question: What risk do samples returned to Earth from spaceflight missions pose, as compared to "the natural influx of material that enters Earth's

5. Clark et al., "Survival of Life," 521.
6. Leslie Orgel, chair, Task Group on Sample Return from Small Solar System Bodies, SSB, *Evaluating the Biological Potential in Samples Returned from Planetary Satellites and Small Solar System Bodies: Framework for Decision Making* (Washington, DC: National Academies Press, 1998), p. 1.

atmosphere as interplanetary dust particles, meteorites, and other small impactors"?[7] To evaluate this risk, the Task Group studied data from Earth, meteorites, the Moon, and astronomical and spacecraft observations to determine how samples returned from small solar system bodies should to be handled as opposed to those returned from a planet such as Mars.

The Task Group needed to identify necessary conditions for the origin and survival of living microorganisms, including dormant as well as metabolically active organisms. The group formulated six questions relevant to the search for life on small bodies:

1. Was there ever liquid water in or on the target body?
2. Were metabolically useful energy sources ever present?
3. Was there ever sufficient organic matter (or CO_2 or carbonates) in or on the target body to support life?
4. After the disappearance of liquid water, was the target body subjected to extreme temperatures (i.e., >160°C)?
5. Was sufficient radiation present to sterilize terrestrial-like life-forms?
6. Is there, or has there been, a natural influx to Earth from the target body of meteorites or material equivalent to a sample returned?[8]

The Task Group answered these questions as best it could, given the existing scientific evidence. By so doing, the group defined the potentials for living entities to exist on a variety of small body types. Subsequent sections of this chapter describe these biological potentials, as well as others identified by NASA or international space organizations, and specify the planetary protection measures that need or will need to be put in place before these small objects are visited.

The Viability and Survivability of Microorganisms

Microorganisms have a far higher probability than multicellular organisms to remain viable on small bodies of the solar system

---

7. Ibid., p. 1.
8. Ibid., p. 2.

because they are able to adapt to a wider spectrum of environmental conditions, either as metabolically active organisms or as dormant forms. Single-celled microbes have successfully invaded every niche of our terrestrial biosphere and make up the majority of Earth's biomass. Microbes grow on land and in fresh and salt water, within larger organisms, and in the extreme environments of deep-sea trenches and radically cold and icy, subzero locations that would kill other forms of life. Such robust microbes often have certain physiological tools facilitating their survival in their particular niche, such as the following:

- Protective layers of material and pigments that shield them from ultraviolet (UV) or ionizing radiation.
- Active, efficient DNA or other cell repair mechanisms.
- Spore-forming capabilities allowing them to remain metabolically frozen.
- Adaptability to desiccation (absence of water). *Tychonema* spp. for instance can survive where there is virtually no water.
- Halophilic capability—able to survive and even thrive in highly saline conditions (such as 30-percent sodium chloride solutions).
- Barophilic capability—able to propagate under high-pressure conditions, such as at the bottom of the deepest ocean trenches—11,000 meters (36,000 feet) below the surface.
- Acidophilic or alkalophilic capability—growth under extremely low or high pH conditions. *Ferroplasma*, for instance, grows at pH 0 in acid mine drainage. These extremophiles use various defenses such as reinforcing their cell membranes and removing hydrogen ions in order to keep their internal pH at a neutral 6.5 to 7.0.
- Autotrophic—able to use carbon dioxide as its source of carbon.
- Able to draw energy from various inorganic chemicals, such as sulfur, ammonia, and iron compounds. *Thermoproteus* spp., for instance, derives its energy from the reduction of elemental sulfur.[9]

---

9. Mindy Richlen, "Microbial Life in Acidic Environments," Microbial Life Educational Resources, Science Education Resource Center at Carleton College, *http://serc.carleton. edu/microbelife/extreme/acidic/index.html* (last updated 17 January 2006, accessed 17 July 2006); Orgel, *Evaluating the Biological Potential*, pp. 11–14.

For all their adaptability, *metabolically active* cells generally require several critical conditions for their survival:

- Presence of liquid water.
- Accessible energy source. These can be photochemical sources, which are available at the body's surface, or various deeply buried sources.
- Temperatures generally no higher than 160°C (there are, however, ocean-vent sulfur-oxidizing bacteria communities that survive in volcanically heated water of 110° to 360°C [230° to 680°F]).
- Carbon source.
- Shielding from high-intensity or long-term exposure to ionizing or UV radiation.[10]

Dormant microbial life-forms do not require such stringent conditions for their survival. The capacity of many organisms to go dormant by forming cysts or spores permits them to survive greater temperature ranges for extended periods of times. Certain microbes show evidence of having survived for thousands of years, and possibly far longer, if they are protected from ionizing radiation fluxes. For instance, *Streptomyces* spp. as well as certain fungi have been isolated from a 5,300-year-old human corpse. In addition, one account reports microbes preserved in arctic and antarctic permafrost for 3 million years;[11] another claims the cultivation of spore-forming *Bacillus sphaericus* taken from the abdomen of a bee entombed in amber that was *25 to 40 million years old.*[12]

---

10. Orgel, *Evaluating the Biological Potential*, p. 16; Margaret S. Race and Michele Bahr, "Conditions for Life," course notes from Planetary Protection: Policies and Practices, sponsored by the NASA Planetary Protection Office and the NASA Astrobiology Institute, Santa Cruz, CA, 19–21 April 2005.
11. D. A. Gilichinsky, "Permafrost as a Microbial Habitat: Extreme for the Earth, Favorable in Space," in *Instruments, Methods, and Missions for the Investigation of Extraterrestrial Microorganisms*, ed. R. B. Hoover (Bellingham, WA: International Society for Optical Engineering, 1997), pp. 472–480, as reported in Orgel, *Evaluating the Biological Potential*, p. 15.
12. R. J. Cano and M. K. Borucki, "Revival and Identification of Bacterial Spores in 25- to 40-Million-Year-Old Dominican Amber," *Science* 268 (1985): 1060–1064.

Surface Environments Hostile to Any Form of Life

Conditions on or near the surfaces of the solar system's small, irregular bodies are typically unfavorable for metabolically active life to survive and may be unfavorable as well for sporulated life-forms. The space environment has characteristics that tend to sterilize the surfaces of small bodies. The three strongest sterilizing factors are as follows:[13]

- Ionizing radiation.
- Thermal inactivation.
- Vacuum.

Each of these factors significantly impacts the survivabilities of microorganisms. In combination, these factors greatly lower the probability of life existing on the surface of a small body.

Ionizing Radiation

Lack of an atmosphere implies intense exposure to damaging forms of radiation. Ionizing radiation from cosmic rays and solar particles consists of protons and ions of sufficient energy to directly destroy chemical bonds and inactivate molecular structures as well as to cause the formation of chemically damaging free radicals. Solar UV radiation, with its short wavelengths and, especially in the inner solar system, its high intensities, also damages molecules.

Types of radiation with the highest energy transfer to a cell are especially destructive. A single high-energy particle, for instance, can cause a double-strand break of a DNA helix. Cosmic rays, in particular, are rich in damaging, high atomic number and energy (HZE) particles.[14]

Ionizing radiation degrades all kinds of matter, including organic molecules.[15] The impact of ionizing radiation on living cells, how-

---

13. Clark et al., "Survival of Life," 521–545.
14. Clark et al., "Survival of Life," 528.
15. Jürgen Kiefer, *Biological Radiation Effects* (Heidelberg: Springer-Verlag, 1990), as reported in Clark et al., "Survival of Life," 525–526; Jürgen Kiefer, "Cellular and Subcellular Effects of Very Heavy Ions," *International Journal of Radiation Biology* 48(6) (December 1985): 873–892.

ever, is far from uniform. While it often causes cell lethality, its effect depends on cell size as well as DNA and RNA content, nuclear size, and ploidy (the number of sets of chromosomes).[16]

Some microbes have developed amazing tolerances to ionizing radiation. *Deinococcus radiodurans,* for instance, was discovered in 1956 at the Oregon Agricultural Experiment Station in a can of ground meat that had spoiled despite exposure to radiation in the megarad range. The name of the organism literally means "strange berry that withstands radiation,"[17] and it is the most DNA damage-tolerant organism ever identified. Its ability to survive extreme environments is attributed to its aptitude for quickly repairing damaged chromosome fragments, usually within 12 to 24 hours. Part of this process involves cutting usable DNA from another molecule and inserting it into the damaged strand. Marvin Frazier, who was the director of the Department of Energy's Microbial Genome Program, which funded a project to sequence all of *Deinococcus radiodurans's* genes, commented that this was the type of organism that might be able to survive interstellar journeys.[18]

Thermal Inactivation

Solar heating in our planetary system causes high temperatures on parts of a celestial body that are not shaded and may lead to thermal inactivation of any microbes on those surfaces. Inactivation comes about through damage to a cell's constituent parts, including its membranes, metabolic equipment, and genomes, with the level of damage increasing rapidly as the temperature rises.

Thermal inactivation of a cell may involve the disruption of a critical chemical bond in an essential molecule. DNA structures are

---

16. Henry S. Kaplan and Lincoln E. Moses, "Biological Complexity and Radiosensitivity," *Science* 145 (3 July 1964): 21.

17. John Travis, "Meet the Superbug," *Science News Online,* 12 December 1998, *http://www.sciencenews.org/pages/sn_arc98/12_12_98/bob1.htm* (accessed 26 January 2011).

18. J. R. Battista, "Against All Odds: the Survival Strategies of Deinococcus Radiodurans," *Annual Review of Microbiology* 51 (1997): 203–224; Michael M. Cox and John R. Battista, "Deinococcus Radiodurans—The Consummate Survivor," *Nature Reviews Microbiology* 3 (November 2005): 882–892.

particularly vulnerable due to their reliance on relatively weak hydrogen bonds. All microorganisms, however, have built-in monitoring and repair mechanisms that can sometimes fix the above structures before they degrade enough that they cease to function.[19]

Most microorganisms that have been observed are quite vulnerable to heat, although there are exceptions to this.

An organism designated "strain 121" was isolated from a water sample collected in an active, 300°C vent named Finn in the Juan de Fuca Ridge of the northeast Pacific Ocean. The microbe exhibited growth at 121°C, the typical temperature of an autoclave, which is employed in hospitals to kill microbes and their heat-resistant spores after an hour or so. Strain 121, in fact, doubled its numbers after 24 hours at 121°C.[20] The upper temperature limit for life is a critical parameter because it defines the boundary on organism survival in hot environments, either terrestrial or extraterrestrial. As mentioned earlier in the chapter, sulfur-oxidizing microbial communities have been found in hydrothermal vents emitting waters as hot as 360°C. Iron-reducing microorganisms have been found in similar environments.

Vacuum

Unlike planets, most small bodies do not have strong enough gravitational fields to retain substantial atmospheres (Saturn's moon Titan is an exception to this), leading to vacuum or near-vacuum conditions on their surfaces. In such environments, microbes cannot biotically employ atmospheric gases such as water vapor; hydrogen; hydrogen sulfide; carbon monoxide; carbon dioxide; and oxides of nitrogen, ammonia, or methane, as they do on Earth. Over an extended time period, high-vacuum conditions cause extreme desiccation of microorganisms, damage to DNA and membranes, and volatization of low molecular weight organic compounds.

Transformation of the microbes exposed to desiccating conditions into metabolically dormant forms—spores or cysts—is generally the

19. Clark et al., "Survival of Life," 528–529, 531.
20. Kazem Kashefi and Derek R. Lovley, "Extending the Upper Temperature Limit for Life," *Science* 301 (15 August 2003): 934.

most successful response for keeping the cell alive. This response also enhances its ability to withstand radiation and temperature extremes. Studies of *Bacillus subtilis* spores in space revealed an ability to survive up to six years if shielded from UV radiation. In fact, one in 10,000 of these spores survived even when unshielded. Although this finding has potential implications for microbe survivabilities on small bodies, it would be unreliable to extrapolate these survivabilities to residence times of thousands or millions of years in space.[21]

Spore formation does not always protect the cells against extremely dry conditions. Research by Dose et al. suggests that at temperatures of around 20°C, dry environments lead to deleterious reactions of dormant cells' structures and DNA, limiting half-lives of the organisms to only a few decades, unless their dormancy is interrupted by a metabolically active period and the presence of liquid water. This is because a transient period of such conditions allows cells to repair their damage. Spore half-lives may, however, reach *geological* time periods when temperatures remain low enough to slow down chemical reaction rates to near zero. This slowdown typically occurs at temperatures of around –130°C (–200°F) or lower.[22]

Initiation of Life on a Small Body: Endogenous and Exogenous Sources

In spite of the hostile conditions discussed above, life may possibly exist on some small bodies of the solar system. If so, it might have arisen through *endogenous* sources—those within the solar system body itself. Life could also have arisen through *exogenous* sources, in that it began elsewhere and was transported to its present location.

21. G. Horneck, H. Bücker, and G. Reitz, "Long-Term Survival of Bacterial Spores in Space," *Advances in Space Research* 14(10) (1994): 41, as reported in, and using additional material from, Clark et al., "Survival of Life," 532.
22. K. Dose, C. Stridde, R. Dillmann, S. Risi, and A. Bieger-Dose, "Biochemical Constraints for Survival Under Martian Conditions," *Advances in Space Research* 15(3) (1995): 207.

Endogenous Sources

The probability of life originating on a small solar system body appears to be far less than the chances of it beginning on a planet. Small body surface areas and volumes are orders of magnitude less than those of planets and thus provide fewer possible microenvironments for prebiotic chemical processes. But more significant, planets (as well as some satellites) have a greater likelihood of providing abundant liquid water and a diversity of environmental conditions that may be critical for the initiation of life. Because of their stronger gravitational fields, planets can hold onto chemically reactive atmospheres with potentially important dynamic processes such as tides, streamflow, turbulence, convection, and variable weather.[23]

Although small bodies are less likely than planets to have harbored the beginnings of life, certain meteorites are rich in organic chemicals and also show signs of having been in contact with liquid water. Studies of carbonaceous meteorites revealed that they once contained moderate- or high-temperature water,[24] while the Murchison meteorite (CM2) had over 600 different organic compounds, including protein-forming amino acids and DNA-forming nucleotide bases.[25] These data suggest that although unlikely, it is at least conceptually possible that at some point in the solar system's past, life may have arisen endogenously in its small bodies.

Scientists have considered whether, if life did arise spontaneously on a small solar system body, but then was frozen and possibly desiccated, could it have survived to the present day? The three major sterilizing factors discussed in the previous section (ionizing radiation, thermal inactivation, and vacuum) would have made this unlikely, but if the organisms spent the eons buried deep enough to be shielded

23. Clark et al., "Survival of Life," 535–536; S. Chang, "Prebiotic Synthesis in Planetary Environments," in The Chemistry of Life's Origins, ed. J. M. Greenberg et al., no. 416, Series C (Dordrecht: Kluwer Acad. Pub., 1993), pp. 259–299, as reported in Clark et al., "Survival of Life."

24. R. E. Grimm and H. Y. McSween, Jr., "Water and the Thermal Evolution of Carbonaceous Chondrite Parent Bodies," Icarus 82 (1989): 244–280.

25. J. R. Cronin and S. Chang, "Organic Matter in Meteorites, Molecular and Isotopic Analysis of the Murchison Meteorite," in The Chemistry of Life's Origins, ed. J. M. Greenberg et al., no. 416, Series C (Dordrecht: Kluwer Acad. Pub., 1993), pp. 209–258, as reported in Clark et al., "Survival of Life."

from ionizing radiation and protected by a layer of ice or other means of encapsulation from the impacts of a hard vacuum, then they might have survived.

If such organisms were buried deep in the body, however, then the possibility of a spacecraft excavating and sampling them would be small indeed. A natural event could have recently (from a geologic perspective) excavated them to a location within a meter or so of the surface. But this would not have been a likely event except, perhaps, in the case of comets. When a comet passes through the inner solar system, ice layers sublimate and a blow-off of materials occurs. Most comets experience high material-loss rates when penetrating the inner solar system—as much as several meters of surface during each perihelion passage. Thus, comets experience a natural means of excavation independent of impacts with other bodies. Endogenous materials, including life-forms that had been protected since the early years of the solar system, could conceivably be exposed during one of a comet's visits to the inner solar system region.[26] Planetary protection measures for spacecraft exploring a comet have typically included the requirement to ensure a very low probability of accidental impact with the body.

Exogenous Sources

It is possible that by means of impacts between celestial bodies, biota were flung into space and traveled through the solar system. It would have been physically easier and perhaps more likely to remove biota from a body with weak, rather than strong, gravity and one with little atmosphere rather than a dense blanket of gases. It would also have required less energy to transport biota inward toward the Sun rather than outward. Biota could have been transported onto and carried by small bodies such as comets, asteroids, and meteoroids.

An important planetary protection consideration that scientists have raised regarding small bodies is that, if some indeed do carry life and if such bodies have already impacted a target planet, then does

---

26. Clark et al., "Survival of Life," 536–537.

this not obviate the need for sterilizing spacecraft headed there? In particular, Gladman et al. examined the issue of whether expensive sterilization activities are justified for lander missions to Mars if meteorites from Earth, formed when some small body impacted our planet, have already found their way to Mars. Studying the orbital details of any terrene (Earthly) meteorites that exist has direct relevance to forward contamination questions such as this.[27]

Cross-contamination of materials among solar system bodies has been appreciable, according to existing data. The analysis of many meteorites, for instance, has revealed that they contain centimeter-sized clasts (constituents, grains, or fragments of sediments or rocks) from foreign parent bodies. Transfer of life-forms is thus conceivable. This possibility of cross-contamination among celestial bodies has "reduced the degree of confidence in the inherent safety of a sample returned"[28] to Earth from a small solar system body.

Missions to the small bodies of the solar system have already been conducted, while others have been proposed and are in the planning stages. Some of the missions involved or will involve sample return, requiring examination of planetary protection issues. Natural sterilizing factors have a range of expected impacts on such missions, depending on the locations, characteristics, and histories of the target bodies and the depths within them from which samples will be taken. Because of the radically different types of bodies and missions to explore them, each mission requires a separate assessment to determine the extent to which forward and backward contamination should be of concern.[29]

---

27. Brett J. Gladman, Joseph A. Burns, Martin Duncan, Pascal Lee, and Harold F. Levison, "The Exchange of Impact Ejecta Between Terrestrial Planets," *Science* 271 (1996): 1392; Clark et al., "Survival of Life."
28. Orgel, *Evaluating the Biological Potential*, pp. 24–25.
29. Clark et al., "Survival of Life," 540.

Our Moon Revisited: The Possibility of Ice Deposits

The elaborate quarantine and analysis procedures carried out at the Lunar Receiving Laboratory after the Apollo 11, 12, and 14 flights did not reveal any indication of life on our Moon. All return samples were certified as safe by the Interagency Committee on Back Contamination (see Chapter 4 on the Apollo program for a fuller discussion of this). Since the release of the lunar samples, "there have been no discernible adverse consequences for researchers or for Earth's ecosystem."[30]

No water was detected in lunar rocks. The SSB Task Group believed, given the existing data, that no special containment was needed for samples returned from the Moon. Hypotheses have been put forth, however, that ice could exist in lunar polar regions, accreted as a result of comet impacts.[31] While most ice delivered by comets would sublimate into water vapor and escape into space due to high surface temperatures during lunar daytime hours, a small fraction of the ice could have reached permanently shadowed "cold traps"[32] and remained stable over geologic time periods. Such traps may have formed near the Moon's poles in craters whose walls perennially block the Sun's light from reaching some sections of the crater floor.

Because solar radiation never directly illuminates cold traps, and because they do not emit much infrared or other types of radiation of their own, they are quite difficult to observe from Earth-based instruments. Frozen volatiles such as ice have a higher reflectivity to radar waves, however, than silicate rocks and also preserve the circular polarization sense of the scattered radar signal, and so under the right conditions ice at the lunar poles can be observable.

The Clementine mission, a joint endeavor between the Strategic Defense Initiative Organization and NASA, conducted numerous observations of the Moon in 1994. The Clementine spacecraft achieved

30. Orgel, *Evaluating the Biological Potential*, p. 27.
31. James R. Arnold, "Ice in the Lunar Polar Regions," *Journal of Geophysical Research* 84(B10) (1979): 5659–5668.
32. S. Nozette, C. L. Lichtenberg, P. Spudis, R. Bonner, W. Ort, E. Malaret, M. Robinson, and E. M. Shoemaker, "The Clementine Bistatic Radar Experiment," *Science* 274 (29 November 1996): 1495–1498.

lunar orbit on 19 February and remained in orbit for 71 days, taking nearly 1 million images of the Moon at 11 different wavelengths, from the ultraviolet to the near-infrared parts of the spectrum.[33] Some of these images yielded important data regarding the lunar polar regions. The lunar south pole is located within the South Pole-Aitken basin, an impact crater, and its relative elevation is likely several kilometers below the average radius of the Moon, resulting in areas that are always in shadow. Data from Clementine indicated that as much as 15,500 square kilometers of south polar terrain, approximately twice the area of Puerto Rico,[34] may be permanently in shadow, as well as a lesser area around the north pole.[35] Clementine also generated data suggestive of polar ice, although Earth-based measurements from the Arecibo observatory[36] raised questions about this.

More recent data from the Lunar Prospector, a NASA Discovery Mission that launched in 1998, provided compelling evidence of between 10 and 300 million tons of water ice in the Moon's polar region (and a greater amount, surprisingly, near the north pole rather than the south pole).[37] The existence of ice in lunar polar regions was by no means universally accepted, however. Based on what he considered more realistic water transport mechanisms and modeling approaches, R. R. Hodges of the Center for Space Sciences at the University of Texas held in a 2002 paper that "the concept of water ice at the lunar poles is insupportable."[38]

In October 2009, NASA's Lunar CRater Observation and Sensing Satellite, or LCROSS, may have "opened a new chapter in our understanding

33. Lunar and Planetary Institute, "The Clementine Mission," 2000, http://www.lpi.usra.edu/expmoon/clementine/clementine.html (accessed 26 January 2011); David R. Williams, "Clementine Project Information," 2005, http://nssdc.gsfc.nasa.gov/planetary/clementine.html (accessed 26 January 2011).
34. Paul D. Spudis, "Ice on the Bone Dry Moon," 21 December 1996, Planetary Science Research Discoveries Web site, http://www.psrd.hawaii.edu/Dec96/IceonMoon.html (accessed 24 July 2006).
35. S. Nozette, C. L. Lichtenberg, P. Spudis, R. Bonner, W. Ort, E. Malaret, M. Robinson, and E. M. Shoemaker, "The Clementine Bistatic Radar Experiment," Science 274 (29 November 1996): 1495–1498.
36. N. J. S. Stacy, D. B. Campbell, and P. G. Ford, "Arecibo Radar Mapping of the Lunar Poles: A Search for Ice Deposits," Science 276 (6 June 1997): 1527–1530.
37. ARC, "Introduction," http://lunar.arc.nasa.gov/project/index.htm, and "Eureka! Ice Found at Lunar Poles," http://lunar.arc.nasa.gov/results/ice/eureka.htm (both were last updated 31 August 2001 and accessed 21 July 2006).
38. R. Richard Hodges, Jr., "Ice in the Lunar Polar Regions Revisited," Journal of Geophysical Research 107(E2) (2002): 5011.

of the moon"[39] when the impact of its Centaur upper stage rocket apparently uncovered water in the Cabeus crater, a permanently shadowed region near the lunar south pole. LCROSS findings indicated that water could be more widespread and in greater quantity than previously suspected. Then in 2010, Spudis et al. published data from NASA radar flying aboard India's Chandrayaan-1 spacecraft that suggested the existence of ice deposits near the lunar north pole. The data, which were taken the year before, are consistent with the existence of more than 40 small craters ranging in diameter from 1 to 9 miles (2 to 15 kilometers) containing water ice. Although the total amount of ice depends on its thickness in each crater, scientists estimated there could be 600 million metric tons or more in the craters.[40]

If lunar polar ice deposits do exist and were delivered by comets or water-rich asteroids, they could have sublimated directly from solid to vapor phase upon impact and then back to solid when they condensed.[41] Liquid-phase water likely never existed on the Moon, with its lack of an atmosphere. Nevertheless, if comet-delivered ice contained dormant microbes and if fragments of this ice reached cold traps on the Moon without sublimating, the possibility exists that such microbes could still be present. If ice deposits are confirmed on the Moon and samples are taken, planetary protection procedures would need to be strongly considered for preventing possible back contamination of Earth. Such regions may be declared by COSPAR to be special regions requiring additional care when studying them.[42] Planetary protection measures may also be applicable if lunar ices were excavated and used as in situ resources for astronauts and mission operations.[43]

---

39. Jonas Dino, "LCROSS Impact Data Indicates Water on Moon," 13 November 2009, *http://www.nasa.gov/mission_pages/LCROSS/main/prelim_water_results.html* (accessed 26 January 2011).

40. P. D. Spudis et al., "Initial results for the north pole of the Moon from Mini-SAR, Chandrayaan-1 mission," *Geophysical Research Letters* 37 (2010): L06204; NASA, "NASA Radar Finds Ice Deposits at Moon's North Pole; Additional Evidence of Water Activity on Moon," *ScienceDaily* (2 March 2010), *http://www.sciencedaily.com/releases/2010/03/100302085214.htm* (accessed 26 January 2011).

41. Orgel, *Evaluating the Biological Potential*, p. 27.

42. Cassie Conley (acting Planetary Protection Officer), interview by *Astrobiology Magazine*, Moffett Field, CA, 5 June 2007, Space Daily Web site, *http://www.spacedaily.com/reports/Restricted_Zone_999.html* (accessed 26 January 2011).

43. Margaret Race, comment to author regarding manuscript of this book, sent 18 September 2007.

●●●

Phobos and Deimos:
The Small, Rocky Satellites of Mars

The two moons of Mars—Phobos and Deimos—are rocky, irregularly shaped, and tiny compared to our Moon. Phobos has a maximum dimension of 27 kilometers (17 miles) and Deimos, only 15 kilometers (9 miles). They are thus similar in size to many asteroids. One or both may in fact be captured asteroids, although such a mechanism would probably not have produced Deimos's near-circular, near-equatorial orbit. A more probable history of the two moons may be that they coformed with Mars out of material left over from the planet's creation.[44]

Phobos and Deimos are not likely locations in which to find biological organisms or evidence of past life. If biological material was at some time formed within the moons themselves, it would probably have been early in the bodies' histories, and such life-forms would long ago have received enough radiation from natural abundances of uranium, potassium, and thorium within the moons "to eliminate the most radiation-resistant microorganisms known."[45] The top several meters of surface would also have received enough cosmic-ray and solar-flare proton flux to ensure its sterility. If, however, voids within the moons were filled with ice that contained microbes, these life-forms could have been effectively shielded from the radioactive element radiation as well as from proton flux.

There are compelling reasons why water ice probably does not exist today in near-surface niches on the Martian satellites. They are dark in color and thus fairly good absorbers of sunlight, and they also lie close enough to the Sun to be relatively warm objects with temperatures near their surfaces that are too high for water ice to persist. Ice could exist at greater depths, but it would be difficult to access for sampling.

Biological material could have been transported to the moons from an exogenous source, such as in the form of ejecta from Mars as a result of impacts that the planet experienced. Viking data indicated,

---

44. Orgel, *Evaluating the Biological Potential*, p. 28; Bill Arnett, "Phobos," Nine Planets Web site, *http://www.nineplanets.org/phobos.html* (last updated 23 June 2006, accessed 25 July 2006).

45. Orgel, *Evaluating the Biological Potential*, pp. 28–29.

however, that much if not most of the Martian surface and near-surface is quite hostile to life-forms. Also, meteorites of possible Martian origin have been extensively sampled and have not been found to contain any life-forms or, in fact, any materials harmful to Earth's ecosystem.

Even if Mars ejecta did contain life and did at some time impact Phobos or Deimos, the odds would be miniscule of capturing such life in a return sample from the satellites, unless the biotic colonies that had been established were widespread. Still, the SSB Task Group recommended conducting remote sensing studies of Phobos and Deimos compositions in order to reduce uncertainties regarding their biological potential.[46]

The Soviet missions to Phobos. The USSR launched two missions in July 1988 meant to land on Phobos. If completed, they might have transported terrestrial organisms to the satellite. Neither mission, however, was successful. The Soviets launched both the Phobos 1 and 2 spacecraft from Baikonur Cosmodrome in Kazakhstan. Both were to study the surface composition of the larger of Mars's moons, but both were lost en route.

A faulty command sequence sent from Earth on 29–30 August 1988 resulted in a shutdown of the Phobos 1 craft. An error in the uploaded software deactivated the craft's attitude thrusters, resulting in the loss of a lock on the Sun. This led to the craft's solar arrays getting pointed away from the Sun, and thus the ship's batteries ran down. Phobos 2 reached the vicinity of Mars on 30 January 1989, but it was lost while maneuvering for an encounter with the moon due to a failure of either the craft's computer or its radio transmitter.[47]

The Icy Moons of Jupiter and Saturn: Life in Submerged Oceans?

Since the Galileo spacecraft sent back its data from its sojourn among the moons of Jupiter, space scientists have realized that

---

46. Orgel, *Evaluating the Biological Potential*, pp. 29, 35–36.
47. National Space Science Data Center (NSSDC), "Phobos Project Information," GSFC, *http://nssdc.gsfc.nasa.gov/planetary/phobos.html* (last updated 18 December 2001, accessed 25 July 2006); *SpaceRef.com*, "Phobos," *http://www.spaceref.com/directory/ exploration_and_missions/robotic_missions/mars/phobos/* (accessed 25 July 2006).

ice-capped oceans of liquid water may be widespread throughout our solar system. Strong evidence points to the existence of a salty ocean beneath the surface ice of Europa, and possibly within two of Jupiter's other Galilean satellites: Ganymede and Callisto. Such oceans may exist on other satellites as well—notably the Saturnian moons Enceladus and Titan.

The energy needed by life-forms may also be available in these oceans, originating from chemical disequilibrium[48] conditions supplying molecular oxygen (O_2) and other oxidants. The processes that result in disequilibria would, however, be very different than those on our planet. On Earth, photosynthesis and organic carbon processes lead to oxidizing surface conditions and chemical disequilibria. Our planet's ocean life depends heavily on the oxidants that filter down from the surface of the ocean. But significant solar energy cannot penetrate the kilometers of ice covering the surfaces of Jupiter's and Saturn's icy moons, and thus cannot drive photosynthesis in the oceans that possibly lie below. Instead, chemical energy could be derived from nonphotosynthetic sources of O_2 and other oxidants available even to subsurface oceans. One pathway for a source of molecular oxygen is for gamma and beta radiation from potassium-40, a naturally occurring isotope of potassium, to decompose water into oxygen and hydrogen.[49]

Jupiter's Galilean Satellites

The four moons of Jupiter that were discovered by Galileo—Io, Europa, Ganymede, and Callisto—are far larger objects than the planets' other satellites. The compositions of the Galilean moons, as well as of Jupiter's other satellites, are functions to some extent of the pressure and temperature environment in the material orbiting the planet during their period of condensation. Immediately after the Galilean

---

48. Viable organisms generally need chemical disequilibrium, an unstable situation that can supply the energy required for survival. Stephen T. Abedon, course concepts for Biology 113 at Ohio State University, *http://www.mansfield.ohio-state.edu/~sabedon/campbl06.htm* (accessed 26 January 2011), derived from "An Introduction to Metabolism," in N. A. Campbell and J. B. Reece, *Biology* (San Francisco: Benjamin/Cummings Publishing Company, 2002), Chap. 6.
49. Christopher F. Chyba, "Life Without Photosynthesis," *Science* 292 (15 June 2001): 2026–2027.

satellites accreted from the primordial Jovian nebula, they were likely composed of some mixture of water ice and chondritic (i.e., having the structure characteristic of the class of meteorites called chondrites) nonicy material.

Assuming a radial temperature gradient in the primordial Jovian nebula, the amount of ice remaining in a satellite after condensation would have depended on the distance from Jupiter at which it formed. Data from observations of the Galilean satellites support such a scenario. Io, the closest Galilean moon to Jupiter, has a density suggesting a rocky composition that lacks even water of hydration, and no spectroscopic evidence of water has been found. Europa, the next distant Galilean moon, has a density consistent with a substantial amount of free water; its surface is also bright with ice. Ganymede and Callisto have densities indicating massive amounts of water—their bulk compositions are nearly 50 percent water by mass.[50]

While the outer three Galilean moons may have environments favorable to life, Io's lack of water and other circumstances make it far less likely to support life. Virtually all of Io's surface material appears to be volcanic, and thus was probably heated at one time to temperatures far too high for organic molecules to remain intact. Io is also subjected to a radiation environment more severe than any other body of its size in the solar system. This flux of high-energy charged particles from Jupiter's magnetosphere powerfully inhibits any biological activity on or near Io's surface.

This is not necessarily the case on other Galilean moons. The particle flux there would be significantly less intense due to their greater distance from Jupiter.[51] If life is to be found in the Jovian system, it will probably be within one or more of those moons.

50. J. S. Lewis, "Satellites of the Outer Planets: Their Physical and Chemical Nature," *Icarus* 15 (1971): 174–185; J. S. Lewis, "Low Temperature Condensation From the Solar Nebula," *Icarus* 16 (1972): 241–252; Orgel, *Evaluating the Biological Potential*, pp. 29–30.
51. Orgel, *Evaluating the Biological Potential*, pp. 30–31.

Figure 6.1 A cutaway view of the possible internal structure of Europa. The rocky interior may be surrounded by a shell of liquid water, which is in turn contained in a shell of ice.

Evidence Suggesting an Ocean Under Europa's Ice

In January 2000, NASA's Galileo spacecraft approached the Jovian moon Europa while space scientists Margaret Kivelson, Krishan Khurana, and their team waited anxiously to measure which way its magnetic field pointed. They knew of two mechanisms that could be generating the field, and the implications for possible life on Europa were enormous. Internal processes and flows deep within the satellite could be generating the magnetic field, much as Earth's molten core does. If such were the case, then the field should point the same direction, regardless of which way the parent planet Jupiter's field was pointing at the time. Alternatively, an external magnetic field such as that of Jupiter might be acting on an electrically conducting layer within Europa and *inducing* a field from it. If the Europan field was indeed induced by its parent planet's massive field, then the direction of the Europan field should be determined by that of the Jovian field.

Europa and the Galileo spacecraft had moved into a region of space in which the Jovian field pointed *opposite* to where it had pointed during previous flybys. If Jupiter was inducing Europa's field, the opposite polarity of the Jovian field should cause an opposite polarity of Europa's field. When Galileo's measurements showed that this was exactly the case, the Kivelson team knew that they were seeing strong evidence of an electrically conducting layer hiding under Europa's icy exterior. The layer's most likely composition was water, warm enough to be liquid, and with dissolved salts in it that would make it a conductor.

Exhaustive data from repeated Europan flybys demonstrated that the observed field, which flipped along with that of Jupiter, could indeed be explained by a subsurface global reservoir of water with a salinity similar to that of Earth's ocean (see figure 6.1). And if liquid water does exist under Europa's ice, so might life.[52]

Galileo's Planetary Protection Approach vs. Those of Voyager and Viking

During early planning for the Galileo mission, which was to make many flybys of Jupiter's icy moons, NASA staff began developing planetary protection protocol. They used the protocol from the Voyager mission, which in 1979 had conducted far more distant flybys of the Jovian system, as a baseline for developing Galileo's plan. Voyager's contamination control program utilized the same class 100,000 cleanrooms as did the Viking Mars lander, whose bioload was reduced to a very low level. But contamination-prevention controls for Voyager, which only flew by planets and did not orbit or land on any, were significantly more relaxed than for Viking. Different contamination control characteristics and requirements of the three missions are compared below:

Particulate contamination. Galileo and Voyager were expected to be less vulnerable to damage from particulate contamination than Viking and its sensitive life-detection experiments. Critical Galileo instruments were purged and covered but not subjected to the rigorous sterilization regimes that Viking's instruments were.

---

52. Margaret G. Kivelson, Krishan K. Khurana, Christopher T. Russell, Martin Volwerk, Raymond J. Walker, and Christophe Zimmer, "Galileo Magnetometer Measurements: A Stronger Case for a Subsurface Ocean at Europa," *Science* 289 (25 August 2000): 1340–1343; Michael Meltzer, *Mission to Jupiter: A History of the Galileo Project* (Washington, DC: NASA SP-2007-4231, 2007).

Alcohol cleaning. Galileo, like Voyager, was not able to be cleaned with alcohol, which would have damaged certain parts.

Personnel control and access restrictions. Galileo's construction operations required fewer controls on personnel clothing than did Viking and did not employ any access restrictions that were exclusively meant for contamination prevention.

Terminal sterilization and encapsulation. Unlike Viking, the Galileo spacecraft was not subjected to terminal sterilization. It was also not encapsulated prior to mating with the Space Shuttle, from which it was launched. Encapsulation would have seriously impacted the spacecraft's weight as well as its mission performance and budget.

Launch-related contamination sources. NASA considered Galileo–Space Shuttle prelaunch and launch environments to be greater sources of contamination than the environments for the Viking and Voyager missions, since those missions employed Titan *robotic* launch vehicles instead of the Shuttle, with its human crew.[53]

Preventing an Accidental Impact Between Galileo and Europa or Another Icy Moon

The Galileo mission had an interesting approach to planetary protection that Jack Barengoltz called "a flexible PP plan."[54] The mission initially had an obligation to generate planetary protection documents that included prelaunch, postlaunch, and end-of-mission reports and to document the contamination control procedures implemented, probabilities of impacts with other bodies, and "the disposition of all launched hardware at the end of the mission."[55] But otherwise, the mission had no hard planetary protection requirements. Galileo faced

---

53. John R. Casani to B. R. McCullar, "NASA Memorandum SBL(RSY:dr), SBL/R. S. Young to SL/B. R. McCullar, 'Jupiter Orbiter/Probe Planetary Protection Requirements,' dated 31 October 1977," John Casani Collection, JPL Archives, JPL 14, folder 41, "Galileo Correspondence 3/78-4/78," box 5 of 6, reference number 230-JRC:db-94, 17 April 1978.

54. J. Barengoltz, "Galileo and Cassini: Planetary Protection Report to COSPAR," COSPAR no. 545, 2004, http://www.cosis.net/abstracts/COSPAR04/00545/COSPAR04-A-00545.pdf (accessed 26 January 2011).

55. J. B. Barengoltz, *Project Galileo Planetary Protection Plan* (draft), NASA-JPL document no. PD 625-14, 28 March 1984, pp. 5-1, 7-1. This is an attachment to John R. Casani to Harry Manheimer, 230-JRC:db-137, 6 April 1984.

the possibility, however, that its investigations would reveal the need to upgrade biological interest in one or more of Jupiter's satellites. With this possibility in mind, the mission's planetary protection plan included an agreement that if Galileo made such discoveries, then the project would modify its end-of-mission planning to reflect the new circumstances.

This is indeed what happened. The science return from the mission indicated that Europa most probably had liquid water under its layer of ice and that Callisto and Ganymede might also have such water layers. And if there were layers of water inside these moons, there was the possibility of life existing within them. The Galileo orbiter had not been sterilized and could have contained terrestrial organisms. Thus, NASA recognized the need to prevent the orbiter from accidentally impacting one of these moons.

Discussions between Galileo mission staff and NASA Planetary Protection Officer John Rummel ensued. In March 2000, Rummel made a verbal request, followed by a letter in April, to the National Academies Space Studies Board Committee on Planetary and Lunar Exploration (COMPLEX) to provide advice on this matter. Specifically, Rummel addressed the final disposition of the Galileo orbiter. In addition, Torrence Johnson, Galileo's project scientist, briefed COMPLEX in March 2000 on the "risks associated with the last phase of Galileo's operational life."[56]

In its deliberations, COMPLEX balanced planetary protection considerations with the unique scientific opportunities provided by various end-of-mission scenarios. A June 2000 letter from COMPLEX to Rummel expressed the following conclusions:

- No planetary protection-related objection existed to the "disposal of Galileo by intentional or unintentional impact with Io or Jupiter."[57]
- Serious objections existed to the intentional or unintentional disposal of Galileo on Europa.

---

56. Claude Canizares and John Wood to John Rummel, "On Scientific Assessment of Options for the Disposal of the Galileo Spacecraft," National Academies SSB and its Committee on Planetary and Lunar Exploration, 28 June 2000, http://www7.nationalacademies.org/ssb/galileoltr.html.
57. Ibid.

- Disposal on Ganymede or Callisto had implications "intermediate in the broad range between those for disposal on Io and for disposal on Europa."[58]

COMPLEX concurred with NASA that the best means available for disposing of Galileo was to send it on a collision course into Jupiter. The scientific benefits of this plan were that it would allow a close flyby of the small Jovian moon Amalthea and possibly one or more polar flybys of Io before the spacecraft's final encounter with Jupiter.

NASA had not developed quantitative estimates for COMPLEX of spacecraft-failure probabilities as functions of time, and thus the committee could not calculate the risks of an accidental Europa impact should additional flybys be planned. COMPLEX recommended that Galileo staff do this calculation. COMPLEX did have the data to analyze the probable effects of additional Jovian radiation absorbed by Galileo during an extra year of operations. Estimating the increase in radiation to be only about 20 percent, and taking into account the redundancy of all of Galileo's essential operating systems and the fact that radiation effects to date had not handicapped the spacecraft, COMPLEX estimated that the probability of total loss of control of the craft during an extra year of operation was relatively small. And if a major Galileo control system did experience trouble, the spacecraft could be quickly retargeted onto a Jupiter-bound trajectory and away from any collision with a moon. Pending NASA's completion of failure-probability calculations, COMPLEX reached a consensus that the best course of action was to defer the destruction of Galileo until after it completed its Io polar flybys. This would maximize the science returns from the mission, especially as related to the study of the Jovian system's magnetic fields and plasma phenomena.[59]

On Sunday, 21 September 2003, after polar passes of Io and a flyby of Amalthea, Galileo plowed into and disintegrated in Jupiter's 60,000-kilometer-thick atmosphere, avoiding any chance that the craft might eventually strike and contaminate Europa or another icy

---

58. Ibid.
59. Ibid.; SSB, "Magnetospheres," in *An Integrated Strategy for the Planetary Sciences 1995–2010* (Washington, DC: National Academies Press, 1994), Chap. 4, http://www.nap.edu/readingroom/books/planet_sci/contents/chap4d.html (accessed 26 January 2011).

moon.[60] This was the first time that a mission had been "purposefully crashed into a planet to protect another solar system body."[61]

Setting an Appropriate Planetary Protection Policy for Future Europa Exploration

For the last 40 years of the U.S. space program, Mars has been the focus of NASA's search for extraterrestrial life. NASA developed forward contamination prevention protocols that reflected its understanding of Mars's biological potential. Although these policies were derived from protocols developed for the 1970s Viking missions to Mars, NASA applied them to other solar system bodies as well. But given the distinctly different features of Europa (as well as other icy moons) from Martian characteristics, NASA eventually recognized that a different set of planetary protection criteria needed to be formulated. Toward this end, in the 1990s NASA's planetary protection officer requested the Space Studies Board (SSB) of the National Academies' National Research Council to analyze present policy and recommend necessary changes. Particular SSB tasks were to include the following:

- Assess the levels of spacecraft cleanliness and sterilization needed to protect Europa from contamination.
- Identify the necessary modifications to existing protocols.
- Identify the scientific investigations required to reduce the uncertainties in the above assessments.

In response to NASA's request, the SSB created the Task Group on the Forward Contamination of Europa, which began its study in April 1999 and drew its members from both academia and the aerospace industry. The core of the Task Group's work was to determine whether present planetary protection procedures, which drew heavily from those originally formulated for the 1970s Viking Mars lander missions, were appropriate for Europa missions.[62]

---

60. Meltzer, *Mission to Jupiter.*
61. Norine E. Noonan to Edward J. Weiler, 26 May 2004, *http://hurricanes.nasa.gov/strategy/ppac/letters/ppac0401.pdf#search=%22may%2026%2C%202004%20Dr.%20Edward%20J.%20Weiler%22* (accessed 31 August 2006).
62. SSB, National Research Council, *Preventing the Forward Contamination of Europa* (Washington, DC: National Academies Press, 2000).

Current contamination prevention techniques included cleaning spacecraft surfaces with isopropyl alcohol and sporicides and sterilizing by means of dry heat. These approaches were derived from Viking procedures, but the current protocol also included sterilization using hydrogen peroxide. In addition, NASA staff envisioned Europa's high-radiation environment contributing to sterilization.

The Task Group concluded that the present procedures were "satisfactory to meet the needs of future space missions to Europa,"[63] although they could be improved in certain areas. For instance, the Task Group recommended that the current spore-based culturing procedures for estimating bioloads be supplemented with screening tests for extremophiles, including radiation-resistant microbes.

The Task Group was also aware of the significant cost penalties and operational complexities of carrying out current procedures. More modern analytical methods such as those based on the polymerase chain reaction (PCR),[64] which is a molecular biology technique for replicating DNA, could cut down on time and add to sensitivity for identifying biocontamination. PCR employs DNA-copying enzymes and is typically used for finding genetic fingerprints and diagnosing infectious diseases.[65]

In addition, the Task Group recommended the following actions for improving planetary protection approaches:

- Analyze the ecology of cleanrooms and spacecraft assembly areas, focusing on extremophile microbes.
- Compare current bioload assay methodologies.
- Investigate the characteristics of desiccation- and radiation-resistant microbes that could contaminate spacecraft during assembly.
- Examine techniques for detecting autotroph contaminants—microbial organisms capable of synthesizing their own food using light or chemical energy.
- Research Europa's surface environment and, in particular, its hydrologic and tectonic cycles.

63. Ibid.
64. Ibid.
65. National Center for Human Genome Research, "Polymerase Chain Reaction—Xeroxing DNA," *New Tools for Tomorrow's Health Research* (1992), National Institutes of Health Web site, *http://www.accessexcellence.org/RC/AB/IE/PCR_Xeroxing_DNA.html* (accessed 26 January 2011).

The Task Group could not agree on the particular planetary protection standards appropriate for Europa, although most members believed that investigating the satellite could be key to understanding the origins of life and that current planetary protection protocols were not easily applied to Europan missions. Scientific understanding of terrestrial organisms' abilities to survive under extreme conditions was very limited, and so a conservative approach to forward contamination prevention was called for. The first lander mission to Europa needed to meet "the highest reasonable level of safeguard"[66] through severe bioload reduction in order to ensure a very low probability that a viable organism would be delivered to the satellite's subsurface ocean. One particular concern was that over geological time scales of 10 million to 100 million years, *any surface contamination* was likely to be carried into Europa's deep ice crust or into its underlying ocean.

NASA's 2005 document, *Planetary Protection Provisions for Robotic Extraterrestrial Missions* (NPR 8020.12C),[67] promulgates mandatory requirements for preventing forward and back contamination on a range of mission types, including those to Europa. Requirements for Europa flyby, orbiter, or lander missions were aimed at reducing the probability of inadvertent contamination of a Europan ocean to less than 1×10^{-4} per mission, and they addressed the following factors:

- Microbial burden at launch.
- Cruise survival for contaminating organisms.
- Organism survival in the radiation environment adjacent to Europa.
- Probability of landing on Europa.
- Mechanisms of transport to the Europan subsurface.
- Organism survival and proliferation before, during, and after subsurface transfer.

Preliminary models suggest that microbial reduction will likely be necessary for Europa orbiters as well as landers. Implementing this will require precision-cleaning of all parts before assembly, cleanroom technology for spacecraft assembly, careful monitoring of processes, and a thorough understanding of the bioload and its microbial diversity,

---

66. SSB, *Preventing the Forward Contamination of Europa*, p. 2.
67. NASA Science Mission Directorate, *Planetary Protection Provisions for Robotic Extraterrestrial Missions*, NPR 8020.12C, 27 April 2005.

including specific problematic species. NPR 8020.12C requirements are discussed further in Chapter 8, which addresses legal and ethical considerations of planetary protection.

Ganymede and Callisto—Abundant Water, But Are the Conditions Right for Life?

Ganymede, which is larger than Earth's Moon, Pluto, or Mercury, and is nearly as large as Mars, would be considered a planet if it orbited around the Sun rather than Jupiter. It has a crust that is probably a thick layer of water ice and that likely covers a salty, liquid water ocean beneath. The presence of a submerged ocean would be the best way to explain the magnetic readings taken by Galileo during flybys in May 2000 and earlier.

In addition, mineral characteristics and infrared reflectance data on Ganymede's surface suggest that salty water may have emerged in the past from below or melted at the surface. Reflectance spectra from Galileo's near-infrared mapping spectrometer (NIMS) provided evidence of possible hydrated materials similar to those found on Europa's surface—in particular, frozen magnesium sulfate brines and possibly sodium sulfate as well, both derived from a subsurface briny layer of fluid.[68]

According to Dave Stevenson, a planetary scientist at the California Institute of Technology, natural radioactivity in Ganymede's interior should provide enough heating to maintain a stable layer of liquid water between two layers of ice, about 150 to 200 kilometers (90 to 120 miles) below the surface. That would be a different situation than found in Europa, where cyclic deformation from the tidal effects of Jupiter's gravity could provide much of the internal heat.[69]

While life-forms may exist within a Ganymede ocean, less evidence has been found than on Europa of biologically important

---

68. Thomas B. McCord, Gary B. Hansen, and Charles A. Hibbitts, "Hydrated Salt Minerals on Ganymede's Surface: Evidence of an Ocean Below," *Science* 292 (25 May 2001): 1523–1525.
69. Guy Webster, "Solar System's Largest Moon Likely Has a Hidden Ocean," JPL Media Relations Office, 16 December 2000; Space Today Online, "Ganymede Is the Largest Solar System Moon," 2004, *http://www.spacetoday.org/SolSys/Jupiter/GanymedeInfo.html* (accessed 26 January 2011); Meltzer, *Mission to Jupiter.*

chemistry. For instance, magnesium salts and other materials significant for exobiology appear to be present on Europa in greater variety and abundance than on Ganymede, according to Ron Greeley, a professor at Arizona State University and head of the NASA Astrobiology Institute's (NAI's) Europa Focus Group.[70] But because of the possibility of life, similar planetary protection precautions will need to be taken for the exploration of Ganymede as for Europa.

Jupiter's second largest moon, Callisto, may also have a liquid water ocean hidden under its icy, cratered crust, based on interpretations of Galileo data. According to Margaret Kivelson, space physics professor at UCLA and principal investigator for Galileo's magnetometer instrument observations, this finding was quite a surprise. She commented that "Until now, we thought Callisto was a dead and boring moon, just a hunk of rock and ice."[71] But the analysis of Galileo data obtained during Callisto flybys in November 1996 and June and September 1997 indicated the presence of a magnetic field fluctuating in time with Jupiter's rotation. Thus, it was most likely a field induced by that of Jupiter.

The question was, where were the induced electric currents flowing to produce Callisto's magnetic field? The moon's icy surface was a poor electrical conductor, and its atmosphere, which lacked charged particles, was not a serious possibility for the medium in which a field could be induced. The best explanation for the observations was, once again, that Jupiter's magnetic field was inducing electric currents within a salty ocean beneath Callisto's icy surface.[72]

Thus, like Europa and Ganymede, exploration of Callisto would have to include planetary protection procedures taking into account the possibility of life within a salty water ocean. This is not to say that the existence of life on Callisto is a likely occurrence. The presence of a Callisto ocean is not sufficient for life to exist or have existed there. Energy to support such life is also needed. An ocean on Callisto would probably be heated almost completely by radioactive elements,

---

70. Astrobiology News, "Ganymede's Liquid Past," NAI Features Archive, NASA Astrobiology Institute, 9 March 2001, http://nai.nasa.gov/news_stories/news_detail.cfm?ID=190 (accessed 26 January 2011).

71. Douglas Isbell and Jane Platt, "Jupiter's Moon Callisto May Hide Salty Ocean," JPL news release 98-192, 21 October 1998, http://www2.jpl.nasa.gov/galileo/news32.html.

72. K. K. Khurana, M. G. Kivelson, D. J. Stevenson, G. Schubert, C. T. Russell, R. J. Walker, and C. Polanskey, "Induced Magnetic Fields as Evidence for Subsurface Oceans in Europa and Callisto," Nature 395 (22 October 1998): 777–780.

whereas Europa has tidal sources of energy as well, due to its greater proximity to Jupiter. Europa may be a more likely satellite to harbor life simply because it is the warmer of the two bodies.[73]

Enceladus: Saturn's Bright, Surprising Moon

The Cassini spacecraft's exploration of Saturn yielded unexpected data regarding its small satellite Enceladus. Space scientists once believed the moon to be too modest in size—only 314 miles in diameter—to be so active, but Cassini data have shown it to be "one of the most geologically dynamic objects in the solar system,"[74] exhibiting a watery, gaseous plume reminiscent of Yellowstone National Park's Old Faithful geyser (although *far* more diffuse),[75] a geothermally heated spot in its southern polar region (20°C hotter than expected), deep canyons, and evidence of thick flows and periodic resurfacing on its exterior. In fact, amorphous and crystalline ice deposits may be only hours to decades old in some places. Partial melting and even a subsurface ocean may also exist within the moon.

According to Robert H. Brown of the University of Arizona, leader of the Cassini visual and infrared mapping spectrometer team, "The kind of geophysical activity we see is quite likely being driven by liquid water below the surface."[76] Tidal heating related to the parent planet's gravitational field is a probable energy source for Enceladus's activity; radioactive sources may play a part as well. The moon has two of the important factors for sustaining life: liquid water and ample energy. It also has organic chemicals, which were detected in formations near the south pole.

Over time—and Enceladus has been around 4.5 billion years, just like Earth and the rest of the solar system—heating a cocktail of simple organics, water, and nitrogen could form some of the most basic

---

73. NASA, "Callisto Makes a Big Splash," Science@NASA Headline News, 23 October 1998, *http://science.nasa.gov/science-news/science-at-nasa/1998/ast22oct98_2/* (accessed 26 January 2011).

74. Jeffrey S. Kargel, "Enceladus: Cosmic Gymnast, Volatile Miniworld," *Science* 311 (10 March 2006): 1389–1391.

75. Mark Dahl, NASA Program Executive, interview by author, Washington, DC, 26 October 2007.

76. Lori Stiles, "Tiny Enceladus May Hold Ingredients of Life," *uanews.org*, University of Arizona, 5 September 2005, *http://uanews.org/?ArticleID=11622* (accessed 26 January 2011).

building blocks of life. Whether that's happened at Enceladus is not clear, but Enceladus, much like Jupiter's moon Europa and the planet Mars, now has to be a place where we eventually search for life.[77]

Planetary protection concerns for Enceladus flybys are similar to those for Europa flybys. Appropriate actions for future flyby missions will likely involve measures such as trajectory biasing for preventing accidental impacts, as well as cleanliness requirements imposed to some level on the spacecraft, in order to lower the probability of contaminating the moon should an accidental collision occur.[78] A mission planning to bring samples from Enceladus's jets back to Earth would of course require additional measures to prevent back contamination, such as quarantining the collected material.

Titan—Organics and Very Cold Temperatures

NASA launched the Cassini spacecraft in 1997, carrying the Huygens probe that would be sent to Titan. Project scientists thought at that time, and still believe, that life is unlikely to exist on the cold moon, whose temperatures have been measured at −160°C (−256°F). Near the atmosphere's stratosphere-troposphere boundary, in fact, temperatures drop to less than −190°C (−310°F).[79] Space scientists often describe Titan as a window on the early Earth, demonstrating what our planet may have looked like before life emerged. But the "harsh chemistry

---

77. Ibid.
78. Margaret S. Race, "Planetary Protection & Enceladus Missions: Don't Leave Home Without It" (presentation to Enceladus Focus Group Meeting, JPL, October 2006), https://encfg. ciclops.org/files/RACE_Oct2006_mtg.ppt.
79. Anne M. Stark, "Cassini-Huygens Observations Show How Titan Compares With the Earth," Lawrence Livermore National Laboratory Public Affairs news release no. NR-05-05-04, 12 May 2005, http://www.llnl.gov/pao/news/news_releases/2005/NR-05-05-04.html; F. M. Flasar, R. K. Achterberg, B. J. Conrath, P. J. Gierasch, V. G. Kunde, C. A. Nixon, G. L. Bjoraker, D. E. Jennings, P. N. Romani, A. A. Simon-Miller, B. Bézard, A. Coustenis, P. G. J. Irwin, N. A. Teanby, J. Brasunas, J. C. Pearl, M. E. Segura, R. C. Carlson, A. Mamoutkine, P. J. Schinder, A. Barucci, R. Courtin, T. Fouchet, D. Gautier, E. Lellouch, A. Marten, R. Prangé, S. Vinatier, D. F. Strobel, S. B. Calcutt, P. L. Read, F. W. Taylor, N. Bowles, R. E. Samuelson, G. S. Orton, L. J. Spilker, T. C. Owen, J. R. Spencer, M. R. Showalter, C. Ferrari, M. M. Abbas, F. Raulin, S. Edgington, P. Ade, and E. H. Wishnow, "Titan's Atmospheric Temperatures, Winds, and Composition," Science 308 (13 May 2005): 975–978.

and frigid temperatures of the moon make many scientists skeptical about the current potential for life on Titan."[80]

Saturn's largest satellite has ample organic material in its thick atmosphere. Methane (CH_4) gas is its second-most abundant atmospheric constituent next to nitrogen, but it has long puzzled scientists how the methane is generated. On Earth, methane is produced by life processes, the degradation of organic material, or geologic processes. Cassini mission scientists concluded that if life does exist on Titan, it cannot be manufacturing all the methane that exists. Geologic activity in Titan's interior rather than biological activity is the most likely source of the chemical.

Nevertheless, if Titan does prove to harbor life, it would most likely be in the interior of the satellite in the presence of liquid water. According to Cassini scientist Francois Raulin of the University of Paris, "You have there, in the interior, what you need to make good prebiotic chemistry."[81] Mission scientists believe that, in fact, Titan may become "a laboratory for studying the organic chemistry that preceded life and provided the building blocks for life on Earth,"[82] even though life may never be found on the satellite.

Planetary protection considerations. Because of the reasons discussed above, the Planetary Protection Officer (PPO), in consultation with the Space Studies Board, decided that the Cassini mission had a low risk of adversely contaminating Titan. The PPO did not deem any spacecraft sterilization procedures necessary, although COSPAR rules did require the spacecraft to be assembled in a class 100,000 cleanroom (that is, with less than 100,000 particles per unit of volume).[83] Cassini was classified by the PPO as Category II, indicating that any contamination on the spacecraft had only a remote chance of jeopardizing future exploration that focused on chemical evolution and the origin of life. The planetary protection requirements included only

---

80. Leslie Mullen, "Titan: Passport to the Early Earth?" *Astrobiology Magazine* (1 December 2005), *http://www.astrobio.net/exclusive/1790/titan-passport-to-the-early-earth* (accessed 26 January 2011).
81. Ibid.
82. Carolina Martinez, "Organic Materials Spotted High Above Titan's Surface," JPL news release 2005-062, 25 April 2005, *http://saturn.jpl.nasa.gov/news/press-release-details.cfm?newsID=561* (accessed 9 September 2006).
83. European Space Agency, "No Bugs, Please, This Is a Clean Planet," 30 July 2002, *http://www.esa.int/esaCP/ESAUB676K3D_Life_2.html* (accessed 26 January 2011); Perry Stabekis, telephone interview by author, 7 September 2007.

simple documentation, such as the preparation of a short planetary protection plan, and assurances that the spacecraft was not likely to accidentally impact the moon.[84]

●●●

Do We Need To Quarantine Samples From Asteroids, Meteoroids, Comets, or Interplanetary Particles?

The space environments surrounding asteroids, meteoroids, comets, and interplanetary particles, which are referred to as "minor bodies" in this section, impose severe limitations on their abilities to harbor viable microbes, especially those exhibiting metabolic activity. Minor body environments differ in several important aspects from most planets and some of their larger satellites (such as Saturn's Titan). Minor bodies cannot sustain atmospheres due to their weak gravitational fields. This eliminates the presence of gaseous metabolites such as water vapor; hydrogen sulfide; carbon monoxide and dioxide; oxides of nitrogen, ammonia, or methane; and others that are exploited for biotic use by Earth organisms.

The lack of an atmospheric blanket on minor bodies makes the existence of life on them unlikely in another way as well. Atmospheres moderate temperature swings and create a gaseous greenhouse effect to help warm cold surfaces. Liquid water, which is needed for life as we know it, cannot exist on the surface of a body without an atmosphere. The water will quickly sublimate or freeze. The ultrahigh vacuum on surfaces without atmospheres—less than 10^{-15} torr—is also challenging to life because it can cause severe desiccation.[85] Furthermore, ionizing UV radiation can reach the surfaces of atmosphere-lacking bodies unimpeded, and this has a sterilizing effect. Above our planet, we have an ozone layer that efficiently absorbs UV radiation and provides

---

84. JPL, "Frequently Asked Questions—Huygens Probe," Cassini-Huygens Web site, *http://saturn.jpl.nasa.gov/faq/huygens.cfm* (last updated 6 April 2005, accessed 9 September 2006).
85. A torr is a unit of pressure equal to that exerted by 1 mm of mercury, or 1/760 of the pressure of our atmosphere at sea level.

important protection for the life-forms below, but minor bodies do not have such an atmospheric filter.[86]

Most minor bodies have diameters less than 100 kilometers and thus are too little to maintain internal heat sources over geologic time scales. As a result, liquid water cores are not probable except within the extremely rare, giant asteroids and comets greater than 300 kilometers in diameter.[87] In spite of these challenges, however, some of the solar system's minor bodies may possibly harbor life, and planetary protection considerations need to be taken into account when exploring them.

Asteroids

Asteroids are typically irregularly shaped bodies composed of rock and metal that travel around the Sun, chiefly between the orbits of Mars and Jupiter (although some pass closer to the Sun than Earth and others much farther from the Sun than Jupiter). Many asteroids are devoid of organic substances and thus do not have one of the basic foundations that scientists believe is necessary for harboring life. While the largest known asteroid—Ceres—is approximately as wide as the state of Texas, and several dozen asteroids are a few hundred miles in diameter (large enough to have conceivably retained liquid water cores over geologic time periods), most have far humbler dimensions. Hundreds of thousands are only a small fraction of a mile in size, and millions are even tinier pieces of rock and metal.[88]

Because of the ionizing radiation that is pervasive in space, Clark et al. expected the top meter or so of asteroid surface to be sterile. For this reason as well as because of the effects of high vacuum and thermal disruption (which, among other impacts, cause outgassing of water vapor from regoliths), Clark et al. suggested that sample return missions excavating asteroidal material only to shallow depths should not require back contamination protection such as quarantining.[89]

---

86. Clark et al., "Survival of Life," 521–545.
87. C. Chyba and G. McDonald, *Annual Review of Earth and Planetary Science* 23 (1995): 215, as reported in Clark et al., "Survival of Life," 523.
88. Courtney Seligman, "Asteroid Size and Mass Distribution," *cseligman.com*, Online Astronomy Text Web site, *http://cseligman.com/text/asteroids/sizedistribution.htm* (accessed 14 July 2006).
89. Clark et al., "Survival of Life," 533, 540, 542.

Grimm and McSween, however, performed calculations indicating that for asteroids 2 astronomical units (AUs)[90] from the Sun with compacted, low-porosity material at their surfaces, water ice (and thus life) could have persisted for the age of the solar system at depths greater than 1 to 40 meters, depending on pore size. For asteroids at distances greater than 3 AUs, ice may exist as close as several tenths of a meter from the surface.[91] Furthermore, the largest population of asteroids is of a distance from the Sun such that their temperatures range from –113° to –23°C (–171° to –9°F),[92] an interval in which the rates of chemical reactions able to damage living cells are quite slow.

The Hayabusa Mission. The Japanese space mission Hayabusa (formerly named MUSES-C), which translates to "peregrine falcon," was a robotic venture led by the Japan Aerospace Exploration Agency (JAXA) for the purpose of returning a sample of material from the near-Earth asteroid 25143 Itokawa. The mission plan was to fire a projectile from the spacecraft into the asteroid and capture some of the ejecta. Hayabusa launched on 9 May 2003 and reached Itokawa in mid-September 2005. After surveying the asteroid surface from a distance of about 20 kilometers, the spacecraft moved closer and then, like the bird it was named for, "swooped down to perch for about 30 minutes on its surface."[93]

Hayabusa performed two such touchdowns on the smooth terrain of the Muses Sea region of Itokawa, carried out on 19 and 25 November 2005.[94] The craft found Itokawa to be a mass of rubble—loosely packed rocks held together by their tenuous gravity. This is quite different than many other asteroids—for instance, Eros, which

---

90. An AU is a unit of distance equal to the average distance between Earth and the Sun and is equal to about 150 million kilometers (93 million miles). An AU is a useful unit for expressing distances in our solar system.

91. R. E. Grimm and H. Y. McSween, Jr., "Water and the Thermal Evolution of Carbonaceous Chondrite Parent Bodies," *Icarus* 82 (1989): 244–280, as reported in Clark et al., "Survival of Life," 533.

92. R. E. Grimm, H. Y. McSween, Jr., "Water and the Thermal Evolution of Carbonaceous Chondrite Parent Bodies," *Icarus* 82 (1989): 244 (temperature estimates given in kelvins), as reported in Clark et al., "Survival of Life," 521–545.

93. Joanne Baker, "The Falcon Has Landed," *Science* 312 (June 2006): 1327.

94. Hajime Yano, T. Kubota, H. Miyamoto, T. Okada, D. Scheeres, Y. Takagi, K. Yoshida, M. Abe, S. Abe, O. Barnouin-Jha, A. Fujiwara, S. Hasegawa, T. Hashimoto, M. Ishiguro, M. Kato, J. Kawaguchi, T. Mukai, J. Saito, S. Sasaki, and M. Yoshikawa, "Touchdown of the Hayabusa Spacecraft at the Muses Sea on Itokawa," *Science* 312 (June 2006): 1350–1353.

is solid. Itokawa's shape suggested that two unequally sized lumps were squashed together in an earlier collision. Surface features gave clues of an early collisional breakup of a parent asteroid, followed by reagglomeration into the rubble-pile observed today.[95]

After mission controllers sent commands to Hayabusa to fire its tiny projectiles at the surface and then sample the resulting spray, the mission staff was initially unable to determine whether the projectiles were released. Subsequent telemetry data indicated that release of the projectiles was unlikely, although bits of dust or pebbles may still have been collected by the craft's sample retrieval system. Earth return maneuvers were threatened due to problems with the spacecraft's engines. JAXA did identify a possible recovery approach, however, that would combine functionalities of two different engines. If it completed its mission successfully, Hayabusa would be the first spacecraft to return an asteroid sample to Earth.[96]

Hayabusa planetary protection considerations. The projectiles that Hayabusa was programmed to fire into the surface of Itokawa were expected to penetrate the asteroid's surface no more than several centimeters, according to estimates reported in Clark et al. that took into account expected surface strengths. Thus, because of ionizing radiation, high vacuum, and thermal disruption, these samples would likely have been sterile and would not have needed quarantining.[95] Nevertheless, Japan requested NASA to review the acceptability of unrestricted Earth return for Hayabusa. NASA studied the mission and agreed that unrestricted Earth return was indeed appropriate. NASA's findings were in turn presented and discussed at a COSPAR workshop, which concurred with NASA's findings.[97]

---

95. A. Fujiwara, J. Kawaguchi, D. K. Yeomans, M. Abe, T. Mukai, T. Okada, J. Saito, H. Yano, M. Yoshikawa, D. J. Scheeres, O. Barnouin-Jha, A. F. Cheng, H. Demura, R. W. Gaskell, N. Hirata, H. Ikeda, T. Kominato, H. Miyamoto, A. M. Nakamura, R. Nakamura, S. Sasaki, and K. Uesugi, "The Rubble-Pile Asteroid Itokawa as Observed by Hayabusa," *Science* 312 (June 2006): 1330–1334.

96. JAXA, "Asteroid Explorer 'HAYABUSA' Ion Engine Anomaly," *http://www.jaxa.jp/press/2009/11/20091109_hayabusa_e.html* (accessed 9 November 2009); JAXA, "Restoration of Asteroid Explorer, HAYABUSA's Return Cruise," *http://www.jaxa.jp/press/2009/11/20091119_hayabusa_e.html* (accessed 19 November 2009); Nancy Atkinson, "Hard-Luck Hayabusa In More Trouble," Universe Today, *http://www.universetoday.com/2009/11/10/hard-luck-hayabusa-in-more-trouble/* (accessed 10 November 2009).

97. UN Office for Outer Space Affairs, *Highlights in Space 2006* (New York: UN E.07.I.92006, 2007), p. 207, *http://www.unoosa.org/pdf/publications/st_space_34E.pdf* (accessed 26 January 2011); Stabekis, telephone interview by author, 7 September 2007.

Although the above reviews were very thorough, there is a conceivable way that the Itokawa asteroid could harbor life. Orgel pointed out that in the past, water may have existed within some types of asteroids. Large pockets of water ice within rubblized asteroids (such as Itokawa) possibly exist today as well. If such pockets of ice do exist, they would have the potential to attenuate the natural radiation field and prevent it from killing any microorganisms that might be present, thus permitting the survival of dormant life within the ice.[98]

In June 2010, Hayabusa successfully delivered its payload back to Earth, dropping a return capsule over Australia. The controlling regulatory agency was Biosecurity Australia, a part of the Australian Department of Agriculture, Fisheries and Forestry. To deal with the sample return, this agency issued a Quarantine Review of the MUSES-C Project in July 2002. This review discussed the importing of surface samples, noting that even though the back contamination risk to Earth might be extremely low to negligible, environmental exposure could conceivably occur.[99] Biosecurity Australia thus recommended that "it is prudent to consider risk management procedures for return samples from small solar system bodies [and] that, on re-entry, the returned sample and all associated equipment potentially contaminated with the sample be immediately placed into secure impervious containers, transported in a safe manner and exported from Australia"[100] to the laboratories that will analyze the material.

According to JAXA, scanning electron microscope analyses of returned samples confirmed that they were definitely from the asteroid Itokawa.[101]

Possible future studies of asteroids. Important data could be generated from spacecraft studies conducted to better assess heterogeneities, including ice pockets, in rubblized meteorites such as Itokawa. Studies could include spectral reflectance mapping of the bodies as

---

98. Orgel, *Evaluating the Biological Potential*, pp. 47, 49.

99. John Matson, "Hayabusa Probe Succeeded in Returning Asteroid Dust to Earth," *Scientific American* (16 November 2010).

100. Biosecurity Australia—Australian Department of Agriculture, Fisheries and Forestry, *Quarantine Review of the MUSES-C Project*, July 2002, p. 6, http://www.affa.gov.au/corporate_docs/publications/pdf/market_access/biosecurity/animal/2002/2002-28a.pdf.

101. Japan Aerospace Exploration Agency (JAXA), "Identification of Origin of Particles Brought Back by Hayabusa," http://www.jaxa.jp/press/2010/11/20101116_hayabusa_e.html, 16 November 2010 (accessed 17 February 2010).

well as x-ray and gamma-ray mapping. Determining the characteristics of an asteroid's heterogeneities would enable estimates of the ranges of exposure to radiation to which the different parts of the body would be subjected to and could thus reduce current uncertainties regarding the need for quarantining return samples. These uncertainties will also be narrowed when we learn more about two key questions:

- To what degree do existing meteorites represent the range of materials in the asteroids believed to be their parent bodies?
- To what level of certainty can we estimate the resemblance of a target asteroid to typical asteroids of its type?[102]

Planetary Protection Protocols for Returned Asteroid Samples

The SSB's 1998 review of the planetary protection aspects of asteroid sample return missions[103] recommended an approach for handling returned asteroid samples that included several aspects. At that time, no special containment and handling was recommended for asteroid types that space scientists believed to have an extremely low potential for harboring life. Included in this group were many C-type (carbonaceous) asteroids as well as S-type undifferentiated metamorphosed asteroids (those with stony, silicaceous compositions). Together, these two types comprise over 90 percent of known asteroids. Scientists believed that except for "possible localized volumes of water ice in C-types,"[104] the interiors of both asteroid types would have received sterilizing quantities of radiation during the last 4.5 billion years from the decay of naturally occurring radioactive materials.

Much less is known about P- and D-type asteroids (reddish, poorly reflective bodies found in the outer asteroid belt and beyond), and thus the SSB urged caution in handling samples returned from them. The SSB recommendations were to implement strict containment procedures since they could not determine the chances of living entities

---

102. Orgel, *Evaluating the Biological Potential*, p. 49.
103. Ibid., pp. 77–79; C. C. Allen and M. M. Lindstrom, "Near-Earth Asteroid Sample Curation," Near-Earth Asteroid Sample Return Workshop, Lunar and Planetary Institute (LPI), Houston, TX, 11–12 December 2000, pp. 1–2, *http://www.lpi.usra.edu/meetings/asteroid2000/pdf/8027. pdf#search=%22meteorite%20planetary%20protection%22* (accessed 26 January 2011).
104. Orgel, *Evaluating the Biological Potential*, p. 77.

being present within the bodies. Handling these asteroids would require a dedicated asteroid sample curation laboratory with characteristics that include the following:

- Cleanroom environment of class 1,000 or better.
- Subfreezing storage and processing capabilities.
- Dedicated processing cabinets for samples from each parent body, with positive-pressure nitrogen atmospheres.
- Continuous monitoring of inorganic, organic, and biological contamination.[105]

Meteoroids and Meteorites

Meteoroids—solid, rocky masses moving through space that are smaller than asteroids but at least the size of a speck of dust—get heated to incandescence by friction when they encounter Earth's atmosphere and generate the bright trails or streaks that we call *meteors*. Some large meteoroids do not burn up in the atmosphere, but reach the surface of our planet as *meteorites*. Interest in meteoroids from a planetary protection perspective stems from findings that, according to some investigators, suggest the presence of microfossils within the bodies. In particular, McKay et al.[106] presented arguments that the carbonate globules within the meteorite ALH84001 were likely formed at temperatures consistent with life and are related to Martian microorganisms. McKay and his colleagues observed elliptical, rope-like, and tubular structures in fractures in the carbonate mineral globules resembling living and fossilized terrestrial bacteria that grew inside volcanic rock similar to that found in the meteorite. ALH84001 is believed to have originated on Mars largely because it resembles another meteorite, EETA79001, which

---

105. C. C. Allen and M. M. Lindstrom, "Near-Earth Asteroid," p. 2.
106. D. S. McKay, E. K., Gibson, Jr., K. L. Thomas-Keprta, H. Vali, C. S. Romanek, S. J. Clemett, X. D. F. Chillier, C. R. Maechling, and R. N. Zare, "Search for Past Life on Mars: Possible Relic Biogenic Activity in Martian Meteorite ALH84001," *Science* 273 (1996): 924–930; Allan H. Treiman, "Current State of Controversy About Traces of Ancient Martian Life in Meteorite ALH84001," Mars Sample Handling Protocol Workshop—Summaries of Key Documents, February 2000, an attachment to John D. Rummel to Steven Dick, 9 March 2000.

has a composition matching that of the current Martian atmosphere as measured by the Viking landers.

An examination of meteorite ALH84001, a softball-sized igneous rock weighing 1.9 kilograms (4.2 pounds) and found in the Allan Hills region of Antarctica, revealed the following in its carbonate globules:

- Hydrocarbons that were the same as broken-down products of dead microorganisms on Earth.
- Mineral phases consistent with byproducts of bacterial activity.
- Evidence of possible microfossils of primitive bacteria.

Based on age-dating of the igneous component of ALH84001, a scenario was proposed that the original igneous rock from which it came solidified within Mars 4.5 billion years ago, or 100 million years after the formation of the planet. Between 3.6 and 4 billion years ago the rock was fractured, perhaps by meteorite impacts. Water then permeated its cracks, depositing carbonate minerals and permitting primitive bacteria to live in the fractures. About 3.6 billion years ago, the bacteria and their byproducts became fossilized in the fractures, and then 16 million years ago, a large meteorite struck Mars, dislodging a section of this rock and ejecting it into space. Thirteen thousand years ago, the meteorite crashed into Antarctica, where it was discovered in 1984.[107]

While the possibility of fossilized extraterrestrial life in the meteorite was, to say the least, intriguing, alternate theories arose holding that nonbiological processes and contamination could explain the findings. In addition, a 1996 letter to the journal *Nature* by Harvey and McSween presented evidence for why life processes could not have formed the carbonate globules. The most likely process of forming the carbonates in the meteorite, according to the researchers, relied on reactions involving hot carbon dioxide–rich fluids whose temperatures exceeded 650°C (1,200°F). If the carbonate globules in the meteorite were indeed this hot, it is quite improbable that their formation could have been due to a biotic process.[108]

---

107. Allan H. Treiman, "Fossil Life in ALH 84001?" Lunar and Planetary Institute Web site, 21 August 1996, *http://www.lpi.usra.edu/lpi/meteorites/life.html* (accessed 13 July 2006).
108. Ralph P. Harvey and Harry Y. McSween, Jr., "A Possible High-Temperature Origin for the Carbonates in the Martian Meteorite ALH84001," *Nature* 382 (4 July 1996): 49–51.

One line of evidence for bacterial life in the meteorite that was difficult to refute focused on microscopic crystals in the meteorite of a mineral called magnetite. According to NASA scientists, the magnetite crystals were of a structure, size, chemical purity, and distinctive shape that only bacteria could have manufactured, rather than an inorganic process. In particular, researchers found that magnetite crystals embedded in the meteorite were ordered in long chains, which they believed had to have been formed by once-living organisms.

These results were reported in the 27 February 2001 proceedings of the National Academy of Sciences.[109] But this claim, too, was assailed, this time by new data and analyses of an Arizona State University team that included Peter Buseck, professor of geological sciences and biochemistry, and Martha McCartney, a scientist at the university's Center for Solid State Science. They presented evidence that the match between the meteoritic crystals and those in bacteria was "inadequate to support the inference of former life on Mars."[110] In particular, tomographic and holographic methods employing a transmission electron microscope showed that the crystals formed by terrestrial bacteria did not uniquely match those in the Martian meteorite.[111]

While the presence of biotic traces in meteorites is still in question, some of the bodies do contain strong evidence, in the form of hydrated minerals, for the past presence of liquid water. "Hydrated" refers to minerals whose crystals are chemically bound to hydroxyl groups (OH) or water molecules. Such hydrated minerals were likely formed in the presence of aqueous fluids.

One critical question related to the existence of extraterrestrial life is whether these hydrated meteorites, which include most types of

---

109. *Space Daily*, "Evidence Of Martian Life In Meteorite Dealt Critical Blow," and "Scientists Find Evidence of Ancient Microbial Life on Mars," 26 February 2001, http://www.spacedaily.com/news/mars-life-01j.html (accessed 13 July 2006).

110. Peter Buseck, Martha McCartney, Rafal Dunin-Borkowski, Paul Midgley, Matthew Weyland, Bertrand Devouard, Richard Frankel, and Mihály Pósfai, "Magnetite Morphology and Life on Mars," *Proceedings of the National Academy of Sciences* 98(24) (20 November 2001): 13490–13495.

111. Tomography employs computer analysis of a series of cross-sectional scans typically taken with x rays to examine a solid object. Holography is a method of producing a three-dimensional image of an object from the interference pattern formed by two coherent beams of light. The transmission electron microscope is a tool that operates similarly to a normal light microscope but uses electron beams instead of light.

carbonaceous chondrites,[112] were altered in an *extraterrestrial location* (rather than on Earth) by the presence of water. Evidence that this was indeed the case includes the following:[113]

- Some of the hydrated meteorites were observed falling to Earth and were recovered too soon thereafter for terrestrial hydration reactions to occur.
- Radioisotope dating of the meteorites' carbonate structures indicated that their formation occurred during a time when Earth was too hot to sustain liquid water (suggesting that the formation had to have occurred in another location).
- Oxygen isotope compositions in hydrated chondrite meteorites are distinct from those in terrestrial materials.[114]
- The ratios of two forms of iron, Fe^{2+}/Fe^{3+}, in the meteorites' secondary minerals (those resulting from the decomposition of another mineral) are noticeably higher than in secondary minerals formed on Earth's surface. Since processes that produce the secondary minerals involve aqueous alteration, this may indicate that aqueous impacts on the meteorites were of nonterrestrial origin.[115]

If direct evidence for meteoritic transfer of living organisms between planets is discovered, questions must be asked as to the ecological impacts that have accrued from this phenomenon, as well as whether alterations in our planetary protection policy are needed.[116] For instance, if we establish that viable microorganisms from Mars have been transported to our planet by means of meteorites

---

112. Carbonaceous chondrites are a type of meteorite that is rich in organic material. Virtual Planetary Laboratory (VPL), "Glossary," VPL home page, Caltech Web site, *http://vpl.ipac.caltech.edu/epo/glossary.html* (accessed 30 August 2006).

113. David Jewitt, Lisa Chizmadia, Robert Grimm, and Dina Prialnik, "Water in the Small Bodies of the Solar System," University of Hawaii Institute for Astronomy preprint IfA-06-026, February 2006, for publication in *Protostars & Planets* V, *http://www.ifa.hawaii.edu/faculty/jewitt/papers/2006/JCGP06.pdf*.

114. Ibid.; Robert N. Clayton, Naoki Onuma, and Toshiko K. Mayeda, "A Classification of Meteorites Based on Oxygen Isotopes," *Earth and Planetary Science Letters* 30(1) (April 1976): 10–18.

115. Jewitt et al., "Water in the Small Bodies"; Michael Zolensky, Ruth Barrett, and Lauren Browning, "Mineralogy and Composition of Matrix and Chondrule Rims in Carbonaceous Chondrites," *Geochimica et Cosmochimica Acta* 57(13) (July 1993): 3123–3148.

116. SSB, National Research Council, "Scientific Investigations That Could Reduce Uncertainty," in *Mars Sample Return: Issues and Recommendations* (Washington, DC: National Academies Press, 1997), Chap. 5.

for many years without apparent harm, how should that alter, if at all, our quarantine policy regarding Martian return samples? It might be argued that even if meteorites did regularly bring Martian organisms to our planet without harm to us, it certainly would not prove that all microbes from Mars would be benign. The best policy in such a case might still be to maintain "a rational program of containment and bio-hazard testing [that] will not only allow us to determine the safety of a returned Mars sample, but . . . will also allow us the best chance of detecting Mars life, if it exists, when a sample is brought to Earth."[117]

The Relationships Between Meteorite and Asteroid Types

Understanding the relationships between meteorite types and asteroid types adds to the knowledge of both types of bodies, and it can shed light on when planetary protection measures may be required in investigating them. This is a currently evolving, multidisciplinary field that draws data from the following:
- Ground-based and space-based remote sensing studies.
- Laboratory analyses of meteorite material.
- Laboratory simulations of space-weathering and other processes that make it difficult to accurately interpret reflectance spectra data.
- Theoretical modeling of collisions and other dynamic processes that liberate meteorites from their parent bodies.[118]

The better we understand the relationships between meteorite and asteroid, the more we can know about a parent asteroid's structure, composition, and potential for harboring life by studying the meteorites generated from it that find their way to Earth.

Comets

Comets contain significant amounts of water ice that scientists believe never had a chance to melt on a large scale. This is because,

---

117. John D. Rummel, "A Case for Caution," *Planetary Report* (November/December 2000).
118. Ibid.

since their creation 4.6 billion years ago in the protoplanetary disc,[119] the bulk of comets' masses have rarely gotten warmer than about −170°C. Their surface layers, however, may have undergone months-long heating during the comets' closest approaches to the Sun to temperatures of the order of 27°C, causing some volatiles to sublimate and escape as vapor.[120] In particular, water sublimates from a comet when it approaches to within about 5 to 6 astronomical units from the Sun (an astronomical unit, or AU, is the mean distance between Earth and Sun). This sublimation leads to formation of the distinctive cometary tail as well as the atmosphere of gas, or coma, surrounding the comet's nucleus.[121]

Cosmic-ray exposure of comets whose orbits take them out of the heliosphere (the region of space through which our Sun's solar wind extends) has been sufficient, according to space scientists, to "destroy any preexisting life in the outermost tens of meters"[122] of the bodies, and the temperatures in that distant region are so low as to preclude the formation of any life. Although some cometary nuclei contain large quantities of organic molecules, this is explainable in terms of abiotic processes. Such molecules have also been observed in the interstellar medium, and these have definitely been generated by abiotic means.

The SSB Task Group on Sample Return from Small Solar System Bodies that was discussed earlier in the chapter considered it extremely unlikely that life could exist on or near the surfaces of comets but could not conclude that the probability was zero. Nevertheless, in its study on the biological potential of small solar system bodies, SSB did not recommend any special containment or handling procedures for comet return samples beyond what was required for the needs of scientific analyses.[123]

The Stardust Mission. NASA's Stardust spacecraft flew within 150 miles of the comet Wild 2 (pronounced "vilt two") in January 2004, trapping particles from the comet in a matrix of low-density silica glass called aerogel and returning them to Earth on 15 January 2006 (see

---

119. The protoplanetary disc was the rotating disk of dense gas surrounding our young, newly formed Sun, from which our solar system's planets arose.
120. Orgel, *Evaluating the Biological Potential*, p. 78.
121. Jewitt et al., "Water in Small Bodies," p. 8.
122. Orgel, *Evaluating the Biological Potential*, pp. 78–79.
123. Ibid.

Figure 6.2 An artist's concept of Stardust sample return capsule (SRC) parachuting down to Earth, bringing samples of comet particles and interstellar dust, including recently discovered dust streaming into the solar system from the direction of Sagittarius.

Figure 6.3 The Stardust sample return capsule was transported by helicopter from its landing site at the U.S. Air Force Utah Test and Training Range. This image shows the SRC inside a protective covering.

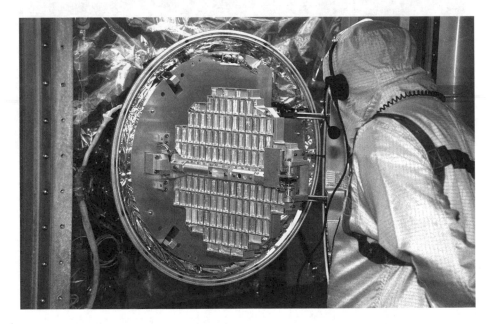

Figure 6.4 Mission staff prepare to remove an aerogel particle collection grid from the Stardust SRC.

Figure 6.5 A closeup view of a cometary impact (center) into aerogel, observed at JSC hours after the Stardust SRC was delivered from the spacecraft's landing site in Utah.

Figure 6.6 An artist's concept of the Genesis spacecraft in its collection mode, opened up to catch and store samples of solar wind particles. The two rectangular solar panels provided electrical energy for spacecraft functions.

figures 6.2 and 6.3). Stardust also collected interstellar dust grains. The aerogel matrices, depicted in figures 6.4 and 6.5, had the capability of slowing down the particles upon impact without destroying them. Most of the thousands of samples returned were smaller than the width of a hair, with only about two dozen larger particles visible to the naked eye. The aerogel matrices were scanned, mapped, dissected, and prepared for distribution in a dedicated class 10 cleanroom[124] at Johnson Space Center (JSC).[125]

NASA deemed Stardust to be a Category V, Unrestricted Earth Return mission. Material brought back to Earth on Stardust did not have to meet any planetary protection requirements because NASA had already decided that it posed no biological risk. Even if the material *did* initially contain microbes which could back contaminate Earth (an extremely small possibility), Stardust's procedure for collecting samples would almost certainly have killed those microbes. The particles collected were traveling at 22,000 kilometers (14,000 miles) per hour—roughly six times the speed of a bullet—when they hit the fluffy aerogel material of Stardust's collector. As a result, the particles would have been heated to temperatures severe enough to biologically sterilize them.[126]

Approximately 40,000 tons of dust particles from extraterrestrial bodies impact our planet every year, and a large fraction of this (one billion particles per second) is cometary material. Comet particles have been bombarding Earth for more than the 3 billion years in which life has existed on this planet. These facts argue strongly for their minimal impact on terrestrial life-forms and support SSB's

---

124. Class 10 indicates that there are fewer than 10 particles per cubic foot within the cleanroom.
125. Susan Watanabe, ed., and Brian Dunbar (NASA official), "Stardust Samples Show Evidence of Fire, Ice," http://www.nasa.gov/mission_pages/stardust/main/index.html (last updated 15 May 2006, accessed 31 August 2006); Allen and Lindstrom, "Near-Earth Asteroid," p. 1; Randy Russell, "The Stardust Mission to a Comet," Windows to the Universe Web site of the University Corporation for Atmospheric Research (UCAR), http://www.windows.ucar.edu/tour/link=/space_missions/comets/stardust.html&edu=elem (last updated 21 March 2006, accessed 31 August 2006).
126. JPL, "Comets & the Question of Life," 12 September 2005, http://stardust.jpl.nasa.gov/science/life.html (accessed 31 August 2006); Glennda Chui, "Chasing Comets: Stardust Mission," San Jose Mercury News, 11 January 2006, http://www.lbl.gov/today/2006/Jan/11-Wed/01-11-2006.html (accessed 26 January 2011).

recommendation for no special containment requirements for sample return missions from comets.

Interplanetary Dust Particles

Interplanetary dust particles, or IDPs, are some of the solar system's most primitive materials. IDPs are extremely small grains, typically 0.001 cm in diameter (about one-tenth the thickness of a human hair) and are a major component of the solar wind. Like meteorites, they fall on all parts of Earth, but their relation to meteorites is not completely understood. Some IDPs have compositions that match known meteorite classes. A 1991 study by Bradley and Brownlee, for instance, identified in an IDP the metallic mineral tochilinite, which is also found in carbonaceous chondrite meteorites. But most IDPs differ from meteorites in crystal chemistry and petrography and may be more primitive. IDP parent bodies probably include asteroids or comets, while some scientists have suggested that a percentage of IDPs come from other solar systems. Tomeoka and Buseck have found evidence that they are possible residue from protostellar clouds.[127]

One astrophysics theory suggests that low-energy protons in the solar wind could have interacted with IDPs containing the carbon compound anthracene to produce quinones. These are a type of molecule that exists in most living organisms. According to Lubomir Gabla and colleagues of Jagiellonian University in Poland, quinones may have been the precursors of life on Earth, carried to our planet by interplanetary dust. Tuleta et al. commented in this regard that the

---

127. Tom Bernatowicz, "Interplanetary Dust Particles," Laboratory for Space Sciences Web site, Washington University in St. Louis, *http://presolar.wustl.edu/work/idp.html* (accessed 31 August 2006); P. H. Benoit, "Interplanetary Dust Particles (IDPs)," Meteoritical Society Web site, University of Arkansas, *http://www.uark.edu/campus-resources/metsoc/idp.htm* (accessed 31 August 2006); J. P. Bradley and D. E. Brownlee, "An Interplanetary Dust Particle Linked Directly to Type CM Meteorites and an Asteroidal Origin," *Science* 251 (1 February 1991): 549; J. P. Bradley and D. E. Brownlee, "Cometary Particles: Thin Sectioning and Electron Beam Analysis," *Science* 231 (28 March 1986): 1542–1544; Kazushige Tomeoka and Peter R. Buseck, "A Carbonate-Rich, Hydrated, Interplanetary Dust Particle: Possible Residue from Protostellar Clouds," *Science* 231 (28 March 1986): 1544; Stanford University News Service, "Scientists Make First Measurements of Organic Molecules on Interplanetary Dust," 17 March 1993, *http://www.stanford.edu/dept/news/pr/93/930317Arc3362.html* (accessed 26 January 2011).

"simple prebiotic molecules [quinones] could play an essential role in the formation of more complex organisms."[128] Astrophysicist Chandra Wickramasinghe of Cardiff University in Wales, however, remained cautious on the connection of such molecules to life, commenting only that "this shows that biochemical monomers can form under solar-system conditions. Although this process must occur, its contribution to the origin of life remains conjectural."[129]

Models of the temperatures attained by IDPs due to frictional heating after entering Earth's atmosphere estimate that only small particles of less than 20 μm (0.0008 inches), which constitute about 10^{-5} the IDP mass incident on our planet, stay cooler than 160°C.[130] This is the temperature that the Microbiology Advisory Committee considered as the minimum necessary for biological sterilization under dry heat conditions.[131] Nevertheless, 10^{-5} of the total IDP mass incident on Earth still constitutes a significant amount of matter—about 400,000 grams per year (roughly half a ton) of potentially unsterilized particles.

This continuing bombardment of Earth by matter possibly containing biotic material, resulting in no observed deleterious effect on our planet's life-forms, must be taken into account when developing planetary protection policy for the miniscule return samples of IDPs. The possibility cannot be ignored, however, that future IDP recovery missions might sample particles "that are not represented in the present-day . . . flux to Earth,"[132] and thus could conceivably contain material harmful to our planet's biota. A mitigating factor is that if the IDPs that were collected had been resident in interplanetary space for geological periods of time (rather than, for instance, protected by the atmosphere of a planet), then they would probably have received sterilizing doses of radiation. For this reason, the SSB recommended that no special containment should be warranted for IDP return samples

128. M. Tuleta, L. Gabla, and J. Madej, "Bioastrophysical Aspects of Low Energy Ion Irradiation of Frozen Anthracene Containing Water," *Physical Review Letters* 87(7) (13 August 2001): 78103.

129. Katie Pennicott, "Did Interplanetary Dust Seed Life on Earth?" PhysicsWeb, 10 August 2001, *http://physicsweb.org/articles/news/5/8/9* (accessed 31 August 2006).

130. Orgel, *Evaluating the Biological Potential*, Chap. 6, "Cosmic Dust."

131. Microbiology Advisory Committee, *Sterilisation, Disinfection and Cleaning of Medical Equipment, Guidance on Decontamination to the Department of Health Medical Devices Directorate* (London: Medical Devices Directorate Publication, 1993).

132. Orgel, *Evaluating the Biological Potential*, p. 67.

taken from interplanetary space. In addition, if the process of collecting IDPs results in the heating of the samples to temperatures considered sufficient for sterilization, as was the case with the Stardust mission, then this collection procedure would also eliminate the need for special sample containment requirements, regardless of the source of the IDPs.

The Genesis Mission: Questions Regarding Safe Transport of Extraterrestrial Material Back to Earth

The Sun and the bodies of our solar system originated, according to current theories, from a cloud of gas, dust, and ice. Examination of the solar wind, the particle streams emitted from the surface of the Sun, may tell us more about this process of solar system formation. To this end, NASA launched its Genesis mission in August 2001 to collect samples from the solar wind (see figure 6.6) and return them to Earth for laboratory analysis. Although hydrogen makes up most of the solar wind, Genesis aimed to collect many different types of matter. The mission had a goal of capturing samples of all naturally occurring elements in the periodic table, if possible, in order to determine in detail the Sun's composition.[133]

After a three-year trip around the Sun, on 8 September 2004, an Earth entry vehicle from the Genesis spacecraft landed in the Utah Test and Training Range at Dugway Proving Ground. On the descent, an essential parachute failed to deploy; the root cause was traced to incorrect drawings that resulted in assemblers installing critical sensors upside down in the vehicle.[134] As a result, the entry vehicle hit Earth at a speed of over 300 kilometers per hour (nearly 200 miles per hour) and cracked open upon impact. The vehicle's outer shell as well as its inner science canister were breached, and some of the collector arrays, containing pristine particles from the solar wind, were shattered.

---

133. Dolores Beasley, William Jeffs, and Susan Killenberg McGinn, "NASA Sends First Genesis Early-Science Sample to Researchers," NASA news release 05-030, 27 January 2005, *http://www.nasa.gov/home/hqnews/2005/jan/HQ_05030_genesis_sample.html* (accessed 7 February 2005).

134. Richard A. Kerr, "Flipped Switch Sealed the Fate of Genesis Spacecraft," *Science* 306 (22 October 2004): 587.

The crash produced major fragmentation of the arrays and loss of collector materials, but much of the material recovered still had utility. Mission planners had expected to receive 275 hexagonal collector parts; instead, they now had greater than 15,000 small but potentially usable pieces (NASA estimated that the smallest usable fragment of collector needed to be at least 3 millimeters in size). The hardest task in recovering science data from the broken collectors was to mitigate surface contamination that included particles, mud, water drop residues, and inorganic aerosols.[135]

Space policy analyst John Logsdon of the George Washington University in Washington, DC, noted that Genesis was designed during the "height of the 'faster, cheaper, better' era"[136] of NASA missions when spacecraft plans were drafted, and vehicles constructed and operated, by fewer personnel in less time than had been the case earlier. Genesis was thus vulnerable to the same kinds of failures as the Mars Polar Lander and Climate Orbiter probes that had run into mission-ending problems. In spite of Genesis's crash landing, however, intensive cleaning and other mitigation efforts by NASA personnel on the spacecraft's collector pieces has resulted in slow but steady recovery of some mission science data. A major cause for optimism in recovery efforts is that the contamination needing to be removed is on the surfaces of collector fragments, while the solar wind particles are buried below those surfaces. Curatorial and research staff are hopeful that the major impact of the crash will prove to be only be a delay in extracting the science data, rather than a loss of those data.[137]

While the contamination of solar wind samples with Earth matter has caused challenging problems in extracting science data from the mission, most space scientists do not believe that the solar wind samples contain any biological material or pose a risk of biologically contaminating our terrestrial environment.[138] NASA classified the mis-

135. D. S. Burnett and the Genesis Science Team, "Genesis Mission: Overview and Status," Lunar and Planetary Science Conference XXXVII, March 2006, http://www.lpi.usra.edu/meetings/lpsc2006/pdf/1848.pdf (accessed 1 September 2006).

136. Kerr, "Flipped Switch Sealed the Fate of Genesis Spacecraft."

137. Mishap Investigation Board, Genesis Mishap Investigation Board Report, Volume I, NASA, 2005, http://www.nasa.gov/pdf/149414main_Genesis_MIB.pdf (accessed 26 January 2011); Burnett et al., "Genesis Mission"; Race comments to author, 2007.

138. John Rummel, "Genesis," Planetary Protection Web site, http://planetaryprotection.nasa.gov/pp/missions/past/genesis.htm (last updated 14 July 2006, accessed 31 August 2006); Amir Alexander, "NASA Assessing Damage to Genesis Capsule from Crash," Planetary

sion as Category V, a grouping that pertains to all missions for which a spacecraft or one of its components returns to Earth. Category V has two subcategories, one of which is termed "unrestricted Earth return" and applies to the exploration of "solar system bodies deemed by scientific opinion to have no indigenous life-forms."[139] This was the subcategory into which Genesis's sample return actions were placed (for a discussion of the evolution of NASA's mission categorization scheme, see Chapter 7 on the return to Mars).

Implications of the Genesis mishap for planetary protection. The Genesis crash did not endanger Earth's biosphere, but it did raise serious questions about *how to safely transport matter containing potential biotic material from space to Earth.* Furthermore, we need to be able to assess the contamination risks incurred if return sample operations go wrong, as they did for Genesis,[140] and develop dependable methodologies for minimizing these risks.

The Genesis crash engendered somber concerns about Mars sample return missions, which could conceivably carry extraterrestrial life-forms. Plans for such a mission at the time of the Genesis mishap called for using a passive Earth entry vehicle that did not need a parachute. It would instead employ atmospheric friction to slow its descent before impacting somewhere in the Utah desert. While such an entry vehicle would not have to rely on the proper opening and functioning of a parachute assembly, it would have to be robust enough for the sample canister to remain unbreached even during a hard landing. Heavily cushioning the sample canister within the entry vehicle would help to attain this goal by minimizing landing impact to the sample canister.[141]

Pete Worden, USAF Brigadier General (retired) and the NASA Ames Research Center Director, supported another approach to Mars sample return that would avoid the problem Genesis ran into. The

Society Web site, *http://www.planetary.org/news/2004/0908_NASA_Assessing_Damage_to_Genesis.html* (last updated 8 September 2004, accessed 31 August 2006).

139. John Rummel, "Categories," Planetary Protection Web site, *http://planetaryprotection.nasa.gov/pp/about/categories.htm#5* (last updated 14 July 2006, accessed 31 August 2006).

140. Alexander, "NASA Assessing Damage."

141. Leonard David, "Biological Mishap Renews Debate About Mars Sample Return," Space News Business Report, 20 September 2004, *http://www.space.com/spacenews/archive04/genesisarch_092004.html* (accessed 31 August 2006).

Moon would be employed as a quarantine zone for material brought back from Mars, avoiding a direct return of Martian samples to Earth.[142] Worden commented that "building a Mars habitat and receiving laboratory on the Moon for Mars samples could allow us to determine its safety—and perhaps compatibility with Earth life."[143]

The space science community has not given lunar quarantine the support that Worden did. For instance, in a 1990 workshop held at NASA Ames Research Center, organized by Donald DeVincenzi, the participants favored an Earth-based facility as opposed to one on the Moon.[144] In a later meeting, John Rummel expressed the thought "that a lunar quarantine would . . . result in a net decrement to safety—especially if one wishes to include the safety of the people setting up and doing the analyses . . . ."[145] He was concerned as well with the effectiveness of such an operation, pointing out that scientists would still have a desire to study the extraterrestrial materials in the most capable laboratories possible, and it would probably not be feasible to build such facilities on the Moon. The samples would most likely have to be transported to Earth to receive the most reliable, in-depth examination.[146]

142. Jeff Foust, "Exploiting the Moon and Saving the Earth," *Space Review* Web site, 7 November 2005, *http://www.thespacereview.com/article/490/1* (accessed 26 January 2011).

143. Simon P. Worden (speech given at the International Space Development Conference, Los Angeles, 7 May 2006, and transcribed and published by *SpaceRef.com*, 9 May 2006), *http://www.spaceref.com/news/viewnews.html?id=1119* (accessed 26 January 2011).

144. D. L. DeVincenzi, H. P. Klein, and J. R. Bagby, eds., *Planetary Protection Issues and Future Mars Missions*, proceedings of the workshop, "Planetary Protection Issues and Human Exploration of Mars," held at ARC, Moffett Field, CA, 7–9 March 1990, NASA Conference Publication 10086, 1991, p. 3.

145. John D. Rummel presentation, "Planetary Protection, NASA, the Science Mission Directorate, and Everything," NASA Advisory Committee (NAC) Science Subcommittee Meetings, NASA Headquarters, Washington, DC, 6 July 2006, *http://science.hq.nasa.gov/strategy/NAC_sci_subcom/planetary_protection/PPS_Jul06_Rummel.pdf*.

146. Leonard David, "Biological Mishap Renews Debate About Mars Sample Return," Space News Business Report, 20 September 2004, *http://www.space.com/spacenews/archive04/genesisarch_092004.html* (accessed 31 August 2006).

Summary of NASA Planetary Protection Requirements for Missions to Small Solar System Bodies

This chapter has discussed various types of past or planned missions to explore a variety of small solar system bodies and has referred several times to planetary protection requirements for such missions. These requirements range from no actions needed for the exploration of bodies not of interest to studies of life or chemical evolution to rather restrictive constraints on mission operations returning samples from bodies that could potentially harbor life. As stipulated by COSPAR planetary protection policy, NASA's document, *Planetary Protection Provisions for Robotic Extraterrestrial Missions,*[147] is formulated around five different categories, with subcategories, to aid in determining appropriate forward and back contamination prevention approaches for various types of space projects. Typically, three of these categories, I, II, and V, apply to small body missions. Constraints on the mission depend on its characteristics and that of the target body or bodies. Summarized below are the constraints of NASA's different planetary protection categories as applied to various types of missions to small bodies.[148]

For further discussion of these planetary protection categories, refer to Chapter 7, "Return to Mars." In particular, table 7.1 provides a summary of the characteristics of all five target body and mission type categories. Also refer to Appendix A of this book, or Appendix A of NASA Procedural Requirement (NPR) 8020.12C, for more information on these categories.[149]

---

147. NASA Science Mission Directorate, *Planetary Protection Provisions for Robotic Extraterrestrial Missions,* NPR 8020.12C, 27 April 2005.

148. John Rummel, "Categories," Planetary Protection Web site, *http://planetaryprotection.nasa.gov/pp/about/categories.htm#5* (last updated 14 July 2006, accessed 31 August 2006); Orgel, *Evaluating the Biological Potential,* Appendix D, "Planetary Protection Policy—NASA and COSPAR"; D. L. DeVincenzi, P. Stabekis, and J. Barengoltz, "Refinement of Planetary Protection Policy for Mars Missions," *Advances in Space Research* 18(1/2) (1996): 311–316.

149. NASA Science Mission Directorate, "Detailed Planetary Protection Requirements," in *Planetary Protection Provisions for Robotic Extraterrestrial Missions,* NPR 8020.12C, 27 April 2005, Appendix A.

Forward Contamination Prevention for Small Bodies: Outbound Issues

Forward contamination prevention requirements are generally not warranted or are fairly limited for missions to the many small bodies of the solar system, except on a case-by-case basis, such as for missions to icy satellites like Europa that could conceivably harbor life. Most small body missions are covered by NPR 8020.12C's Category I or II requirements,[150] whose defining characteristics are summarized as follows:[151]

Category I: Missions that are not of direct interest for understanding the processes of chemical evolution or origin of life. No planetary protection requirements are imposed on these missions. Category I applies to outbound missions (or the outbound part of sample return missions) to the Sun, Mercury, and our Moon because biological contamination of these bodies is not considered a risk, given their extreme environmental conditions. Evidence that strongly supports this position for the Moon is given by Apollo mission results. None of the more than 2,000 samples of lunar material returned to Earth by six Apollo missions yielded any evidence of past or present lunar biological activity. Consequently, missions to and from the Moon are not required to have formal planetary protection plans in place.

Category II: Missions that have significant interest for understanding chemical evolution and origin of life processes, but only a remote chance that contamination carried by spacecraft could jeopardize future exploration. Category II missions must provide documentation of mission preparations and operations to the Planetary Protection Officer and must implement procedures to minimize the likelihood of accidental impact with their target planets or satellites, as the Galileo orbiter did by plunging into Jupiter at the end of its life. Both the Galileo orbiter mission to Jupiter and the Cassini orbiter mission to Saturn were considered Category II endeavors. So was the comet exploratory phase of the European Space Agency's Rosetta

---

150. NASA Science Mission Directorate, *Planetary Protection Provisions.*
151. Rummel, "Categories."

mission.[152] The probes and landers of a Category II mission, however, such as the Huygens probe to Saturn's moon Titan, may require a higher level of planetary protection, such as contamination control measures, that their mother ships do not.

Back Contamination Prevention: Inbound Issues

Category V: All missions for which the spacecraft, or a spacecraft component, returns to Earth. The concern for these missions is the protection of Earth from back contamination resulting from the return of extraterrestrial samples (usually soils, dusts, and rocks). A subcategory called *Unrestricted Earth Return* is defined for missions to bodies that the scientific community deems to have no indigenous life-forms. The Genesis spacecraft, which sampled solar wind particles, and Stardust, which sampled particles in the Wild 2 comet's tail as well as interstellar dust sweeping through the solar system, were both classified as Category V—Unrestricted Earth Return. Missions in the unrestricted return subcategory may have planetary protection requirements imposed, but only on their outbound (Earth to target body) phase. The appropriate classification for the outbound phase to many, if not most, small bodies is either Category I (no requirements imposed) or Category II (requirements for documenting mission activities and avoiding accidental impact with the target).

The second subcategory of Category V missions is *Restricted Earth Return*. For this classification, which applies to spacecraft that have targeted bodies with biological potential, the highest degree of back contamination prevention concern is expressed by requiring the following:

- Containment throughout the return phase of all unsterilized material from the target body and returning hardware which directly contacted the target body.
- No destructive impact of the spacecraft with Earth that could release unsterilized material from the target body.

---

152. G. Schwehm, J. Ellwood, H. Scheuerle, S. Ulamec, D. Moura, and A. Debus, "Planetary Protection Implementation for the Rosetta Mission" (presented at the 34th COSPAR Scientific Assembly, Second World Space Congress, 10–19 October 2002, Houston, TX, 2002); NASA NPR 8020.12C.

- Timely postmission analyses of the returned unsterilized samples, under strict containment, and using the most sensitive techniques.
- Continued containment of the return sample, if any evidence of a nonterrestrial replicating organism is detected, until a reliable sterilization procedure is applied.

A mission returning samples from the Jovian moon Europa would fit into this category because of the possibility of life being present within those samples.[153]

For a summary of planetary protection methodologies used on missions to small solar system bodies as well as other missions, see Appendix F.

---

153. NASA Science Mission Directorate, "Detailed Planetary Protection Requirements."

RETURN TO MARS

7

We may discover resources on . . . Mars that will boggle the imagination, that will test our limits to dream. And the fascination generated by further exploration will inspire our young people to study math and science and engineering and create a new generation of innovators and pioneers.

—President George W. Bush,
speech at NASA Headquarters, 14 January 2004[1]

Although the Viking mission greatly expanded our knowledge of Mars, it was seriously limited in its search for organic chemicals and the existence of life by the two landing sites that NASA chose "more for spacecraft safety than for scientific interest."[2] Viking found no evidence of life in the vicinity of the landing sites or of any organic substances in the soil.[3] But the mission's data could not be reliably extrapolated to the whole planet and thus could not answer many questions about possible Martian prebiotic chemistries or the past or present existence of life.

1. Office of the Press Secretary, "President Bush Announces New Vision for Space Exploration Program," the White House Web site, 14 January 2004, *http://www.whitehouse.gov/news/releases/2004/01/20040114-3.html* (accessed 7 April 2008).
2. D. L. DeVincenzi, H. P. Klein, and J. R. Bagby, eds., "Executive Summary," *Planetary Protection Issues and Future Mars Missions*, NASA Conference Publication 10086, proceedings of a workshop held at ARC, 7–9 March 1990, p. iii, obtained from LaRC Library.
3. SSB, National Research Council, *Mars Sample Return: Issues and Recommendations* (Washington, DC: National Academies Press, 1997), p. 11.

Biological studies since Viking have highlighted the durability of extremophile life-forms on Earth—those that thrive in harsh environments previously thought incapable of supporting an ecology. Some of these microorganisms abound in very cold biospheres such as in polar ice caps, glaciers, and permafrost,[4] and even within rocks in the dry, frigid valleys of the Antarctic.[5] Other terrestrial microbes, the anhydrobiotes, have adapted to states of extreme dehydration, allowing them to live in very dry environments.[6] These findings raise the question of whether similar extremophiles might also exist in the cold, desiccated Martian environment.

The Martian surface environment may once have been far less challenging to life than it is now. The planet's average temperature was probably at one time considerably warmer and, according to geological evidence, liquid water may have flowed on the surface in past eras. It is possible that life arose in the planet's distant past, about the time that this was happening on Earth, and may still exist in localized niches.[7]

Since Viking, the technology for detecting microorganisms has advanced remarkably, and this has understandably elicited scientific interest in reexamining the Martian surface. The many new methods of bioassessment include epifluorescent microscopic techniques[8] that

---

4. NASA, "Search for Life on Mars Will Start in Siberia: NASA Funds Permafrost Study to Support Astrobiology Research," Science@NASA Web site, http://science.nasa.gov/newhome/headlines/ast27may99_1.htm (accessed 26 September 2006).

5. Task Group on Planetary Protection, SSB, National Research Council, Biological Contamination of Mars: Issues and Recommendations (Washington, DC: National Academies Press, 1992), p. 5, http://www7.nationalacademies.org/ssb/bcmarsmenu.html. Also available from Steven Dick Unprocessed Collection, NASA Historical Reference Collection.

6. K. Dose, C. Stridde, R. Dillmann, S. Risi, and A. Bieger-Dose, "Biochemical Constraints for Survival Under Martian Conditions," Advances in Space Research 15(3) (1995): 203.

7. NASA Exobiology Program Office, "The Present State of Knowledge," in An Exobiological Strategy for Mars Exploration, http://cmex-www.arc.nasa.gov/Exo_Strat/exo_strat.html, January 1995 (accessed 26 September 2006); SSB, Mars Sample Return: Issues and Recommendations, pp. 10–14.

8. Epifluorescent microscopy relies on the excitation of susceptible molecules in a sample with short-wavelength, high-energy light, followed by observation of the emitted lower-energy light, or fluorescence, of those molecules. Susceptible molecules include chlorophyll as well as substances that bind to specific cellular compounds. Epifluorescent microscopy enables scientists to observe cellular components not detectable with conventional light microscopy and is sensitive enough to detect a single molecule. From "Expanding Our Knowledge of the Limits of Life on Earth," in Committee on Preventing the Forward Contamination of Mars, SSB, National Research Council, Preventing the Forward Contamination of Mars (Washington, DC: National Academies Press, 2005), Chap. 5.

directly count viable cells, as well as the polymerase chain reaction that enzymatically amplifies biomarkers of even a single cell to levels that can be detected.[9]

The above considerations have led to plans for new Mars missions that will visit different locations than Viking investigated and employ far more sensitive technologies than that project had access to. The renewed possibility of finding extant or past life on Mars, as well as on other celestial bodies, underlines the need for continuing to incorporate strict planetary protection concerns into all phases of such exploration. The following section outlines alterations to the planetary protection approach that have been made since Viking in order to better meet the evolving needs of missions to Mars and other solar system bodies while reliably guarding against forward and backward contamination dangers.

●●●

Changes in the Planetary Protection Approach Since Viking: Categorizing by Target Body and Type of Mission

The planetary protection procedures for the Viking mission's two spacecraft were governed by the quantitative criterion that the probability of contaminating a planet of biological interest must not exceed one in 1,000 (1×10^{-3}) during the period of biological interest. This was the period when space scientists would be searching for evidence of chemical evolution and the origin of life.[10]

Approximately 7,000 surface assays of Viking Lander Capsules, Orbiters, and shrouds taken during assembly and testing indicated that these components had initially carried significant bioloads. To meet the probability-of-contamination suballocation for each of the landers of roughly 10^{-4}, reliable bioload reduction measures had to be incorporated during Viking fabrication processes. Experimental testing of various techniques led NASA to choose dry-heat sterilization of the

---

9. Task Group, *Biological Contamination*, p. 5.
10. D. L. DeVincenzi, P. D. Stabekis, and J. B. Barengoltz, "A Proposed New Policy for Planetary Protection," *Advances in Space Research* 3(8) (1983): 15.

landers as the major method for effectively reducing microbial counts. Because Viking landers contained sensitive metabolic assay technologies, the Viking biology team imposed additional heat treatment in order to greatly lower the possibility of terrestrial organisms contaminating their instrument.[11] These approaches to bioload reduction were discussed in detail in Chapter 5.

Post-Viking planetary missions had a different, evolving set of planetary protection criteria than the Viking mission. An influential post-Viking SSB study was published in 1978 that identified the need for such new criteria, entitled *Recommendations on Quarantine Policy for Mars, Jupiter, Saturn, Uranus, Neptune, and Titan*. As it pertained to Mars, this study concluded that the probability of terrestrial organism growth on that planet was so low that landers conducting initial exploratory visits to a subpolar region did not require terminal heat sterilization.[12]

While the 1978 SSB study made clear the need for alterations to planetary protection policy, the guidelines that had governed Viking flights did not begin to change until the 1980s. This occurred when the quantitative modeling basis for the policy was reexamined in light of data obtained from a decade of planetary exploration, as well as changes to, or uncertainties in, some of the parameters employed in the existing policy. This quantitative modeling approach had inherent weaknesses in its application to Martian exploration (as well as to other target bodies). For instance, there were gross uncertainties in probability estimates of terrestrial microbe growth at different Martian surface locations. NASA also did not know how many future landings would be made on Mars, but both of these data needed to be input into the quantitative model, and thus its capability to accurately

---

11. Donald L. DeVincenzi, Margaret S. Race, and Harold P. Klein, "Planetary Protection, Sample Return Missions and Mars Exploration: History, Status, and Future Needs," *Journal of Geophysical Research* 103(E12) (25 November 1998): 28, 577–528, 585; J. R. Puleo, N. D. Fields, S. L. Bergstrom, G. S. Oxborrow, P. D. Stabekis, and R. Koukol, "Microbiological Profiles of the Viking Spacecraft," *Applied and Environmental Microbiology* 33(2) (February 1977): 379–384.

12. Space Science Board, Committee on Planetary Biology and Chemical Evolution, *Recommendations on Quarantine Policy for Mars, Jupiter, Saturn, Uranus, Neptune, and Titan* (Washington, DC: National Academies Press, 1978): 12–13, folder 006697, "Quarantine/Sterilization," NASA Historical Reference Collection.

predict the probability of contaminating Mars or other target bodies was severely strained.

In light of these issues, NASA's planetary protection policy was reevaluated during the early 1980s, and a new approach was proposed,[13] whose key features were the following:

- The quantitative, probabilistic methods employed for Viking would be deemphasized.
- Most planetary protection requirements for the majority of space missions would be eliminated. This recommendation was supported by "the wealth of data obtained by planetary exploration during the past decade [that was] consistently negative on the existence of indigenous life-forms on other planets."[14]
- The missions requiring planetary protection measures, as well as the particular protection measures that were warranted, would be identified by the characteristics of the target body and of the mission itself—for instance, whether exploration was to take place using a flyby, orbiter, probe, or lander.

This approach better reflected the space science community's degree of concern for protecting a specific target body, which was dependent on the characteristics of the body as well as on the type of mission. The scientists proposing the new method considered it to be superior to one depending on highly uncertain probabilistic calculations. Five categories corresponding to the various target planet and mission type combinations, as well as each category's suggested planetary protection requirements, were proposed:

Category I included any mission to a target body not of direct interest for understanding the process of chemical evolution. No protection of such planets was warranted and thus no planetary protection requirements were recommended.

Category II missions comprised those to target bodies where there was significant interest in chemical evolution processes, but where there was only a remote chance that contamination carried by a

---

13. J. B. Barengoltz, S. L. Bergstrom, G. L. Hobby, and P. D. Stabekis, JPL Publication 81-90, 1982, as reported in DeVincenzi et al., "A Proposed New Policy for Planetary Protection," 14–15, 20.
14. DeVincenzi et al., "A Proposed New Policy for Planetary Protection," 14.

spacecraft could jeopardize future exploration goals. Recommended requirements were for simple documentation, such as a short planetary protection plan outlining intended or potential impact targets, brief pre- and postlaunch analyses giving details of impact avoidance strategies, and postencounter and end-of-mission reports providing the impact location if such an event occurred.

Category III covered "no direct contact" missions—flybys and orbiters—to bodies of exobiological interest for which contamination would likely jeopardize future biological exploration. Requirements consisted of more involved documentation than Category II and the implementation of some procedures that included the following:

- Trajectory biasing. The probabilities of accidental impact could not exceed 10^{-5} for launch vehicles and 10^{-3} for flybys.
- Use of cleanrooms during spacecraft assembly and testing. In particular, all orbiter and flyby vehicles were to be assembled in cleanrooms of class 100,000 or better.
- Bioburden reduction.

Category IV embraced "direct contact" missions—probes and landers—to target bodies of chemical-evolution and origin-of-life interest for which there was a significant chance of contamination that could jeopardize future biological experiments. Recommended requirements to be imposed entailed detailed documentation more severe than Category III, including a bioassay to quantify the bioburden, a probability-of-contamination analysis for non-nominal events such as a crash or equipment failure, and a bulk constituent organics inventory. The implementation of an increased number of planetary protection procedures was recommended as well, such as trajectory biasing, the use of cleanrooms, bioload reduction, the possible partial sterilization of direct-contact hardware, and the installation of a bioshield for that hardware. The requirements were similar to those for Viking, with the probable exception of complete lander and probe sterilization.

Category V comprised all Earth return missions and was concerned with back contamination prevention of the terrestrial system—Earth *and* the Moon. By protecting both bodies, travel between them would not require planetary protection procedures.[15]

---

15. DeVincenzi et al., "A Proposed New Policy for Planetary Protection," 17.

For the set of target bodies that the space science community did not believe had indigenous life-forms, a special subcategory of Category V, *safe for Earth return*, was to be designated. Missions in this subcategory were to have planetary protection requirements on their outbound phase only, corresponding to the appropriate category for that phase—typically Category I (as for the Genesis mission that brought back samples from the solar wind) or Category II (as for the Stardust mission to a comet).

The very highest degree of contamination concern was to be given to other Category V missions to targets that scientists believed could harbor life. These missions were termed *restricted Earth return* and required the sterilization of returned spacecraft hardware that directly contacted the target planet as well as the quarantining of any unsterilized samples. Category V recommendations included a range of requirements that encompassed those of Category IV plus the monitoring of project activities and conducting of preproject studies such as those for remote sterilization and containment techniques.

Table 7.1 summarizes the characteristics of each of the above target body and mission type characteristics. Also refer to Appendix A of NASA Procedural Requirement (NPR) 8020.12C,[16] which can be found in the appendices of this book.

The major provisions of the recommended policy were reviewed by NASA, the SSB, and members of the U.S. and international space science communities, and they were also presented at the July 1984 COSPAR meeting in Graz, Austria. Following minor changes, COSPAR unanimously accepted a resolution approving the revised planetary protection policy as a replacement for existing COSPAR guidelines.[17]

Still, the question of whether to include terminal sterilization in the planetary protection protocol was not yet answered. Meanwhile, the aerospace industry was changing and making it more difficult to employ a terminal sterilization regime without damaging spacecraft components. Don Larson, a planetary protection lead at Lockheed Martin, gave some

---

16. NASA Science Mission Directorate, "Detailed Planetary Protection Requirements," in *Planetary Protection Provisions for Robotic Extraterrestrial Missions*, NPR 8020.12C, 27 April 2005, Appendix A.

17. D. L. DeVincenzi and P. D. Stabekis, "Revised Planetary Protection Policy for Solar System Exploration," *Advances in Space Research* 4(12) (1984): 291–295.

Table 7.1 Summary of target body/mission type categories.

	Category I	Category II	Category III	Category IVa
Type of mission	Flyby, orbiter, and lander.	Flyby, orbiter, and lander.	No direct contact (flybys and orbiters).	Direct contact (landers and probes).
Subcategory				IVa: Spacecraft not carrying instruments for the investigation of extant Martian life.
Target examples	The Moon; Mercury; undifferentiated, metamorphosed asteroids.	Venus, Jupiter (exclusive of its icy moons), Saturn, Titan, Uranus, Neptune, Triton, Pluto/Charon, Kuiper belt objects, comets, carbonaceous chondrite asteroids.	Mars, Europa, Ganymede, Callisto.	Same as Category III.
Defining characteristic of mission	Not of direct interest for understanding process of chemical evolution or origin of life.	Significant interest in process of chemical evolution and origin of life, but only a remote chance that contamination carried by spacecraft could jeopardize future exploration.	A target body of chemical-evolution or origin-of-life interest that if contaminated, could jeopardize future biological exploration.	A target body of chemical-evolution or origin-of-life interest that if contaminated, could jeopardize future biological exploration; surface microbial burden level limits are based on average Viking lander burden levels before sterilization.

Category IVb	Category IVc	Category V Unrestricted	Category V Restricted	
Same.	Same.	Earth return (missions making contact with another body and returning to Earth).	Same.	**Type of mission**
IVb: Spacecraft carrying instruments designed to investigate extant Martian life.	IVc: Investigation of Martian special regions—those within which terrestrial organisms are likely to propagate, or regions which have high potential for existence of extant Martian life-forms.	Unrestricted Earth return of samples.	Restricted Earth return of samples.	**Subcategory**
Same.	Same.	The Moon (Apollo); undifferentiated, metamorphosed asteroids; comets (Stardust); solar wind (Genesis).	Mars, Europa.	**Target examples**
Same.	Same.	No concern about back contamination resulting from return of extraterrestrial samples (usually soil and rock).	Concern about back contamination of the terrestrial system (Earth and the Moon) resulting from return of extraterrestrial samples (usually soil and rock).	**Defining characteristic of mission**

	Category I	Category II	Category III	Category IVa
Degree of planetary protection concern	None.	Documentation of procedures and activities.	Passive bioload control.	Active bioload control: Total surface microbial burden no greater than Viking lander *preterminal* sterilization levels.
Representative planetary protection procedures	None.	Impact avoidance strategies.	Documentation more detailed than Category II, contamination control, organics inventory as necessary, microbial burden limits or impact or orbital lifetime requirements.	Detailed documentation surpassing that for Category III, microbial assay and reduction plans, organics inventory.

Category IVb	Category IVc	Category V Unrestricted	Category V Restricted	
Possible need for sterilization to Viking lander *postterminal* sterilization levels.	Possible need for sterilization to Viking lander *postterminal* sterilization levels.	No procedures needed for back contamination prevention.	Prevention of back contamination is necessary.	**Degree of planetary protection concern**
Either entire landed system must be sterilized to especially stringent microbial burden levels or subsystems involved in acquisition, delivery, and analysis of samples used for life detection must be sterilized and a method of preventing recontamination implemented.	Landing site within special region: sterilize entire landed system. Special region accessed through horizontal or vertical mobility— either sterilize entire landed system or sterilize subsystems directly contacting special region and prevent their recontamination. If off-nominal condition such as a hard landing would cause high probability of special region biocontamination, entire landed system must be sterilized to Viking poststerilization microbial burden levels.	Outbound leg: depends on category of target object (Category I, II, etc.). Inbound leg: No contamination control procedures needed for back contamination prevention.	Outbound leg: Category IVb requirements unless specifically exempted. Inbound leg: breaking chain of contact with target planet; postreturn sample containment (quarantine) and biohazard testing in receiving laboratory.	**Representative planetary protection procedures**

Source: NASA Science Mission Directorate, "Detailed Planetary Protection Requirements," in *Planetary Protection Provisions for Robotic Extraterrestrial Missions*, NASA Procedural Requirement (NPR) 8020.12C, 27 April 2005, Chapter 2 and Appendix A.

details of this changing situation. In the late 1960s and early 1970s, aerospace-specific parts were generally manufactured with the government as client. When the space industry shifted to more commercial ventures, such as the launches of numerous telecommunications satellites, vendors and manufacturers had to become increasingly focused on cost-efficient solutions rather than catering their operations to government space projects with deeper pockets. This evolution of the aerospace industry was accompanied by a major increase in the use of microcircuitry, which contained electrical conductors that were extremely close together and delicate and could more easily be damaged than many of the old printed circuit boards. The result of these factors was that many post-Viking parts used in spacecraft could no longer withstand the terminal sterilization environment.[18]

Human Mars Missions: The 1990 NASA Ames Policy Development Workshop

Policies for exploring Mars continued to develop, including those related to human exploration. A March 1990 Ames Research Center workshop that focused on planetary protection issues for both robotic and human exploration of Mars[19] drew scientists from the United States and the international community with expertise in astrophysics, atmospheric chemistry, biology and microbiology, ecology, environmental science, instrument design, legal issues, and other fields. Their object was to analyze the unique circumstances that arise during human Mars exploration and to understand how planetary contamination resulting from such missions might be prevented. The conference proposed guidelines that included the following elements:

---

18. Donald C. Larson, Phoenix Flight Program Planetary Protection Lead, Lockheed Martin Space Systems Company, Denver, CO, interview by author, 15 August 2006.
19. D. L. DeVincenzi, H. P. Klein, and J. R. Bagby, eds., *Planetary Protection Issues and Future Mars Missions*, proceedings of the workshop "Planetary Protection Issues and Human Exploration of Mars," held at ARC, Moffett Field, CA, 7–9 March 1990, NASA Conference Publication 10086, 1991, pp. i, 3–4.

- Planetary protection questions needed to be thoroughly investigated during the precursor robotic mission phase of exploration, before any human contact took place with the planet.
- To prevent forward contamination of Mars from orbiting vehicles,
  - all orbiters needed to be assembled in cleanrooms to minimize their initial bioloads,
  - their trajectories needed to be biased to avoid unplanned impacts with and releases of microbes on Mars, and
  - minimum orbital lifetime requirements needed to be met to ensure that no impacts occurred during the period of biological exploration.
- To prevent forward contamination from landers (including roving vehicles, surface penetrators, and surface stations), they needed to be
  - assembled in cleanrooms,
  - subjected to procedures for minimizing bioloads to prescribed levels, and
  - enclosed in bioshields to prevent recontamination.
- To prevent back contamination of Earth,
  - samples returned from Mars needed to be enclosed in hermetically sealed containers,
  - the contact chain between the Martian surface and the return vehicle needed to be broken to prevent transfer of unwanted substances to Earth, and
  - the samples needed to be subjected to a thorough quarantine protocol on Earth to investigate the possible presence of harmful species or other constituents.

The return of Martian samples to either a lunar or orbital quarantine facility was *not* advised by the workshop because the scientific community thought that such facilities could not provide both the required containment conditions and the thorough capabilities for assaying the samples.[20] Proper containment of potentially dangerous life-forms required a level of sophistication, including remote handling technology, airflow control and filtering, and high-integrity seals, that only existed on Earth in facilities designed for handling highly infectious and virulent

---

20. Ibid., pp. 3–4.

microbes. Replicating such facilities, and the experience necessary to operate them, was unlikely to be achieved either on the Moon or in orbit.

●●●

These interim planetary protection recommendations, suggested by the international science community but not yet vetted and approved for implementation, were developed to help mission designers and planners.[21] The vetting process for updating both human and robotic exploration policy was formally begun in July 1990 with a request from NASA Planetary Protection Officer John Rummel to the National Research Council's Space Studies Board (SSB), which had served as NASA's primary adviser on planetary quarantine matters for many years. Rummel asked SSB to give its advice on planetary protection policy and its application to upcoming Mars missions. Rummel also mentioned that if an initial SSB report on Mars planetary protection issues could be given to NASA prior to the 1992 COSPAR meeting, it would enable "a robust U.S. position"[22] in the analysis of Russia's envisioned planetary protection measures for its Mars-94 mission.

●●●

The 1992 SSB Study and the Policy Refinements That Followed

As requested by PPO John Rummel, SSB reviewed current planetary protection policy and published its findings in a 1992 study, *Biological Contamination of Mars: Issues and Recommendations.* The SSB study concluded that the contamination of Mars by the propagation of terrestrial microbes was an extremely unlikely event, largely due to "the high levels of ultraviolet radiation, the thin atmosphere, the extremely

---

21. DeVincenzi et al., *Planetary Protection Issues*, pp. 2–4.
22. John D. Rummel to Louis Lanzerotti, 16 July 1990, in Task Group on Planetary Protection, SSB, National Research Council, *Biological Contamination of Mars: Issues and Recommendations* (Washington, DC: National Academies Press, 1992), Appendix A, *http://www7.nationalacademies.org/ssb/bcmarsmenu.html.* Also available from Steven Dick Unprocessed Collection, NASA Historical Reference Collection.

low maximum temperatures, and the absence of liquid water on the surface."[23] However, terrestrial contaminants did present a significant risk *to the integrity of experiments designed to search for extant or fossil Martian microbes.* These experiments used extremely sensitive technologies, and small amounts of biotic contamination from Earth could cause false-positive results. Therefore, SSB recommended that minimization of bioloads still needed to be carried out on all lander missions to Mars but that the level of contamination control should be determined by specific mission objectives as follows:

1. Landers carrying instrumentation for in situ investigation of extant Martian life should be subject to *at least Viking-level sterilization procedures.* Specific sterilization methods were to be driven by the nature and sensitivity of the particular experiments, although terminal dry-heat sterilization, as had been implemented for Viking, was a likely candidate. The objective of this requirement was the reduction, to the greatest feasible extent, of contamination by terrestrial organic matter and/or microorganisms at the landing site.[24]

2. Space vehicles (including orbiters) without biological experiments should be subject to at least Viking-level *presterilization* procedures for bioload reduction, such as cleanroom assembly and cleaning of all components, but such vehicles need not be sterilized.

By requiring stricter control of bioburden for landers carrying search-for-life instrumentation, SSB appeared to "shift the emphasis of planetary protection away from concerns about contaminating the planet to protection of exobiological science."[25] This change raised concerns that to avoid the extra restrictions and expenses, future missions might be biased away from experiments and target sites of exobiological interest.

---

23. SSB Task Group on Planetary Protection, *Biological Contamination of Mars: Issues and Recommendations* (Washington, DC: National Academies Press, 1992), http://www7.nationalacademies.org/ssb/bcmarsmenu.html.

24. SSB, *Biological Contamination of Mars.*

25. R. Howell and D. L. DeVincenzi, *Planetary Protection Implementation on Future Mars Lander Missions,* proceedings of the joint U.S.-Russian Workshop on Planetary Protection Implementation for Future Mars Lander Missions, Palo Alto, CA, 13–15 July 1992, NASA Conference Publication 3216, 1993, p. 14.

DeVincenzi, Stabekis, and Barengoltz[26] examined the 1984 planetary protection guidelines in light of SSB's 1992 recommendations and proposed several refinements to the policy. The first was to split Category IV into Categories IVa and IVb in order to make the important distinction between missions without and with specific life-detection objectives. *Category IVa* applied to *landers without life-detection instruments.* Such landers would not require terminal sterilization, but would require procedures for reducing bioburdens to levels no greater than those found on Viking landers *just prior to terminal sterilization.*

The average presterilization spore burden on the Viking landers of ≤300 spores per square meter of surface and total spore burden level of 3×10^5 had been determined using data from over 2,000 bioassays performed during Viking assembly and processing operations. DeVincenzi et al. thus proposed that the surface burden for each future Mars lander without life-detection experiments be set at an average of ≤300 spores per square meter of surface, and the total lander burden at $\leq 3 \times 10^5$ at the time of launch, using the same measurement protocol as for Viking. SSB, however, recommended that Viking protocols for bioload assessment be upgraded to include state-of-the-art technologies. Using more modern instrumentation had the potential for requiring changes to surface burden specifications.

Category IVb related to *spacecraft carrying search-for-life experiments.* The driver for defining Category IVb, according to former Planetary Protection Officer John Rummel, was "to avoid false positives in looking for signs of life on Mars, not to protect Mars from contamination."[27] This meant that the search-for-life instruments had to be exceptionally clean, as had been the case on Viking. Category IVb thus required spacecraft to receive at least Viking-level sterilization procedures. DeVincenzi et al. interpreted this to mean that such vehicles should have bioburden levels no greater than was present on the Viking landers *following terminal sterilization.* The

---

26. D. L. DeVincenzi, P. Stabekis, and J. Barengoltz, "Refinement of Planetary Protection Policy for Mars Missions," *Advances in Space Research* 18(1/2) (1996): 311–316.

27. John D. Rummel (executive secretary) and Norine E. Noonan (chair), "Planetary Protection Advisory Committee Meeting Report," JPL, 12–13 January 2004, p. 5, *http://science.hq.nasa. gov/strategy/ppac/minutes/PPACmin0401.pdf* (accessed 31 October 2006).

two Viking Lander Capsules (VLCs), remember, were each subjected to a terminal dry-heat sterilization cycle. These cycles were such that each lander was exposed to a temperature of 111.7°C for 30 hours, *after* the coldest contaminated point reached that temperature. Thus, some parts of the lander experienced slightly higher temperatures. It was Viking project management that chose the target temperature (111.7°C), with the concurrence of NASA's Planetary Quarantine Office.[28]

Category III missions were those *without direct contact* of the space vehicle with the target object (Mars), and they included orbiters and flybys. Category III orbiters had to meet either one of two requirements: a cleanliness requirement specifying a maximum total spore burden of 5×10^5 (including *surface, mated,* and *encapsulated* locations) *or* an orbital lifetime requirement specifying that the probability the craft would remain in orbit around Mars for 20 years must be at least 10^{-2}, and for 50 years, at least 5×10^{-2}. Thus, if an orbiter could attain a reliably stable orbit, it did not have to meet a cleanliness requirement.

Category III *flybys* had a much simpler requirement—to have a probability of impact with Mars that did not exceed 10^{-3}. *Launch vehicles* in Category III missions also had a requirement imposed on them—a probability of impact with Mars that did not exceed 10^{-5}.[29]

An important caveat to the cleanliness requirement for Category III orbiters was that it be placed *on the vehicle as a whole,* not just on its surface (as was the case for Category IVa landers). This was because if an orbiter's or flyby's microbes did contaminate Mars, it would likely be through an accidental impact in which the vehicle broke open and microbes from inside got loose.[30]

---

28. DeVincenzi et al., "Refinement of Planetary Protection Policy for Mars Missions," 312–314; Perry Stabekis to author, 8 October 2007.

29. SSB, *Biological Contamination of Mars*; SSB, National Research Council, *Preventing the Forward Contamination of Europa* (Washington, DC: National Academies Press, 2000); DeVincenzi et al., "Refinement of Planetary Protection Policy," 314–315; Stabekis, telephone interview by author, 7 September 2007.

30. DeVincenzi et al., "Refinement of Planetary Protection Policy," 315.

Piloted vs. Unpiloted Missions

Although the presence of astronauts may help in the detection of extant or past life, such as through the wide-ranging collection of samples, a human presence on the Martian surface will cause unavoidable contamination. SSB thus recognized the need for identifying local Martian environments, such as hydrothermal areas, in which life could exist, and to explore such areas with robotic vehicles before allowing humans onto these sites.[31] NASA eventually developed explicit planetary protection policies and categories for such "special" areas; these will be discussed later in the chapter.

Applying the New Policies
During Missions of the 1990s

The Mars projects of the 1990s that were conducted by NASA as well as by the space agencies of other countries provided a range of examples of how planetary protection measures were woven into mission operations and how scientific findings could impact policies for future missions.

NASA's Mars Observer

The robotic orbiter Mars Observer, whose goal was to examine and map the planet's atmosphere, surface, interior, and magnetic field for one Martian year,[32] was the first mission to the planet to comply with the refined 1984 planetary protection guidelines discussed above. NASA designated the Mars Observer as Category III (no direct contact of the space vehicle with the target object).[33] The mission carried out planetary

---

31. SSB, *Biological Contamination of Mars*, pp. 54–55.

32. JPL, "Fact Sheet: Mars Observer," JPL Office of Public Information, 23 July 1992, p. 1, Mars Observer File 16614, NASA Historical Reference Collection.

33. R. Howell and D. L. DeVincenzi, *Planetary Protection Implementation on Future Mars Lander Missions*, proceedings of the joint U.S.-Russian Workshop on Planetary Protection Implementation for Future Mars Lander Missions, Palo Alto, CA, 13–15 July 1992, NASA Conference Publication 3216, N94-13183, 1993, pp. 5–6.

Table 7.2 Planetary protection activities during different Mars Observer mission phases.

Mission Phase	Time Period	Pollution Prevention Implementations
Launch	Liftoff until spacecraft separation.	Design components so as to prevent contaminating spacecraft with hot gases from, or collision with, upper stage during its separation procedures. Bias upper stage trajectory to avoid impact with Mars.
Cruise	Spacecraft separation until start of Mars orbit insertion (11 months).	Set spacecraft aimpoint so as to avoid accidental impact with Mars. Implement increased tracking 90 days before orbit insertion to ensure accurate targeting at Mars.
Orbit insertion	Mars orbit insertion until mapping orbit achieved (8–9/93 until 12/93).	Design orbital insertion maneuver to minimize probability of accidental spacecraft impact, either during engine firing, on first periapsis after firing, or during intermediate orbits between orbital insertion and achievement of mapping orbit.
Mapping and quarantine orbits	Mapping orbit: Insertion until raising of spacecraft to quarantine orbit (12/93 to 2/96).	Design mapping and quarantine orbits to have adequate lifetimes. Design quarantine raise maneuver to reliably raise spacecraft to desired orbit.

protection activities during each of its phases: launch, cruise, orbit insertion, mapping orbit, and quarantine orbit. These activities are described below and summarized in table 7.2, and they are indicative of the care that has to be taken to prevent forward contamination of a planet, even from spacecraft never meant to come in contact with the body.

Launch phase. During the launch phase, the trajectory of the launch vehicle's upper stage was carefully biased so as to prevent its eventual impact with Mars, which would have resulted in contamination of the planet. As indicated in table 7.3, NASA required that the probability the launch vehicle would impact Mars had to be less than 1×10^{-5} (one chance in 100,000). The particular trajectory modification that was needed to attain this probability varied depending on the date that the mission launched.

The upper stage eventually separated from the spacecraft using a maneuver designed to minimize the probability that the spacecraft

Table 7.3 Mars Observer impact probability estimates and requirements.

Mission Phase	Estimated Probability of Impact with Mars	NASA's Maximum Allowable Probability of Impact
Launch vehicle	$< 1 \times 10^{-5}$	$< 1 \times 10^{-5}$
Spacecraft		
Injection and cruise	1.2×10^{-5}	
Orbit insertion	1.2×10^{-5}	
Mapping orbit and quarantine raise	2×10^{-5} (until 2009) 8.3×10^{-3} (until 2039)	
Spacecraft total	4.4×10^{-5} (until 2009) 8.3×10^{-3} (until 2039)	1×10^{-2} (until 2009) 5×10^{-2} (until 2039)

would get contaminated, either through a collision with the upper stage or through materials given off by it. This maneuver had the secondary purpose of further biasing the upper stage away from a Mars impact trajectory. It was critical that the mission took these measures *before* the upper stage used all its fuel, since the upper stage would not be capable after that point of trajectory modifications. The spacecraft, on the other hand, had engines and fuel aboard that allowed it to make later trajectory corrections if needed.[34]

Cruise phase. The aimpoint of the spacecraft's trajectory during its cruise phase was biased so as to be suitable for orbit insertion but not lead to an impact with Mars if mission control lost control of the craft. Approximately 90 days before orbit insertion at Mars, NASA conducted increased tracking efforts in order to reduce navigation delivery errors and ensure as accurate targeting of the spacecraft as possible.[35]

Orbit insertion. During the first part of Mars orbit insertion, the spacecraft entered a three-day capture orbit. This maneuver needed to be performed with care because at two points during it, the spacecraft engine firing and the first periapsis[36] after the burn, there was a small probability of impact with and subsequent contamination of the Martian surface. All intermediate orbits between the insertion

---

34. J. B. Barengoltz, *Mars Observer Planetary Protection Plan*, Supplement 1: Prelaunch Report, identification numbers given include: 642-32, Supplement 1, and JPL D-4481, Supplement 1, July 1992, pp. 3-6 to 3-8, folder 16614, "Mars Observer File," NASA Historical Reference Collection.
35. Barengoltz, *Mars Observer*, p. 3-8.
36. The spacecraft's periapsis was the point in its orbit when it was closest to Mars.

maneuver and the eventual mapping orbit that the craft attained needed to be rigorously analyzed to make sure that their lifetimes were of sufficient length for the maneuver to be completed. The principal factor that decayed these orbits and shortened their lifetimes was atmospheric drag. Another important factor was the nonsphericity of the Martian gravitational field. Contributions to uncertainties in the prediction of orbital lifetimes included variances in Mars's atmospheric density and solar activity.[37]

Mapping and quarantine orbit phases. Planetary protection procedures implemented during the mapping phase of the mission included designing the mapping orbit to have an adequate lifetime for the phase to be completed. They also included designing a quarantine raise procedure that would be as trustworthy as possible in maneuvering the spacecraft into a higher terminal orbit, circular in shape and 405 kilometers (252 miles) above the planetary surface. To meet NASA planetary protection requirements, this "quarantine orbit"[38] needed to be stable enough such that the probability of spacecraft impact with Mars would not exceed 1×10^{-2} by the year 2009 and 5×10^{-2} by the year 2039.[39] Table 7.3 includes estimates of the actual probabilities of impact at different phases of the mission as well the maximum allowable impact probabilities. Note that estimated probabilities never exceed NASA's required limits.

NASA's Mars Surveyor Program

After NASA launched the Mars Observer, it continued plans to explore Mars in fiscal year 1995, but with some changes to its operational strategy. The program was conducted within a fairly constrained cost ceiling of approximately $100 million per year.[40] The Mars Surveyor program called for the development of a small orbiter to be launched in November 1996 to conduct surface studies of Mars. The

---

37. Barengoltz, *Mars Observer*, pp. 3-8 to 3-9, 5-3 to 5-4.
38. Ibid., p. 3-9.
39. Ibid., p. 4-3.
40. Donald L. Savage and Diane Ainsworth, "NASA Begins Development of New Mars Exploration Program," Mars Surveyor Mission press release 94-20, 7 February 1994, *http://tes.asu.edu/MARS_SURVEYOR/mars_surveyor_info.2_8_94.html* (accessed 26 January 2011).

orbiter, which was small enough to be launched on a Delta expendable launch vehicle and carried only half the science payload of the Mars Observer, aimed to "lay the foundation for a series of missions to Mars in a decade-long program of Mars exploration."[41] The orbiter—the Mars Global Surveyor (MGS)—was the first in a series of low-cost, lightweight spacecraft bound for Mars. It was built by Martin Marietta Technologies of Denver, in part because of the company's success in developing the Magellan Venus radar mapping mission and the Viking Mars landers.[42]

The Mars Surveyor missions took advantage of launch opportunities that occurred approximately every two years as Mars came into alignment with Earth. The orbiter designed for the 1998 launch was even smaller than the initial MGS and carried the remainder of the Mars Observer science instruments. It was meant to act as a communications relay satellite for a companion lander, launched the same year, and other landers in the future, such as the Russian Mars-96 lander.

A major challenge of the Mars Surveyor program, with its emphasis on fast development and low cost, was to build in effective planetary protection measures without driving up the budget or delaying the aggressive schedule of launches every two years. As with other programs, trajectory course maneuvers (TCMs) were employed during the voyage to Mars[43] to better aim the craft toward its destination as well as to minimize the chances of an accidental impact with, and contamination of, the planet (this strategy for MGS is described in more detail in the next section). The MGS spacecraft (consisting of its payload of science instruments and its bus, which contained its propulsion system, flight hardware, and other subsystems) was also required to have the capability of achieving a near-circular quarantine

---

41. "NASA Plans for 1996 Mars Orbiter Mission," from NASA news release 94-20, February 1994, http://barsoom.msss.com/mars/global_surveyor/mgs_project_releases/96press.html (accessed 27 October 2006).

42. Diane Ainsworth, "Mars Global Surveyor To Be Built By Martin Marietta of Denver, Colorado," JPL Press Release, 8 July 1994, http://tes.asu.edu/MARS_SURVEYOR/mgs_press_rel.7_8_94.html (accessed 30 October 2006).

43. Glenn E. Cunningham, "Mars Global Surveyor Is On Its Way to Mars," Martian Chronicle 8 (November 1996), http://mpfwww.jpl.nasa.gov/MPF/martianchronicle/martianchron8/mgs.html (accessed 23 January 2007).

orbit with an average altitude of 405 kilometers (252 miles) above Mars if necessary.[44]

The MGS spacecraft, which was an orbiter, was not sterilized. The requirement on its outside was only that "The spacecraft external surfaces shall be visibly clean at payload encapsulation. Visibly clean is defined as the absence of all particulate and nonparticulate matter visible to the normal unaided . . . eye . . . ."[45]

NASA's Mars Global Surveyor—Discoverer of Potential Special Regions

The MGS spacecraft, the first of the Mars Surveyor program, launched on 7 November 1996, bound for a Mars orbit that enabled it to gather data on surface morphology, topography, composition, gravity, atmospheric dynamics, and the magnetic field.[46] MGS began its primary mapping mission in March 1999.[47] Although its primary mission was to be two years (or approximately one Martian year), it kept operating for far longer, last communicating with Earth on 2 November 2006. A review of the mission determined that the loss of the spacecraft was related to a computer error made five months before the craft's apparent battery failure. If not for that error, MGS might have kept operating even longer.[48]

44. NASA, "Mission Background," in *Mars Global Surveyor Mission Plan* (draft), no. 542-405, 30 June 1995, section 2, *http://www.msss.com/mars/global_surveyor/mgs_msn_plan/section2/section2.html#RTFToC8* (accessed 23 January 2007).

45. JPL, "Introduction," "System Requirements," and "Payload Description and Accommodation," in *Mars Global Surveyor Spacecraft Requirements*, Revision A, JPL D-11509, 10 September 1996, sections 1, 3, and 4, *http://mars.jpl.nasa.gov/mgs/scsys/e3/e30.html* (accessed 23 January 2007).

46. Mars Now Team and the California Space Institute, "What Was the Purpose of Mars Global Surveyor?" University of California at San Diego, California Space Institute site, *http://calspace.ucsd.edu/Mars99/docs/library/mars_exploration/robotic_missions/orbiters/mars_global_surveyor/purpose1.html* (accessed 29 October 2006).

47. JPL, "Current Mission—Mars Global Surveyor," *http://www.jpl.nasa.gov/missions/current/marsglobalsurveyor.html* (accessed 29 October 2006).

48. Guy Webster and Dwayne Brown, "Report Reveals Likely Causes of Mars Spacecraft Loss," NASA news release 2007-040, 13 April 2007, *http://marsprogram.jpl.nasa.gov/mgs/newsroom/20070413a.html* (accessed 7 September 2009); Malin Space Science Systems, 2006, *http://www.msss.com/* (accessed 7 November 2009).

Among its key science findings related to planetary protection, MGS imaged gullies and debris-flow features suggesting the presence of current sources of liquid water, similar to an aquifer, at or near the Martian surface. These areas could be Martian special regions, an important designation discussed later in the chapter.[49]

As with other NASA Mars missions, a major consideration in planning the maneuvers of the spacecraft was to meet all NASA planetary protection quarantine requirements, which it did in part by using the various biasing strategies that other spacecraft performed as well. Each TCM had to be designed to satisfy the requirements while at the same time minimizing the necessary change in velocity (and fuel usage) of the craft.

The aimpoint at the time of injection of the launch vehicle upper stage and spacecraft onto a Mars trajectory had to be selected to ensure that the probability of the upper stage and spacecraft accidentally impacting Mars was no more than 1×10^{-5}. During the cruise phase of the mission, when the spacecraft journeyed to the vicinity of Mars, several small TCMs were implemented to remove the trajectory bias given for planetary protection purposes to the craft at the beginning of the phase. The TCMs also were performed to control the path of the spacecraft as it approached Mars, in part to meet planetary protection objectives. The navigation design had to ensure that the probability of accidental Martian impact did not exceed 1×10^{-2} within 20 years after launch and 5×10^{-2} for an additional 30 years.[50]

Several aspects of MGS's orbits around Mars also related to planetary protection. The spacecraft's mapping orbit was elliptical, with a minimum height of 354 kilometers (220 miles) over the southern polar region and 409 kilometers (254 miles) over the northern region.[51] This orbit was carefully selected to be low enough so that the MGS instruments could make close-range observations of the Martian surface, but

49. JPL, "Mars Global Surveyor," NASA Facts, http://www.jpl.nasa.gov/news/fact_sheets/mgs.pdf (accessed 29 October 2006).

50. JPL, Mars Global Surveyor Project—Mission Requirements Document, 542-400, JPL D-11956, September 1995, pp. 2-2, 3-1, 3-9, http://mars.sgi.com/mgs/pdf/400.PDF; P. B. Esposito, Mars Global Surveyor Navigation Plan, Rev. B, 542-406, JPL D-12002, August 1996, pp. 6-2, 6-7, 6-8, http://mars.sgi.com/mgs/pdf/406.PDF.

51. NASA, "Mission Background," in Mars Global Surveyor Mission Plan, section 2.

not so low that friction from the planet's atmosphere would slow the spacecraft down enough for an impact.

The MGS orbit was nearly polar, which meant that it was able to progressively observe all of Mars as the planet rotated below it. The orbit was also Sun-synchronous, in that the spacecraft passed over a given part of Mars at the same time each day. For instance, at about 2 p.m. local Mars time every orbit, the spacecraft crossed the equator on the planet's daytime side. This timing was essential for certain measurements because it allowed scientists to separate local daily variations from the longer term seasonal and annual trends.[52]

When the spacecraft completed its mapping work, its propulsion system was designed to raise the average altitude of the orbit in order to greatly reduce the frictional drag of the Martian atmosphere. This "quarantine orbit" practice conformed with international planetary protection agreements for minimizing, at least for several decades, the chances that the craft would crash and potentially contaminate Mars with terrestrial organisms. As discussed above, this strategy was intended to provide a contamination-free period in which to conduct missions examining the biotic nature of the planet.[53]

Russia's Mars-94/96

The planetary protection approach taken in Russia's Mars-94/96 mission was somewhat different from that carried out in NASA's missions, and it is thus interesting to examine. Mars-94/96 was an attempt to accomplish the first landings on Mars since the 1976 Viking craft. In 1990, Russia decided to split this mission into two launches—1994 and 1996. The Mars-94 segment was to include an orbiter, two small autonomous stations on the surface of Mars, and two penetrators digging into the planet's surface. Scientists from 20 countries participated in the effort, employing instruments built by a range of international

---

52. JPL, "Destination: Mars," Mars Global Surveyor site, *http://mars.jpl.nasa.gov/mgs/sci/oldstuff/mgsbrocb.html* (last updated 20 October 2006, accessed 29 October 2006).
53. JPL, "Destination: Mars."

partners. The Mars-96 mission was to be configured similarly to Mars-94 but was to include one or two balloons and small rovers.[54]

Russia claimed that its Mars missions were in accord with COSPAR planetary protection policy, with two main foci: protecting planetary environments from influence or contamination by terrestrial microorganisms and protecting experiments searching for Martian life.[55] Past Soviet planetary protection performance had been questionable, however. While there was evidence that some precautions were taken on past missions, there was also reason to believe, as noted earlier in the book, that Soviet space exploration may have already contaminated Mars and possibly Venus as well by landing or crashing insufficiently sterilized spacecraft on the planets.[56]

Searches of available scientific literature and news articles indicated that by 1990, the Soviets had attempted 17 missions to Mars. According to DeVincenzi et al., three of the missions probably landed on the surface of Mars.[57] The Soviet scientific community consistently maintained that their spacecraft complied with COSPAR guidelines through decontamination methods that included radiation, chemicals, and heat, either singly or in combination, but no literature is available on the details of how these technologies were applied to specific missions. However, in a report to COSPAR in 1973, Soviet scientists provided general information on bioburden reduction for their Mars series of spacecraft, stating that subcomponents were first treated by heat, gas, chemical, or radiation, and the final spacecraft was then assembled in an "ultra-clean" room.

The Soviets had the policy of building three identical spacecraft and disassembling one of them, pulverizing its subcomponents and

---

54. International Scientific Council, "Mars-94 Unmanned Spacecraft Mission to Mars," briefing material for International Scientific Council Meeting, Moscow, 1–3 December 1992, p. 55, as reported in "Mars 94/96," NASA Center for Mars Exploration Web site, *http://cmex.ihmc.us/MarsTools/Mars_Cat/Part_2/mar.html* (accessed 27 October 2006).

55. G. Rogovski, V. Bogomolov, M. Ivanov, J. Runavot, A. Debus, A. Victorov, J. C. Darbord, "Planetary Protection Program for Mars 94/96 Mission," *Advances in Space Research* 18(1-2) (1996): 323–332.

56. D. L. DeVincenzi, H. P. Klein, and J. R. Bagby, eds., *Planetary Protection Issues and Future Mars Missions*, NASA Conference Publication 10086, proceedings of a workshop held at ARC, 7–9 March 1990, p. 9.

57. E. Burgess, *Return to the Red Planet* (New York: Columbia University Press, 1990), as reported in DeVincenzi et al., *Planetary Protection Issues*, p. 10.

assaying them before launching the other two craft, in order to verify that the appropriate level of decontamination had been achieved. If these methods were indeed followed, they would have complied with the intent of the COSPAR guidelines. But it is known that although these approaches might have reduced spacecraft bioburdens sufficiently before transport to the launch site, liftoff operations were carried out under military authority, over which the Soviet planetary protection team had no jurisdiction.[58]

Russia's envisioned Mars-94 launch began running into problems in 1993. By the end of the year, there was significant concern as to whether the launch schedule could be achieved. By April 1994, six months before the planned launch, the Russian Space Agency postponed Mars-94 until the next flight opportunity in 1996. The reason given by the Russian Academy of Sciences was "because of national economic difficulties."[59] The new flight date, however, did not have as favorable a launch window and required a reduction in payload. The original Mars-96 craft's launch was delayed until 1998, but it was eventually canceled.

By the end of 1994, concerns arose that a complete cancellation of the project might be necessary.[60] The spacecraft did take off on 16 November 1996, but its fourth-stage launch vehicle ran into problems, resulting in the spacecraft crashing in the vicinity of the Chilean coast and the fourth stage impacting the Pacific Ocean near Easter Island.[61] These events elicited some concerns regarding the contamination of Earth by 0.44 pounds of plutonium[62] aboard the spacecraft. The

58. V. I. Vashkov, N. V. Ramkova, G. V. Scheglova, L. Z. Skala, and A. G. Nekhorosheva, "Verification of the Efficiency of Spacecraft Sterilization," *Life Sciences and Space Research* XII (1974): 119–202, as reported in DeVincenzi et al., *Planetary Protection Issues*, pp. 10–11.

59. A. Zakharov, ed., "Robotic Spacecraft Mission to Mars," Space Research Institute, Russian Academy of Sciences, 1996, *http://www.iki.rssi.ru/mars96/* (accessed 24 January 2007).

60. Federation of American Scientists (FAS), "Mars," FAS Space Policy Project-World Space Guide, *http://www.fas.org/spp/guide/russia/science/solarsystem/mars.htm* (accessed 24 January 2007).

61. Roger Bourke, "Mars 96: Failure and Aftermath," JPL *Martian Chronicle* 8 (1997), *http://mpfwww.jpl.nasa.gov/MPF/martianchronicle/martianchron8/mars96.html* (accessed 24 January 2007); "Russian Planetary Science Sinks With Mars '96," *ScienceNOW: Daily News Archive* (18 November 1996).

62. Robert M. Bowman, "Mars 96 Failure Fuels Cassini Protest," Space and Security News home page, 10 January 2006, *http://www.rmbowman.com/ssn/cassini.htm* (accessed 24 January 2007).

Russian Academy of Sciences announced that the spacecraft fell into the ocean,[63] although some scientists thought it might have actually crashed on land.

NASA's Mars Pathfinder: Implementing Planetary Protection into a Low-Cost Lander Mission

Since the Pathfinder lander and its Sojourner rover were intended to make contact with the surface of Mars, this mission was classed as Category IVa for landers without life-detection instruments. As described earlier in this chapter, Category IVa requires detailed documentation, bioassays to quantify the burden on the spacecraft, a probability-of-contamination analysis, an inventory of bulk constituent organics, trajectory biasing, the use of class 100,000 or better cleanrooms during spacecraft assembly and testing, and bioload reduction to levels of the average spore burden on the Viking landers before they were subjected to terminal sterilization: < 300 spores per square meter and a total spore burden no higher than 3×10^5.

NASA personnel achieved bioburden requirements by applying cleaning procedures such as those using repeated alcohol solution wiping of the spacecraft during its development and fabrication. Large surface area components such as the craft's parachute as well as its airbags (which deployed around the lander and cushioned it as it hit the Martian surface) were heated for approximately 50 hours to a temperature of 110°C (230°F). Bioassays were taken of spacecraft components at many points during their fabrication in order to determine bioburden levels.[64]

Mars Pathfinder, which launched in December 1996, had ambitious objectives. It was to be developed in only three years using largely off-the-shelf parts, for a cost to place a science payload on the surface of Mars of only 7 percent of the Viking price tag.[65] Pathfinder, like the spacecraft in the Mars Surveyor program, was meant to demonstrate

---

63. A. Zakharov, ed., "Robotic Spacecraft Mission to Mars."

64. NASA, *1996 Mars Missions Press Kit*, doc. no. 96-207, November 1996, p. 46, folder 16263, "Mars Pathfinder, December 1996," NASA Historical Reference Collection.

65. "Mars Pathfinder Mission Objectives," JPL Mars Pathfinder Web site, *http://mars.sgi. com/mpf/mission_obj.html* (accessed 18 October 2006).

NASA's commitment to low-cost planetary exploration. The mission was also intended to showcase the mobility and usefulness of a roving vehicle on the Pathfinder, which needed to achieve its objectives without sacrificing planetary protection requirements, even though the mission budget was modest. The probability that any part of Pathfinder's unsterilized launch vehicle upper stage would impact the Martian surface had to be less than 1×10^{-4}.[66] The upper stage of the launch vehicle could accidentally be carried to the vicinity of Mars along with the Pathfinder spacecraft if, after the launch vehicle's final burn placing the spacecraft on target to Mars, there was a failure in the separation process. For this reason, the aimpoint of the upper stage and spacecraft was biased away from Mars just enough to meet the upper bound of the impact probability. This implied that, after separation from the upper stage, the Pathfinder spacecraft needed to use its propulsion system to change its trajectory toward one that would allow it to eventually enter an orbit around Mars.

The first two trajectory correction maneuvers (TCMs 1 and 2) for accomplishing this changed the path of the spacecraft from the biased aimpoint given it when it was injected onto its cruise trajectory to a still slightly biased Mars arrival trajectory. The reason for maintaining this slight bias was to satisfy another planetary protection requirement—to make sure that the probability of the spacecraft impacting Mars at a speed greater than 1,000 feet per second was no more than 1×10^{-3}. This requirement was implemented to minimize the contamination of Mars from parts of the vehicle in the event of a crash landing.[67] The trajectory after TCMs 1 and 2 was such that if control of the spacecraft was subsequently lost, it would enter the Martian atmosphere at a shallow enough entry angle for atmospheric braking to slow down the craft to below 1,000 feet per second at impact, even if the craft's parachute did not deploy.[68] Only near the end of the cruise to Mars, if control of the spacecraft and its systems had not been lost, were more correction maneuvers (TCM 3, 4, and, if necessary, 5) to

---

66. NASA, "Planetary Protection Specification Sheets," in NPR 8020.12C, *Planetary Protection Provisions for Robotic Extraterrestrial Missions*, 27 April 2005, Appendix B.

67. Pieter Kallemeyn, answer to a query in "Pre-Launch, Launch and Cruise," Mars Pathfinder Frequently Asked Questions, JPL, 11 February 1997, *http://mpfwww.jpl.nasa.gov/MPF/mpf/faqs_launch.html#1000* (accessed 26 January 2011).

68. Robin Vaughan, "Mars Pathfinder Navigation," 29 June 1997, *http://mars.jpl.nasa.gov/MPF/mpf/mpfnavpr.html* (accessed 18 October 2006).

be executed to direct Pathfinder to precisely the desired Mars atmospheric entry conditions for an optimal landing.

Pathfinder successfully touched down on Mars on 4 July 1997 as planned. Only four trajectory correction maneuvers were needed to bring the craft in on target; TCM 5 was not required. Pathfinder landed in an ancient flood plain in the northern hemisphere known as Ares Vallis, among the rockiest parts of the planet. Mission scientists chose this plain because they believed it to be a relatively safe surface to land on and one containing a wide variety of rocks that may have been transported during a catastrophic flood from different geologic regions of the planet.[69]

Japan's Nozomi—Fears of Contaminating Mars Through Accidental Impact

The Nozomi spacecraft, Japan's first interplanetary exploration, was to achieve a highly elliptical Mars orbit taking it as low as 150 kilometers above the surface of the planet and enabling it to take valuable measurements of Mars's upper atmosphere.[70] Launched on 4 July 1998, its mission plan called for it to reach Mars in October 1999. But an early thruster problem prevented Nozomi from attaining the speed desired during a swing-by of Earth, which led to the scripting of an emergency flight plan. Then a burst of solar flares damaged the craft's heating system and caused a loss of communication with Earth. As the craft finally approached Mars in 2003, it had very little fuel left and also suffered from a short circuit in its electrical system, possibly caused by the solar flares. Mission control crew sent various commands to the spacecraft in an attempt to bypass the short circuit. These repeated setbacks delayed

---

69. Richard A. Cook and Anthony J. Spear, "Back to Mars: The Mars Pathfinder Mission," *Acta Astronautica* 41 (1997): 599–608; NASA, "Mars Pathfinder," *http://www.nasa.gov/mission_pages/mars-pathfinder/* (last updated 23 November 2007, accessed 7 September 2009); JPL, "Mars Pathfinder Trajectory Data," 16 August 1997, *http://marsprogram.jpl.nasa.gov/MPF/mpfwwwimages/mpffootp.html* (accessed 26 January 2011).

70. Paul Kallender, "Japan's Nozomi Mars Mission Faces Critical Dec. 2 Deadline," *Space.com* site, 2 December 2003, *http://www.space.com/spacenews/archive03/marsarch_120203.html* (accessed 30 October 2006).

Nozomi's arrival and raised fears in the space science community of a possible crash landing on, and contamination of, the Red Planet.[71]

Ichiro Nakatani, Nozomi's spacecraft manager, explained that the electrical problem absolutely had to be fixed before Nozomi's engines were fired for an attempt at Mars orbital insertion. Otherwise, mission staff would command Nozomi to fly by Mars rather than orbit it. If the electrical short was fixed, Nozomi's thrusters would be fired during a 9 December 2003 TCM whose purpose was to realign the spacecraft and prepare it for orbit insertion on 14 December. The 9 December TCM itself had a certain risk of failure, however, that could lead to contamination of the planet. Nakatani commented that "We have a probability factor for Nozomi hitting Mars at about or less than one percent if we fail the December 9 maneuver."[72]

An impact with Mars could have had serious planetary protection implications because, according to Nozomi project manager Hajime Hayakawa, "no sterilization has been done before launch."[73] Some space scientists were not overly concerned about such an incident, arguing that even if Nozomi did impact Mars, it was not likely to present a biocontamination hazard. The spacecraft had been subjected to years of radiation exposure in space—several years more than planned. In addition, an accidental impact would occur only after the craft was intensely heated during its flight through the Martian atmosphere. The radiation and atmospheric heating would, the scientists believed, probably eradicate most microbes on the spacecraft.

Others in the science community were more conservative and stressed the importance of Japan's acknowledging the possibility of contamination and then demonstrating why an impact was extremely unlikely. If it could not do this, then it should never insert Nozomi into Mars orbit but should make sure it flew by the planet. John Rummel, NASA Planetary Protection Officer at the time, stressed that the Japanese should insert their craft into Mars orbit only if they "put

71. Leonard David, "Japan's Nozomi Mars Probe Stirs Contamination Qualms," *Space. com* site, 3 July 2003, *http://www.space.com/missionlaunches/nozomi_fears_030703.html* (accessed 30 October 2006).
72. Kallender, "Japan's Nozomi."
73. Govert Schilling, "Japan's Lost Hope," *Science Now* site, 17 November 2003, *http://sciencenow.sciencemag.org/cgi/content/full/2003/1117/3* (accessed 30 October 2006).

extra effort into extending Nozomi's orbital lifetime to achieve the non-contamination objectives required by international consensus."[74]

On 9 December 2003, mission scientists abandoned plans for orbit insertion. They had failed to fix the electrical problem and thus could not fire the craft's main engines.[75] Mission scientists instead sent commands to the craft to fire small alternate thrusters to minimize the possibility of a Mars impact.

NASA's Mars Climate Orbiter—What Was Its Fate, and Did It Generate Planetary Protection Concerns?

The Mars Climate Orbiter (MCO) lifted off on 11 December 1998 atop a Delta II launch vehicle and began its Mars orbit insertion in September 1999. The robotic spacecraft, whose assembly is depicted in figure 7.1, was to orbit the planet as its first weather satellite. MCO was also to relay communications from the Mars Polar Lander (MPL) when it arrived, which was scheduled for December 1999. MCO was lost, however, during its orbit insertion maneuver, at some point after it entered Mars occultation.

NASA received the last carrier signal from the spacecraft on 23 September 1999. The root cause for the loss of MCO was NASA's failure to convert from English to metric units in the coding of a software file employed in trajectory models. Small errors introduced in trajectory estimates over the course of the 9-month journey from Earth to Mars became more important by the time of the Mars orbit insertion, when the spacecraft trajectory was approximately 170 kilometers (106 miles) lower than had been planned. As a result, MCO either burned up in the Martian atmosphere or left the atmosphere and reentered interplanetary space.[76] NASA could not find the spacecraft, even though the mission team employed the 70-meter-diameter

---

74. David, "Japan's Nozomi Mars Probe."

75. Kenji Hall, "Hope Lost, Japan Abandons Mars Probe," Associated Press, 9 December 2003, http://www.space.com/missionlaunches/nozomi_done_031209.html (accessed 30 October 2006); BBC, "Japanese Mars Mission 'Abandoned,'" BBC News site, 9 December 2003, http://news.bbc.co.uk/1/hi/sci/tech/3304131.stm (accessed 30 October 2006).

76. Arthur G. Stephenson (chair), Mars Climate Orbiter Mishap Investigation Board Phase I Report, 10 November 1999, ftp://ftp.hq.nasa.gov/pub/pao/reports/1999/MCO_report.pdf (accessed 26 January 2011).

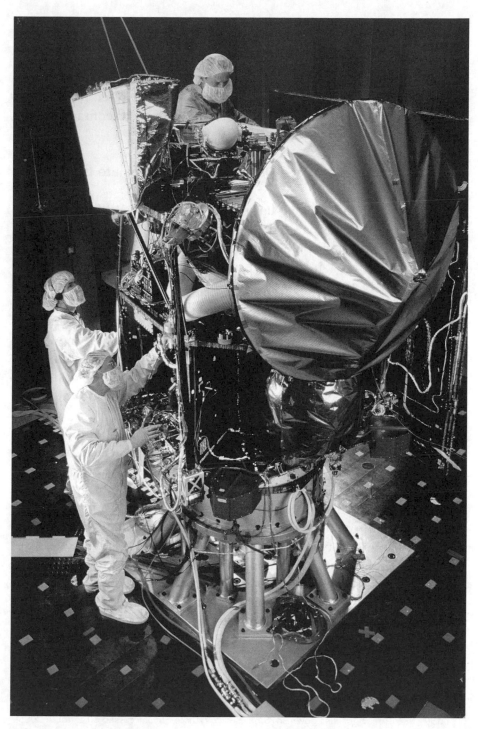

Figure 7.1 Assembly of MCO.

(230-foot) antennas of the Deep Space Network in an attempt to regain contact with it.[77]

From a planetary protection perspective, no contamination of the Martian surface apparently occurred, although under slightly different circumstances, this type of accident could have resulted in the orbiter's impact with the planet, leading to serious forward contamination.

NASA's Mars Polar Lander—Did Its Crash Contaminate the Planet?

The Mars Polar Lander launched 3 January 1999 from Cape Canaveral Air Station, atop a Delta II launch vehicle and carrying equipment to dig for traces of water beneath Mars's surface. The capability to burrow into the Martian surface was provided by the lander's robotic arm as well as by two basketball-sized microprobes, which were to be released as the lander approached Mars. The microprobes were to dive toward the planet's surface and penetrate as much as 1 meter underground, where they would test 10 new technologies, including a water-ice detector. The MPL and MCO missions were the second phase of NASA's long-term program of robotic exploration of Mars, which was initiated with the 1996 launches of the Mars Global Surveyor and the Pathfinder.[78]

The lander itself was to touch down on 3 December 1999 on the cold, windy, barren plain near the edge of Mars's southern polar cap, a few weeks after the seasonal carbon dioxide frosts had disappeared. It was to then use its instruments to analyze surface substances, frost, weather patterns, and surface-atmospheric interactions. Its objectives included examining how the Martian climate has changed over time and identifying where water can be found on Mars today. Scientists believe that water once flowed on Mars, and they want to know where it went. Evidence may be found in the geologic record and, in particular, in the layered terrain of the polar regions. Their alternating bands of color may contain different mixtures of dust and ice and may, like a tree's growth rings, help us understand climate change on Mars.

---

77. JPL Media Relations Office, "Mars Climate Orbiter Mission Status," 24 September 1999, http://mars.jpl.nasa.gov/msp98/news/mco990924.html (accessed 30 October 2006).
78. Stephenson, Mars Climate Orbiter Mishap, p. 9.

Scientists want to know whether it was caused by catastrophic events and whether the changes were episodic in nature or a gradual evolution in the planet's environment.[79]

Similar to Pathfinder, MPL was to dive directly into the Martian atmosphere and slow its initial descent by means of an aeroshell and parachute. These were scaled down from Pathfinder's design since MPL was considerably smaller. Unlike Pathfinder, however, MPL was not going to use airbags, but would instead make a soft landing by means of onboard guidance and retrorockets. The microprobes were to be released approximately 10 minutes before the lander touched down.[80]

MPL did not have specific life-detection experiments on board. Planetary protection thus required the same level of cleaning as for the Mars Pathfinder lander. MPL was to be cleaned to the level of the Viking landers before they received terminal sterilization. MPL surfaces could contain no more than 300 spores per square meter (about 250 per square yard) and 300,000 spores total.

To meet this bioreduction goal, technicians at Lockheed Martin Astronautics (formally Martin Marietta) in Denver repeatedly cleaned MPL with alcohol wipes throughout its fabrication. Large surface areas, such as the thermal blankets and parachute, were baked for about 50 hours at about 230°F (110°C). The spacecraft was bioassayed during processing at Lockheed Martin, and its level of cleanliness was verified just before encapsulation in its aeroshell. A final inspection and spore count was then performed at KSC before the spacecraft was integrated with the Delta II launch vehicle, assuring that the lander contained fewer than 300,000 spores total.[81]

Several factors in the MPL mission plan made planetary protection considerations especially critical. Both the lander's robotic arm and the two microprobes were meant to pierce the planet's surface and possibly contact water deposits. Furthermore, the microprobes were to do this by smashing into the Martian surface at up to 400 miles per hour (200 meters per second), after which their aeroshells would shatter and the

---

79. Douglas Isbell, Franklin O'Donnell, Mary Hardin, Harlan Lebo, Stuart Wolpert, and Susan Lendroth, *Mars Polar Lander/Deep Space 2*, NASA press kit, December 1999, pp. 3–4, *http://www.jpl.nasa.gov/news/press_kits/mplds2hq.pdf* (accessed 30 October 2006).
80. Isbell et al., *Mars Polar Lander*, p. 4.
81. Isbell et al., *Mars Polar Lander*, p. 34.

probe inside would split into two parts. The microprobes' electrical and mechanical systems were supposed to be tough enough to withstand this severe impact, but such an operation had not been tried before.[82] Any terrestrial bacteria within the microprobes or the lander's robotic arm might have opportunities to propagate on Mars that were not available to microbes on either the Viking or Pathfinder craft. Thus, it was especially important on the MPL mission that Lockheed Martin's bioload reduction procedures were carried out effectively.

There were certain challenges to doing this. MPL had a lack of flat surfaces and was thus particularly challenging to clean. Many of its surfaces, in fact, were not allowed to be touched, either for cleaning or for bioassay operations. One reason for making certain areas off-limits was that they contained extremely delicate sensors the size of human hairs. In addition to these areas, other locations on the spacecraft were inaccessible to cleaning. Thus, certain levels of bioburden had to remain in those places.[83]

Unfortunately, MPL did not touch down safely on Mars. As the lander descended, its legs were deployed in preparation for contact with the planet's surface. But sensors on the legs gave erroneous signals that the spacecraft was already on Mars, and this shut down the engine. In fact, the lander was 132 feet (40 meters) above the ground at this point. It crashed into the surface at 50 miles per hour (80 kilometers per hour)—five times faster than intended.

The premature engine shutdown was most likely caused by a software error related to the landing leg touchdown sensors.[84] John Casani, former head of flight programs at JPL and head of the Mars Polar Lander Failure Review Board, speculated on what happened to the MPL when it impacted the planet, commenting that it "probably would have pancaked. The legs would have broken off, it might have

82. Ibid., pp. 35–36.
83. Dave Tisdale, "Lessons Learned: The Engineer's Perspective," *Planetary Protection: Policies and Practices*, author's notes from a NASA course in planetary protection held in Santa Cruz, CA, 19–21 April 2005.
84. John Casani (chair), *Report on the Loss of the Mars Polar Lander and Deep Space 2 Missions*, JPL Special Review Board, JPL D-18709, 22 March 2000, p. 20, *http://spaceflight. nasa.gov/spacenews/releases/2000/mpl/mpl_report_1.pdf* (accessed 31 October 2006); Roberta L. Gross, NASA Inspector General to the Honorable F. James Sensenbrenner, Jr., Chairman, Committee on Science, U.S. House of Representatives, 5 June 2000, *http://oig. nasa.gov/old/inspections_assessments/MPL.pdf* (accessed 31 October 2006).

cart-wheeled or started rolling. You had two propellant tanks that were half full and those would have broken up, though I don't think there would have been a fire or an explosion. In the end, you'd have a lot of pieces to pick up."[85] The question remains, what happened to any microbes that were on the surfaces exposed by the MPL breakup?

NASA's Mars Exploration Rovers: A New, 21st Century Focus for Planetary Protection

NASA's two Mars Exploration Rover (MER) spacecraft lifted off Earth on 10 June and 7 July 2003, with an objective to explore the history of water on the Red Planet. They landed on Mars on 3 January 2004 and 24 January 2004 Pacific standard time (PST). The MER mission was part of NASA's Mars Exploration Program, a long-term robotic exploration effort.[86] Like the Pathfinder mission, MER was classed as Category IVa for landers without life-detection instruments.

The MER mission typified the evolving approach to planetary protection in that it was largely focused on protecting the integrity of the hardware that would make critical biologically related measurements as well as on contamination prevention of the planetary surface. In the words of Laura Newlin, JPL engineer and planetary protection lead for MER, "Keeping the spacecraft as clean as possible before, during and after launch is very important for any science instruments searching for organic compounds on the surface of other planets."[87] In other words, the umbrella of planetary protection now covered not only contamination prevention of the target body during the period of biological interest, but also of the spacecraft instruments themselves in order that

---

85. Paul Hoversten and Andrew Bridges, "Engine Cutoff Doomed Polar Lander," *Space. com* site, 28 March 2000, *http://www.space.com/scienceastronomy/solarsystem/mpl_ report_000328.html* (accessed 31 October 2006).

86. JPL, "Summary," Mars Exploration Rover Mission Web site, *http://marsrovers.nasa.gov/ overview/* (last updated 20 September 2006, accessed 19 October 2006).

87. *MarsToday.com*, "Mars Exploration Rover Spacecraft Undergo Biological Testing and Cleaning Prior to June Launches," SpaceRef Interactive Inc., 23 May 2003, *http://www. marstoday.com/viewpr.html?pid=11592* (accessed 19 October 2006).

they could make more reliable measurements. This broad definition of planetary protection was beginning to develop even during the Viking mission, when extra bioload reduction measures were taken for the life-detection instrument so that it would not generate false-positive indications of biotic processes.

Planetary protection measures were applied not only to the spacecraft itself, but also to the facilities in which it was handled. Before the MER spacecraft were shipped to KSC, for instance, the highbay facility and ground support equipment that would come into contact with the craft "were cleaned, sampled and recleaned to reduce further biological contamination when the spacecraft arrived."[88]

After the two spacecraft arrived at KSC, they were disassembled and cleaned to remove any contaminants received during their cross-country trip. During reassembly, JPL planetary protection team members bioassayed spacecraft surfaces to determine microbial levels. Mission staff cultured the bioassay samples, working in KSC life sciences labs and using equipment either from JPL or provided by KSC. This equipment included sonicators; water baths; incubators; microscopes; biosafety hoods; and a large, magnified colony counter.

Dry heat microbial reduction (DHMR) was performed on large components that could tolerate an environment of 125°C (257°F) for 5 hours and that were difficult to adequately wipe clean. These components included thermal blankets, airbags, honeycombed parts (such as the core structures of the rovers), and parachutes packed in cans. Spacecraft propellant lines were precision-cleaned. High-efficiency particulate air (HEPA) filters[89] were employed on the vents of each rover's core box, which contained computers as well as other critical electronics. The aim was to keep an estimated 99.97 percent of particles 0.3 microns or larger inside. These HEPA filters were also designed to help

---

88. *MarsToday.com*, "Mars Exploration Rover."
89. "High-efficiency particulate air" is the definition of the acronym "HEPA" used by the U.S. Department of Energy to describe a dry-type filter exhibiting a minimum efficiency of 99.97 percent when tested at an aerosol diameter of 0.3 micrometers aerodynamic diameter (U.S. Department of Energy, *DOE HEPA Filter Test Program*, DOE-STD-3022-98, May 1998, p. 2, *http://www.eh.doe.gov/techstds/standard/std3022/std3022.pdf* [accessed 19 October 2006]). HEPA has also been defined as "high-efficiency particulate arrestor" (*MarsToday.com*, "Mars Exploration Rover") and "high-efficiency particulate aerosol" (Dru Sahai, "HEPA Filters," *http://www.csao.org/UploadFiles/Magazine/vol15num2/hepa.htm* [accessed 19 October 2006]).

filter out Martian dust later in the mission. The planetary protection team worked closely with spacecraft design engineers to set contamination reduction strategies and to determine which pieces of hardware needed to be cleaned and which required other approaches such as heat sterilization.[90]

Each Mars Exploration Rover had to comply with requirements to carry a total of no more than 300,000 bacterial spores on surfaces from which the spores could enter the Martian environment.[91] An issue arose with spores detected in the acoustic foam backing of the fairing[92] enclosing the MER-1 spacecraft. At a 2004 Planetary Protection Advisory Committee[93] meeting after the MERs' launches, possible improvements to the fairing's foam layer manufacturing process were discussed. The Committee reached the conclusion that cleanliness and bioburden requirements had to be included in launch vehicle (as well as spacecraft) manufacturing specifications.[94]

As with other missions, the trajectories of various components had to be carefully designed to prevent unwanted impacts. In particular, when the third stage of MER's Delta launch vehicle (which had not been cleaned sufficiently to allow it to impact with Mars) separated from the spacecraft, the two objects were on nearly identical paths. This shared course had been set so that both objects would miss Mars. Approximately 10 days later, the spacecraft performed its first trajectory correction maneuver (TCM) in order to readjust its path to arrive at Mars.[95]

---

90. Donald Savage, Guy Webster, and David Brand, "Mars Exploration Rover Landings," NASA press kit, January 2004, p. 37, folder 16897, "Mars Rover 2003," NASA Historical Reference Collection; MarsToday.com, "Mars Exploration Rover."
91. JPL, "In-situ Exploration and Sample Return: Planetary Protection Technologies," http://mars5.jpl.nasa.gov/technology/is_planetary_protection.html (last updated 12 July 2007).
92. The fairing refers to the nose cone—the forwardmost, separable section of the launch vehicle rocket that was streamlined to present minimum aerodynamic resistance.
93. The Planetary Protection Advisory Committee (now called the PP Subcommittee) was an interagency group established in February 2001 to advise the NASA Administrator on policy and technical matters.
94. Rummel and Noonan, "Planetary Protection Advisory Committee Meeting Report," 12–13 January 2004, pp. 2–3.
95. Savage et al., "Mars Exploration Rover Landings," p. 37.

●●●

Reducing Bioload and Conducting Bioassays on the Phoenix Mars Lander and Other Mars Spacecraft

Contamination—foreign matter that does not belong on the spacecraft—can be grouped into two types: molecular and particulate. Molecular contamination includes the cumulative buildup of individual molecules, such as greases and oils. Particulate contamination refers to the deposition of visible, typically micrometer-sized conglomerations of matter, such as dust. These particulates can fall from the air onto exposed surfaces. Even miniscule quantities of contaminants can seriously degrade the performance of spacecraft hardware. So can outgassing, such as from various organic contaminants.[96]

The presence of organic contamination can be especially troublesome for Mars missions with instruments built to measure extremely small amounts of organic substances, in particular, biotic material. Mars lander equipment capable of detecting even one single spore can generate erroneous results if terrestrial microorganisms are also present. Thus, building clean spacecraft is absolutely critical for Mars as well as other missions concerned with the possible presence of life.

Typical Bioload Reduction Procedures

The procedures used to reduce and assay bioloads on the Phoenix Mars Lander provide a good example of the typical post-Viking approach for lander spacecraft bound for Mars. These procedures were applied during Phoenix's assembly and test operations beginning in early 2006 at Lockheed Martin Space Systems' Denver, Colorado, facility.[97] The operations were quite similar to those used earlier on MPL,

---

96. A. C. Tribble, B. Boyadijian, J. Davis, J. Haffner, and E. McCullough, *Contamination Control Engineering Design Guidelines for the Aerospace Community*, NASA Contractor Report 4740, May 1996, p. 4-1.

97. Gary Napier, "Lockheed Martin Delivers Phoenix Mars Lander Spacecraft to NASA," Lockheed Martin Corporation, 8 May 2007, *http://www.lockheedmartin.com/news/press_releases/2007/LockheedMartinDeliversPhoenixMarsLa.html* (accessed 26 January 2011).

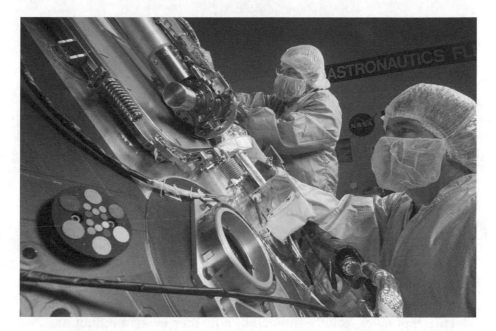

Figure 7.2 Phoenix spacecraft assembly, with Lockheed Martin Space Systems technicians working on the science deck of the Phoenix Mars Lander. Assembly operations took place in a class 100,000 cleanroom with NASA's planetary protection practices being strictly followed to prevent organics from being transported to Mars. The lander's robotic arm, built by JPL, is at the top.

MER, and Pathfinder. Assembly of the Phoenix spacecraft is depicted in figure 7.2.

To begin with, the Phoenix parts delivered to Lockheed Martin for inclusion in the spacecraft arrived already somewhat clean. The project's many subcontractors and vendors were characteristically suppliers to the aerospace industry and understood the requirements of planetary protection. They generally followed precision-cleaning protocols and assembly of components in relatively clean areas.[98] Typical precision-cleaning procedures targeted the removal of both particles and molecular films such as oils and greases and consisted of wiping gross contamination off the article, followed by an application of organic solvents and aqueous rinses with or without ultrasonic treatment. The article was then sometimes subjected to vapor

---

98. Larson interview, 15 August 2006.

degreasing, followed by a final rinse typically conducted with an iso-propyl or other alcohol solution.[99]

The Phoenix parts were usually delivered to Lockheed Martin containing less than 10,000 spores per square meter. The parachute, however, contained less than an estimated 10 spores per square meter because it had been subjected to dry heat microbial reduction involving a bakeout at temperatures above 110°C. This procedure was necessary because of the difficulty of adequately cleaning the large component. To prevent subsequent contamination, the parachute was double bagged before transport and packed with sample strips, which were then archived upon arrival at Lockheed Martin, available if needed to confirm parachute cleanliness. If there was a subsequent indication of life on Mars, it would be compared with the strips to hopefully prove that the life-forms had not been carried to Mars on the parachutes.[100]

After Phoenix parts other than the parachute arrived at Lockheed Martin, they were usually subjected to further bioload reduction. The basic tools for this were cloth wipes saturated with a 100 percent iso-propyl alcohol (IPA) solution. The cleaning action of these wipes was due mainly to the IPA's solvent properties, but a frictional action was applied as well in order to lift contamination off the parts' surfaces. The cloth wipes were repeatedly folded to keep only clean sections touching the parts' surfaces.

Biocides such as sodium hypochlorite (chlorine bleach) and hydrogen peroxide, which can also lower bioloads on a surface, were generally not approved for use on flight hardware since they are strong oxidants that can tarnish metals and break down coatings. Ethylene oxide gas was another biocide that had been considered in the past, but its benefits were too specific—it killed some microbes but not others—and it was also toxic to humans.[101]

---

99. P. R. Mahaffy, D. W. Beaty, M. Anderson, G. Aveni, J. Bada, S. Clemett, D. Des Marais, S. Douglas, J. Dworkin, R. Kern, D. Papanastassiou, F. Palluconi, J. Simmonds, A. Steele, H. Waite, and A. Zent, "Science Priorities Related to the Organic Contamination of Martian Landers," JPL, unpublished white paper posted November 2004 by the Mars Exploration Program Analysis Group (MEPAG) at http://mepag.jpl.nasa.gov/reports/index.html (accessed 26 January 2011).
100. Larson interview, 15 August 2006.
101. Ibid.

In past missions, Lockheed Martin subjected additional parts besides the parachute to DHMR bakeouts when their bioloads would have been difficult to reduce adequately in other ways due to such factors as their sizes, shapes, and materials. The aeroshells for the MER landers, for instance, which had been built by Lockheed Martin, were subjected to DHMR. The Phoenix aeroshell did not receive such treatment, however. It had been fabricated in the 1999 to 2000 time period in accordance with Pathfinder protocols. While its composite structures saw high enough temperatures during their curing processes to demonstrate that they could withstand the bakeout temperature cycle, mission staff feared that the adhesives on the fittings inside the aeroshell may not have been sufficiently temperature resistant to withstand the DHMR regime.[102]

The Bioassay Program

Bioassays of Mars spacecraft were performed typically using two basic sampling tools: absorbent cotton swabs mounted on wood applicator sticks and polyester wipe cloths.

Wet swab sampling. The swabs were purchased individually prepacked in test tubes for protection, then sterilized by autoclaving at a minimum of 121°C for at least 20 minutes. After this, they were checked regularly for sterility. Before sampling, the swab was removed from its container and moistened with sterile distilled or deionized-distilled water. The swab was used to sample a 2" × 2" (25 cm²) sample area of the part's surface. The swab was rubbed slowly and thoroughly over this surface three times, reversing direction between successive strokes and slowly rotating the head of the swab throughout the procedure. The head of the swab was then broken off and dropped back into its test tube into the sample recovery liquid consisting of sterile distilled or deionized-distilled water.[103]

---

102. Ibid.; Donald C. Larson, Phoenix Flight Program Planetary Protection Lead, Lockheed Martin Space Systems Company, Denver, CO, telephone interview by author, 2 February 2007.
103. Lockheed Martin, "Microbiological Assay Procedure" for Phoenix Mars Lander flight hardware, document PHX-312-606.

Wet wipe sampling. The wipes were 100 percent polyester, 9" × 9" (23 cm × 23 cm) cleanroom cloths. These wipes were sterilized by autoclaving them under the same conditions as for the swabs. For sampling purposes, the wipes were premoistened with sterile distilled water. The wipes were folded in quarters and placed flat on the surface to be sampled, then rubbed over the surface using a firm, steady pressure. They were then refolded so that the contaminated surface was inside the new folded configuration. The wipe was rubbed over the sample area three times, rotating the direction of motion 90° after each complete coverage of the sample area. The wipes were rinsed in buffered distilled water, the assay recovery liquid that would extract the contaminants removed from the surface. This recovery liquid contained 0.02 percent of a surfactant (potassium dihydrogen phosphate) to break up the surface tension of the liquid.[104]

●●●

Exploring the Special Areas of Mars

By 2002, exploration data had been received from the Mars Global Surveyor and Mars Odyssey orbiting spacecraft that strongly suggested the need for a new designation for certain sections of the Martian surface. These were the locales that were especially interesting for extant life investigations while at the same time were potentially quite vulnerable to terrestrial biotic contamination. Recognition of this situation led COSPAR to introduce the term "special region" in 2002 as part of the Mars planetary protection policy. Prior to this, forward contamination prevention requirements for spacecraft landing on the Martian surface fell into one of two categories distinguished by the mission's purpose:

- Category IVa—Landers without extant life-detection investigations.
- Category IVb—Landers with extant life-detection investigations.

The new special regions designation differed from those above in that it was for Martian environments that needed a high degree of

---

104. Bruce Keller, Phoenix systems test engineer, Lockheed Martin Space Systems Company, Denver, CO, interview by author, 15 August 2006.

protection, regardless of the mission purpose or category.[105] COSPAR defined a special region as "a region within which terrestrial organisms are likely to propagate, or a region which is interpreted to have a high potential for the existence of extant martian life-forms,"[106] and designated *all* missions that might come into contact with such regions as *Category IVc*. This designation applied to missions without as well as with life-detection instrumentation. Special regions were thought to be those *where liquid water was present or might occur*, and they included the following areas:

- Subsurface areas with access to a depth where the presence of liquid water was probable.
- Penetrations into polar caps.
- Areas with hydrothermal activity.[107]

Modifications of Planetary Protection Policy

NASA incorporated the Mars special regions concept into its own planetary protection policy in 2005 by modifying NASA Procedural Requirement 8020.12C—*Planetary Protection Provisions for Robotic Extraterrestrial Missions*. NASA identified some hurdles, however, to incorporating the policy into its present Mars exploration operations. In 2005, a National Research Council committee referred to as NRC PREVCOM completed the NASA-requested study, *Preventing the Forward Contamination of Mars*, and found that "there is at this time insufficient data to distinguish with confidence 'special regions' from regions that are not special."[108] NRC PREVCOM identified a key issue related to current limits on the resolution that can be obtained from Mars observations. The problem stems from the fact that signifi-

---

105. MEPAG Special Regions Science Analysis Group, "Findings of the Special Regions Science Analysis Group," unpublished white paper, p. 66, posted April 2006 by the NASA-JPL MEPAG, p. 4, *http://mepag.jpl.nasa.gov/reports/index.html* (accessed 26 January 2011).
106. COSPAR Planetary Protection Policy, 20 October 2002 (amended 24 March 2005), approved by the bureau and council, World Space Congress, Houston, TX, *http://www.cosparhq.org/scistr/PPPolicy.htm*.
107. Ibid.
108. Committee on Preventing the Forward Contamination of Mars, SSB, National Research Council, *Preventing the Forward Contamination of Mars* (unedited prepublication copy) (Washington, DC: National Academies Press, 2005).

cant horizontal and spatial diversity occurs on Mars in kilometer and even centimeter spatial scales, but relevant observational data often have spatial resolutions no better than ~3 × 10^5 km². NRC PREVCOM thus recommended an interim policy designating all of Mars a special region, until such time as more data could warrant more a precise designation.[109]

To discern between special and nonspecial regions, NASA made a request of the Mars Exploration Program Analysis Group (MEPAG), the space agency's forum for providing science input to Mars projects over the next several decades. NASA asked MEPAG to charter a Special Regions Science Analysis Group (SR-SAG)[110] to determine whether it was indeed necessary to designate *all* of Mars as special, or just parts of the planet, based on careful studies of the limits to life of known Earth organisms considered in light of known Martian environmental conditions. The answer to this question would strongly affect the conduct of future Mars missions that were already in the planning stage. The major issue, however, was whether even with the newest data, sufficient scientific information existed to construct more precise designations.[111]

The Period of Biological Interest

Space scientists have found from recent orbiter and rover data that Mars is far more diverse than indicated earlier. As a result, the scientific community has identified a very large number of sites for future exploration, and many researchers anticipate that the period of Martian biological interest will span the 21st century. Based on input from the NASA Planetary Protection Officer, the MEPAG SR-SAG study conducted to examine special region designations[112] employed a 100-year timeframe "over which the existence of martian special regions would be considered and could be encountered by

109. MEPAG, *Findings*, p. 5.
110. David W. Beaty, Karen L. Buxbaum, and Michael A. Meyer, "Findings of the Mars Special Regions Science Analysis Group," *Astrobiology* 6(5) (2006): 677–678.
111. MEPAG, *Findings*, p. 2; Margaret Race comments, 18 September 2007.
112. Margaret Race, comments to authors regarding manuscript of this book, sent 18 September 2007.

any given mission."[113] The MEPAG analysis took into account past and present climate and included consideration of current special regions as well as the possibility that a locale could become a special region within a century after a spacecraft's arrival, due to a natural event such as a volcanic eruption. Observational data backed this up—the southern polar cap appears to be able to change within a 100-year time scale. Noticeable changes in the carbon dioxide ice cover[114] have occurred from one year to another,[115] and modifications on the scale of decades have occurred to the polar cap outline. In addition, observational data revealed that where carbon dioxide ice is disappearing, water ice is being exposed.[116]

How Far Below the Surface of Mars Is Planetary Protection Relevant?

The MEPAG SR-SAG study put an important bound on the maximum depth beneath the Martian surface to which planetary protection interest extended. The study used formulas estimating the expected crater magnitudes generated by different-sized landers accidentally crashing at various velocities into a range of surface materials and concluded that only the outermost 5 meters or so of the Martian crust was likely to be contaminated by such impacts. Regions of the planet that lie deeper were not of relevance for planetary protection considerations.[117]

---

113. MEPAG, *Findings*, p. 13.
114. P. C. Thomas, M. C. Malin, P. B. James, B. A. Cantor, R. M. E. Williams, and P. Gierasch, "South Polar Residual Cap of Mars: Features, Stratigraphy, and Changes," *Icarus* 174 (2005): 535–559, as reported in MEPAG, *Findings*, p. 8.
115. Michael C. Malin, Michael A. Caplinger, and Scott D. Davis, "Observational Evidence for an Active Surface Reservoir of Solid Carbon Dioxide on Mars," *Science* 294 (7 December 2001): 2146–2148.
116. Timothy N. Titus, Hugh H. Kieffer, and Phillip R. Christensen, "Exposed Water Ice Discovered Near the South Pole of Mars," *Science* 299 (14 February 2003): 1048–1051; S. Douté, Y. Langevin, F. Schmidt, B. Schmitt, M. Vincendon, J. P. Bibring, F. Poulet, E. Deforas, and B. Gondet, "Monitoring and Physical Characterization of the South Seasonal Cap of Mars from Omega Observations" (Fourth Mars Polar Science Conference, Davos, Switzerland, 2–6 October 2006), p. 8030, http://www.lpi.usra.edu/meetings/polar2006/pdf/download/alpha_a-g.pdf (accessed 26 January 2011).
117. MEPAG, *Findings*, pp. 10–11.

The Minimum Temperature Likely To Support Propagation of Life

Microbes exist on Earth that can stay alive at temperatures well below the normal freezing point of water in environments such as within glacial ice, sea ice, and permafrost. Various mechanisms make this possible. Certain impurities such as mineral acids or salts can reduce the freezing point of water resulting in unfrozen intergranular veins that transport nutrients to and waste products from the microbes. Some cells can themselves resist freezing, while others synthesize stress proteins, reduce their size, enter dormant or spore states, adapt their components (such as making changes in their fatty acid and lipid composition), or alter the liquids in their cytoplasm. The Mars special regions designation, however, is concerned not only with cells' abilities to survive, but also with their abilities to *propagate.*

Many researchers have demonstrated some cellular metabolic activity at temperatures as low as −20°C, although at the lowest temperatures, activity was insufficient to support cell replication and was not sustained beyond a few weeks. According to MEPAG, no one has conclusively demonstrated cell replication at or below −15°C. Adding a safety factor of −5°C, the MEPAG SR-SAG study thus proposed that based on current data, an area needed to experience temperatures above −20°C to be designated a special region. Previously unknown extremophiles are always being discovered, however. If Earth organisms were found that were able to replicate at temperatures at or below −20°C, the MEPAG recommendation would need to be reevaluated.[118]

---

118. H. A. Thieringer, P. G. Jones, and M. Inouye, "Cold Shock and Adaptation," *BioEssays* 20 (1998): 49–57; C. A. Bakermans, A. I. Tsapin, V. Souza-Egipsy, D. A. Gilichinsky, and K. H. Nealson, "Reproduction and Metabolism at −10°C of Bacteria Isolated from Siberian Permafrost," *Environmental Microbiology* 5(4) (April 2003): 321–326; Blaire Steven, Richard Léveillé, Wayne H. Pollard, and Lyle G. Whyte, "Microbial Ecology and Biodiversity in Permafrost," *Extremophiles* 10(4) (August 2006): 259–267; J. N. Breezee, N. Cady, and J. T. Staley, "Subfreezing Growth of the Sea Ice Bacterium *Psychromonas ingrahamii,*" *Microbial Ecology* 47(3) (April 2004): 300–304; MEPAG, *Findings,* pp. 12–14.

The Minimum Water Activity Likely To Support Propagation of Life

While many terrestrial spores can survive extreme desiccation, microbes require liquid water in order to multiply and increase their biomass. Reasons for needing the presence of liquid water include making nutrients available to the cells and transporting waste products away from them. The MEPAG SR-SAG report thus proposed a threshold for the minimum amount of liquid water activity that needs to be present to classify an area as a special region. Water activity is a quantitative measure of the extent to which water is available in a region. It is defined as the actual vapor pressure of the water that is present divided by the vapor pressure that pure water would have at the same temperature. As an example, if pure water was present, it would have a water activity of exactly one.

The MEPAG SR-SAG report noted that based on current data, terrestrial organisms are not known to reproduce at water activities less than 0.62. Adding a safety margin, a water activity threshold of at least 0.5 was proposed when considering special region designation for an area.[119]

Thermodynamic Disequilibrium and Martian Special Regions

MEPAG SR-SAG study results indicated that in regions of the Mars surface and shallow subsurface that are at or close to long-term thermodynamic equilibrium,[120] conditions are not right for the propagation of terrestrial microbes because the temperatures and water activities are considerably below the thresholds required. Regions may exist on Mars, however, in long-term disequilibrium, "where water and temperature were in equilibrium . . . at an earlier time, but for which conditions have changed."[121] In such regions, the propagation of life may be more strongly supported, justifying their classification as special

---

119. MEPAG, *Findings*, pp. 14–16.
120. When a system is in thermodynamic equilibrium, all parts of it have attained a uniform temperature that is the same as that of the system's surroundings.
121. John D. Rummel, "Special Regions in Mars Exploration: Problems and Potential," *Acta Astronautica* 64 (2009): 1293–1297.

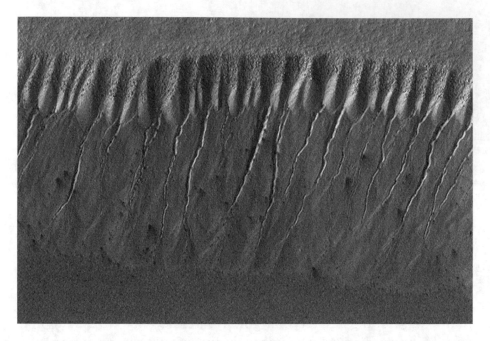

Figure 7.3 Mars Global Surveyor evidence of recent liquid water: gully landforms possibly caused by geologically recent seepage and runoff.

regions. Natural long-term disequilibrium conditions on the Martian surface would have developed as a result of geological phenomena such as geothermal vents or volcanoes.[122]

Martian Gullies: Are Some of Them Special Regions?

The genesis of gullies on the Martian surface may have involved liquid water (see figures 7.3 and 7.4). What's more, the features adjacent to some of the gullies suggest that they might be so young as to be sites at which liquid water could occur today or in the near future, at least for brief periods, which would classify these sites as special regions. It is by no means sure, however, that it was liquid water that formed these features, and this contributes to the difficulty of knowing which Martian surface areas should be designated as special regions.

---

122. MEPAG, *Findings*, pp. 26–27.

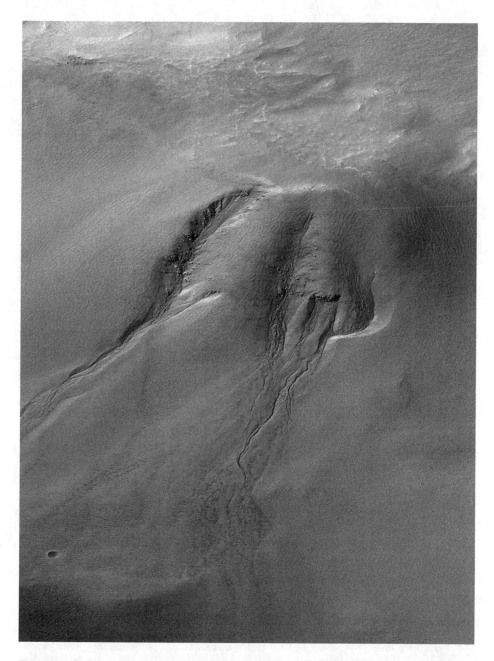

Figure 7.4 Gullies eroded into the wall of a meteor impact crater. This picture taken by the Mars Global Surveyor's Mars Orbiter Camera shows channels and associated aprons of debris that could have formed by groundwater seepage, surface runoff, and debris flow. The lack of small craters superimposed on the channels and apron deposits suggests that these features are geologically young. It is possible that these gullies are indicative of liquid water present within the Martian subsurface today.

Alternate hypotheses consider gully-forming mechanisms such as flows of liquid carbon dioxide released from below-ground reservoirs sealed with dry-ice barriers[123] or dry, granular flows not employing a volatile substance such as water or carbon dioxide. Dry flows have been modeled and appear plausible for cases when individual particle settling is slower than characteristic terrestrial debris flow speeds, as would be the situation with Mars's reduced gravity.[124]

The MEPAG SR-SAG study recommended that those Martian gullies and gully-forming regions in which liquid water could possibly surface within the next 100 years be classified as special regions, but it is problematic to identify such areas. A clue can be given by the gully's age. A gully formed in the recent past may be more likely to have continued activity than a gully that has remained static for a long period. Samplings of Martian gully data suggest that all of them are young relative to Martian geological time scales, and that liquid water sources may exist at shallow depths beneath some of them.[125] No known geomorphic criteria exist, however, that can predict which of the tens of thousands of gullies on Mars are likely to become active in the next century, although gullies with the following characteristics are being examined more closely:

- Young ages.
- Channels within preexisting gully complexes.
- Bright material visible, possibly ice or salts.

For the purposes of planetary protection, the best strategy at present regarding gully-forming areas is probably to extend special region status to all of them, rather than only to specific preexisting gully features. As of yet, however, the scales of Mars's gully-forming areas have not been well defined.[126]

---

123. D. S. Musselwhite, T. D. Swindle, and J. I. Lunine, "Liquid CO_2 Breakout and the Formation of Recent Small Gullies on Mars," *Geophysical Research Letters* 28(7) (2001): 1283–1285; N. Hoffman, "Active Polar Gullies on Mars and the Role of Carbon Dioxide," *Astrobiology* 2(3) (2002): 313–323.

124. Allan H. Treiman, "Geologic Settings of Martian Gullies: Implications for Their Origins," *Journal of Geophysical Research* 108(E4) (8 March 2003): 8031; T. Shinbrot, N. H. Duong, L. Kwan, and M. M. Alvarez, "Dry Granular Flows Can Generate Surface Features Resembling Those Seen in Martian Gullies," *Proceedings of the National Academy of Sciences* 101(23) (8 June 2004): 8542–8546; MEPAG, *Findings*, pp. 26–27.

125. Michael C. Malin and Kenneth S. Edgett, "Evidence for Recent Groundwater Seepage and Surface Runoff on Mars," *Science* 288 (30 June 2000): 2330–2335.

126. MEPAG, *Findings*, pp. 29–30.

Midlatitude Indications of Snow and Ice

The middle Martian latitudes exhibit a range of features that some researchers have interpreted as evidence of ice-bearing materials. These possible special regions may have been formed during a period of different climatic conditions in Mars's past, when much of the midlatitudes may have been mantled by ice-rich materials. Two remnants of these materials are particularly relevant for the discussion of special regions:

1. *Midlatitude mantle* covering major sections of Mars between 30° and 60° latitude in both hemispheres.[127]
2. Materials called *pasted-on terrain* that are most commonly found on poleward-facing midlatitude slopes such as crater walls and massifs.

Midlatitude mantle is a layered deposit that scientists estimate from Mars data to be 1 to 10 meters thick. Mustard et al. proposed that the roughening observed on sections of the mantle resulted from sublimation of ice out of the mixture of ice and dust that originally produced the mantle.[128] The midlatitude mantle varies noticeably as a function of latitude. Lower latitudes of ~30° to 45° characteristically have areas of smooth, intact mantle next to areas where the mantle has been completely stripped from the surface. Higher latitudes of ~45° to 55° typically have a knobby surface texture indicating incomplete removal of the mantle material.[129] Latitudes higher than ~55° exhibit the least removal of mantle material, which suggests that deposits at these lat-

---

127. M. A. Kreslavsky and J. W. Head, "Kilometer-Scale Roughness of Mars: Results from MOLA Data Analysis," *Journal of Geophysical Research* 105 (2000): 26695–626712, http://www.planetary.brown.edu/planetary/documents/2447.pdf (accessed 26 January 2011); M. A. Kreslavsky and J. W. Head, "Fate of Outflow Channel Effluents in the Northern Lowlands of Mars: The Vastitas Borealis Formation as a Sublimation Residue from Frozen Ponded Bodies of Water," *Journal of Geophysical Research* 107(E12) (2002): 5121, http://www.planetary.brown.edu/planetary/documents/2686.pdf (accessed 26 January 2011); M. A. Kreslavsky and J. W. Head III, "Mars: Nature and Evolution of Young Latitude-Dependent Water-Ice-Rich Mantle," *Geophysical Research Letters* 29(15) (2002), http://www.planetary.brown.edu/planetary/documents/2756.pdf (accessed 26 January 2011).

128. J. F. Mustard, C. D. Cooper, and M. K. Rifkin, "Evidence for Recent Climate Change on Mars from the Identification of Youthful Near-Surface Ground Ice," *Nature* (2001): 411–414, http://www.planetary.brown.edu/planetary/documents/2610.pdf (accessed 26 January 2011).

129. R. E. Milliken, J. F. Mustard, and D. L. Goldsby, "Viscous Flow Features on the Surface of Mars: Observations from High Resolution Mars Orbiter Camera (MOC) Images," *Journal of Geophysical Research* 108(E6) (2003): 5057.

itudes have experienced less erosion and thus "may still be ice-rich beneath a thin layer of ice-free dust."[130] Mars Odyssey data supported the possibility of near-surface water ice by detecting an increased abundance of hydrogen within the upper meter of the surface at latitudes greater than about 55°. There is, however, no evidence for melting over much of this region. The mantle material was likely emplaced between 2 and 0.5 million years ago, and it has been undergoing sublimation and desiccation for the last half million years. Scientists do not predict active layers of water flow.

Pasted-on terrain consists of apparently young mantle deposits that have been preferentially preserved on poleward-facing slopes in the midlatitudes of both Martian hemispheres. They were initially noted by Malin and Edgett,[131] who thought they resembled accumulations of snow on colder, more frequently shadowed surfaces, even though the materials are not light-toned like snow. Christensen, using Mars Odyssey images, also noted these accumulations and speculated that they might be remnants of old snow accumulations.[132] The deposits appear to be covered with a residue of dust that could be protecting snow layers from further sublimation.[133]

Both Christensen and Milliken et al.[134] proposed that the pasted-on terrain might be the source of water that created the midlatitude gullies discussed above. While the temperatures at the Martian midlatitudes are so cold that the presence of liquid surface water is unlikely, Christensen theorized that since the surface of a snow deposit had mixed with dust prevalent in the Martian atmosphere, this created a conglomeration darker than the snow alone that readily absorbed solar energy. This top layer then transmitted the energy, in the form of heat, to the pristine layers of snow several centimeters beneath. In roughly 5,000 years, according to his calculations, this heat melted enough of

---

130. MEPAG, *Findings*, p. 32.
131. M. C. Malin and K. S. Edgett, "Mars Global Surveyor Mars Orbiter Camera: Interplanetary Cruise Through Primary Mission," *Journal of Geophysical Research* 106 (October 2001): 23429–23570.
132. P. R. Christensen, "Formation of Recent Martian Gullies Through Melting of Extensive Water-Rich Snow Deposits," *Nature* 422 (6 March 2003): 45–48.
133. MEPAG, *Findings*, p. 33.
134. Milliken et al., "Viscous Flow."

the subsurface snow to erode underlying rock and carve gullies.[135] Some of the pasted-on mantle areas thus may constitute special regions.[136]

Polar Caps

In COSPAR's 2005 planetary protection policy statement regarding special regions,[137] the polar caps were mentioned as an example of a special region. But the results of the MEPAG SR-SAG report suggested that the polar caps do not actually fit the special regions designation.[138] Jakosky et al.[139] (and others) have calculated that at present, maximum summer temperatures reach only about –70°C in the Martian polar regions, too cold for them to be naturally occurring special regions. Resolution of this question will have to wait until additional data on the polar regions are received and interpreted.

Creating a Special Region with a Spacecraft

MEPAG's analysis revealed that it was possible for spacecraft to *induce* conditions that favored biological propagation within a localized region, thus transforming an area that was not a special region into one. For instance, spacecraft generate heat, which can enhance the conditions for the propagation of life. In regions where ice is present, thermal emissions from spacecraft could result in the localized elevation of temperatures and the melting of ice deposits, leading to the formation of

135. Ron Cowen, "Martian Gullies: Carved by Melting Snow?" *Science News* 163(8) (22 February 2003): 116, *http://www.sciencenews.org/articles/20030222/fob3.asp*.
136. MEPAG, *Findings*, p. 34; Rummel, "Special Regions in Mars Exploration: Problems and Potential," 1296.
137. COSPAR Planetary Protection Policy, 20 October 2002 (amended 24 March 2005), approved by the bureau and council, World Space Congress, Houston, TX, *http://www.cosparhq.org/scistr/PPPolicy.htm*.
138. MEPAG SR-SAG, "Findings of the Mars Special Regions Science Analysis Group," *Astrobiology* 6, no. 5 (2006): 718.
139. B. M. Jakosky, B. G. Henderson, and M. T. Mellon, "The Mars Water Cycle at Other Epochs: Recent History of the Polar Caps and Layered Terrain," *Icarus* 102 (1993): 286–297; MEPAG, "Findings," unpublished white paper, p. 47.

liquid water reservoirs. Microbes either initially present in such an area or transported there by the spacecraft could possibly replicate.

Many spacecraft employ radioisotope thermal generators (RTGs) to supply electrical energy. These devices, which are typically powered by the radioactive decay of plutonium, are a perennial heat source. Their use necessitates an in-depth analysis to determine whether they will create temperatures local to the RTG above the threshold needed for biotic propagation.[140]

A key factor in the impact of spacecraft-induced environmental changes is the duration of the new conditions and whether it is of sufficient length to permit microbial propagation. The highest growth rates documented by the MEPAG SR-SAG study at the temperatures of interest on Mars (–15°C to +5°C) indicated that significant replication of terrestrial microorganisms would not occur if the duration of a maximum surface temperature of –5°C did not exceed 22 hours, a maximum of 0°C did not exceed 3 hours, or a maximum of 5°C did not exceed 1 hour.[141] Since the boiling point of water at a Martian surface atmospheric pressure of 8.6 millibars[142] is 5°C, there was no need to extend this analysis to higher temperatures.

●●●

NASA and COSPAR continue to work toward reaching a consensus within the space science community of what constitutes a Mars special region. One such effort toward this end was the COSPAR colloquium held in Rome, Italy, 18–20 September 2007. John Rummel, the former NASA PPO and (at the time of this writing) current Senior Scientist for Astrobiology in the Science Mission Directorate at NASA

---

140. MEPAG, *Findings*, pp. 53–54.
141. D. A. Ratkowsky, J. Olley, T. A. McMeekin, and A. Ball, "Relationship Between Temperature and Growth Rate of Bacterial Cultures," *Journal of Bacteriology* (1982): 1–5; D. A. Ratkowsky, R. K. Lowry, T. A. McMeekin, A. N. Stokes, and R. E. Chandler, "Model for Bacterial Culture Growth Rate Throughout the Entire Biokinetic Temperature Range," *Journal of Bacteriology* (1983): 1222–1226; R. B. Haines, "The Influence of Temperature on the Rate of Growth of *Sporotrichum Carnis* from –10 to 30°C." *Journal of Experimental Biology* (1931): 379–388. All papers were cited in MEPAG, *Findings*, p. 55.
142. By means of comparison, surface atmospheric pressure on Earth is approximately 120 times higher.

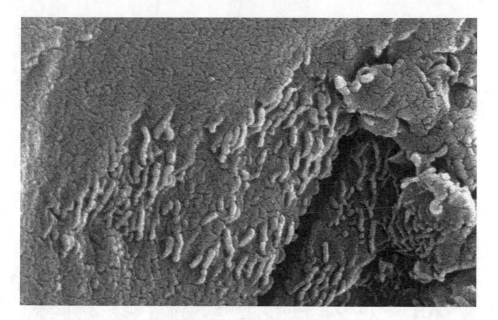

Figure 7.5 Scientists have speculated that the tube-like structures in these electron microscope images may be microscopic fossils of primitive, bacteria-like organisms that lived on Mars more than 3.6 billion years ago. A two-year NASA investigation found organic molecules, mineral features characteristic of biological activity, and possible microscopic fossils such as these inside an ancient Martian rock that fell to Earth as a meteorite. The largest of these structures are less than 1/100th the diameter of a human hair, while most are 10 times smaller.

Headquarters, is very active in this undertaking, which seeks to define the appropriate use of the special regions designation within COSPAR and NASA planetary protection categories. The results of this work will impact many nations' plans for future missions to Mars.

Mars Sample Return

The interest in bringing back sample materials from solar system bodies, especially Mars, increased measurably in 1996 when the analysis of a Martian meteorite found on Earth revealed possible evidence of life (see figure 7.5) as well as the past presence of liquid water on the Red Planet. This find, coupled with the discovery that even under extremely harsh conditions on Earth, life can be found wherever there is liquid water and energy, reinforced the hypothesis that life may have

Figure 7.6 An artist's concept of the launch of a Martian sample back toward Earth.

emerged on Mars at some point in the planet's past. Although considerable data suggest that the Martian surface is hostile to life as we know it, plausible scenarios have been presented for the existence of microbes in special locations such as possible hydrothermal oases or subsurface regions.

Uncertainties regarding life on Mars may be resolved or at least reduced through the examination of returned Martian samples.[143] Sample return missions are also relevant to other objectives, such as the following:

- Understanding the evolution of the Martian surface and interior via local and global studies of its chemistry, lithology, and morphology.
- Characterizing the dynamics and chemistry of the Martian atmosphere.

---

143. SSB Task Group on Issues in Sample Return, *Mars Sample Return—Issues and Recommendations* (Washington, DC: National Academies Press, 1997), pp. vii, 2–3.

- Determining the degree to which climatic conditions have evolved over time.
- Searching the planet not only for extinct or extant life (or remnants thereof), but also for evidence of prebiotic organic compounds and prebiotic chemical evolution.
- Exploring the interaction of Mars with the solar wind and the extent to which this interaction affects the state of the planet's upper atmosphere and ionosphere.[144]

While NASA at one time envisioned Mars sample return missions taking place early in the 21st century, this target has been delayed, largely because of significant technical challenges. One approach for such an effort would be to use a robotic lander spacecraft that included a rocket capable of taking off from Mars to send samples of rocks, soils, and atmosphere back to Earth for detailed chemical and physical analysis. Using Earth laboratories, scientists could measure sample characteristics much more precisely than they could remotely. In addition, for decades after the samples were returned, they could yield new information as future emerging technologies were employed to study them.[145]

Figure 7.6 is an artist's concept of a spacecraft lifting off from Mars to return the samples it collected back to Earth.

The Mars Science Laboratory (MSL) mission, scheduled for launch in 2011, actually had a sample return component under study. To practice caching Mars specimens, the mission team considered including a hockey-puck-sized sample cache to store Martian material until some future mission could bring it back to Earth. The objective of installing this caching capability in MSL, according to Alan Stern, NASA's former Science Mission Directorate Associate Administrator, was to "build the foundation of support for future Mars sample return activities, not only in scientific and public circles, but also in Congress and the White

144. SSB, "On NASA Mars Sample Return Mission Options," National Academies Web site, *http://www7.nationalacademies.org/ssb/msrrep.html* (last updated 10 February 2000, accessed 3 November 2006).
145. JPL, "Missions to Mars," NASA's Mars Exploration Program Web site, *http://marsprogram. jpl.nasa.gov/missions/future/futureMissions.html* (last updated 22 March 2006, accessed 1 November 2006).

House Office of Management and Budget."[146] The sample cache idea was not limited to the MSL project. The European Space Agency also considered including it in its ExoMars program.

Although the execution of Mars sample return missions is in the future, it is interesting to briefly examine the considerable scientific and engineering efforts that have already gone into conceptualizing and planning for such missions.

Past Efforts in the Design of Sample Return Missions

In response to NASA's 1995 request, the National Research Council's Space Studies Board (SSB) specifically convened the Task Group on Issues in Sample Return to examine challenges related to bringing back material from other solar system bodies.[147] The primary driver for this action was a Mars sample return mission that, in the 1990s, had been envisioned for a 2005 launch, although the conclusions and recommendations in the Task Group's study applied to return missions from "any solar system body with a comparable potential for harboring life."[148]

Work on sample return concepts focused on issues of potential back contamination—i.e., how to protect Earth from contamination by exotic biota. This involved assessments of the potential for the following:

1. A viable biological entity to be in a return sample.
2. Large-scale impacts if that entity entered the biosphere of Earth.

The small sizes of any samples returned would preclude concern about their direct toxicity to life on Earth, due to either their chemistry or their radioactivity. These potential dangers would also be mitigated through laboratory protocols. *Replicating biohazards*, however, are serious planetary protection concerns. Even small return samples

146. Leonard David, "Mars Sample Return Proposal Stirs Excitement, Controversy," *Space.com*, 26 July 2007, *http://www.space.com/4116-mars-sample-return-proposal-stirs-excitement-controversy.html* (accessed 26 January 2011).
147. Wesley T. Huntress to Claude R. Canizares, 20 October 1995, in SSB, *Mars Sample Return*, Appendix, p. 46.
148. SSB, *Mars Sample Return*, p. 8.

could conceivably harm our biosphere if they were able to propagate within it.[149]

One of the key questions pertaining to Mars that the Task Group examined was what scientific analyses needed to be performed to reduce uncertainty regarding the existence of life on the Red Planet. The Task Group also assessed the efficacy of existing technology for preventing the return of uncontained and unsterilized Martian material and made recommendations for the appropriate distribution of the samples.

Factors Regarding Possible Extant Life on Mars

The Task Group examined characteristics of the Martian climate, especially what it might have been like during the planet's early history (prior to 3.5 billion years ago). At that time, geological evidence indicates that conditions were possibly warmer and friendlier to life, with liquid water flowing on the surface. Hydrothermal environments may also have been common on the early Mars.

Studies of Earth bacteria suggest that life may first have arisen in hot-spring environments or that during its evolution it resided for some time under such conditions. Thus it was deemed plausible that Martian bacteria may once have existed that followed a similar evolutionary path.[150]

The Task Group also analyzed mechanisms by which Martian life-forms, either of independent origin or having traveled to the planet from Earth by means of meteorite impacts, could have survived in localized niches and as spores until today. The Viking mission's experiments convinced many scientists at the time that Mars was probably devoid of life. But the Task Group was aware of the limitations of that mission's technology, which was only able to test for a few of the possible mechanisms by which Martian organisms could have obtained energy and had access to water. Alternatively, Viking may not have found life simply because it carried out its testing in the wrong locations.

---

149. Ibid., pp. 8–9.
150. C. R. Woese, "Bacterial Evolution," *Microbiology and Molecular Biology Reviews* 51(2) (1987): 221–271; SSB, *Mars Sample Return*, pp. 12, 14.

Niches that Martian life could potentially occupy include the interiors of rocks in which water is present as well as in other "isolated oases where liquid water exists,"[151] such as in recent volcanic vents that Viking would have missed. Including these possibilities in future missions will impact the types and locations of samples that are collected, as well as the technologies for examining and quarantining them.

Protecting Against Large-Scale Impacts from Martian Biota in Earth's Biosphere

A fact that the Task Group considered in its analysis is that pathogenesis, the ability to induce disease, is rare in terrestrial microbes. Only a tiny fraction produce adverse effects in their host organisms. The capacity for pathogenesis is typically generated through selective evolutionary pressure, and such a process of selection for pathogenesis to Earth organisms would have no reason to exist on Mars. What's more, exotic microbes introduced into new biospheres usually do not alter them in any significant manner. Consider that a Martian organism able to survive in Earth's environment would encounter formidable competition from terrestrial biota that have had millions of years to adapt to the same environmental niche. Thus, the chances are extremely low that a Martian organism would be able to cause widespread damage after it was introduced into Earth's ecosystem. Nevertheless, because there is a nonzero possibility that such biota could generate environmental disruption on Earth, the Task Group recommended that return samples from Mars be kept in physical and biological isolation and treated as though they were hazardous until proved otherwise.

The need for sterilization. The Task Group took the position that no spacecraft surfaces exposed to the Martian environment could be returned to our planet unless sterilized. Furthermore, if dependable sample containment could not be verified on the return trip to Earth, the sample would need to be either sterilized or not returned. The containment technology had to remain reliable during the spacecraft's reentry through Earth's atmosphere and landing as well as the sample's

---

151. SSB, *Mars Sample Return*, p. 15.

subsequent transport to its receiving facility. Finally, sample distribution can only be undertaken if rigorous examination determined that the extraterrestrial material was not biologically hazardous.[152]

Breaking the chain of contact with Mars. A key factor in protecting Earth is to design sample return missions and spacecraft so as to "break the chain of contact" with Mars. This means that no uncontained hardware that came into contact with the Martian environment, directly or indirectly, can be returned to Earth. Uncontained hardware that will be returned to Earth needs to be isolated from anything exposed to the Martian environment during all phases of the mission, including during any in-flight transfer operations.[153]

Research Approaches for Reducing Uncertainty Regarding the Existence of Martian Life

The Task Group identified several categories of research aimed at better understanding the possibilities for the existence of Martian life. These included the following:

- Observations conducted from Mars orbiting craft, robot landers, and rovers.
- Examinations of Martian meteorites.
- Studies of terrestrial extremophiles.
- Analyses of Mars return samples.

Several key research questions that needed to be answered based on data generated by the above investigations as well as past work are included below.

Are there present regions of Mars where life-forms could conceivably exist? The search for these niches is closely related to the search for evidence of water, active volcanism, and the presence of nonequilibrium gases.

---

152. SSB, *Mars Sample Return*, pp. 19–20.
153. NASA Science Mission Directorate, "Detailed Planetary Protection Requirements," in *Planetary Protection Provisions for Robotic Extraterrestrial Missions*, NPR 8020.12C, 27 April 2005, Appendix A, *http://nodis3.gsfc.nasa.gov/displayDir.cfm?Internal_ID=N_PR_8020_012C_&page_name=AppendixA* (accessed 26 January 2011).

Are there regions that are sterile? If Mars exploration could identify regions with physical and chemical characteristics that preclude the existence of life, planetary protection guidelines for those locales could be relaxed, opening the door to a wider spectrum of analyses.

Are meteorites able to carry living organisms between planets? To date, 12 meteorites believed to be from Mars have been found on Earth, although scientists estimate that roughly five Martian meteorites fall somewhere on Earth (including into the oceans) each year. If some of these meteorites can be shown to have transported viable microbes to Earth, can any evidence be found that they propagated in or impacted our biosphere in any way?

Transport of terrestrial meteorites to Mars is also believed to have occurred, although considerably less often due to the relative strengths of the two planets' gravitational fields as well as to their different orbital characteristics. The question then emerges whether terrestrial organisms could have survived ejection from Earth, transport through space and through the Martian atmosphere, and impact with the planet. If viable microbe transport from Earth was possible, it could conceivably have resulted in the microbes propagating and evolving on Mars.

The Mars Sample Handling Protocol Workshops

In response to the Task Group's findings and recommendations, NASA undertook a series of workshops to prepare for the receiving, handling, testing, distributing, and archiving of return samples from Mars. The overall objective of the workshop series was to develop a draft protocol guiding the assessment of returned samples for biohazards and evidence of extant or extinct life while at the same time preventing sample contamination with terrestrial materials.[154] Emerging from these workshops were important research questions that needed to be pursued. These questions are clearly relevant to return samples from other target bodies in addition to Mars. Discussions of some of these questions are included below.

---

154. Margaret S. Race, Donald L. DeVincenzi, John D. Rummel, and Sara E. Acevedo, eds., *Workshop 4 Final Report and Proceedings—Arlington, Virginia, June 5–7, 2001*, NASA/CP-2002-211841, Mars Sample Handling Protocol Workshop Series, June 2002, pp. 1, 71–132.

What criteria must be satisfied to demonstrate that a Martian material does not present a biohazard? Answering this was to involve testing for adverse effects on a range of organisms, from unicellular organisms to animals and plants, as well as tests specifically for identifying mutagenesis and DNA damage. Testing on multi-organism populations was also identified as important in order to gauge the potential for ecosystem disruption.[155]

What are the facilities needed for making reliable assessments? To determine this, an important issue that needed to be resolved was whether all sample receiving, processing, characterization, life, and biohazard detection should take place at one sample receiving facility (SRF), or were these functions better addressed by distributing them among several labs? Also, the best equipment to perform all these functions needed to be identified. These were issues similar to those that NASA faced in developing concepts for the Apollo Lunar Receiving Laboratory (see Chapter 4).

What sample sizes are required? These depend on such factors as the relevance of the dose of a material being examined, limitations to the dose that the biological model system can receive, the particular sample preparation procedure being used, the number of tests that need to be conducted, and the total biohazard testing time required.

What total duration of biohazard testing is needed? To estimate this, NASA had to determine the durations of the slowest tests and add a safety margin to these times. Workshop participants estimated that three months would be too short a time for biohazard testing and considered four to six months more preferable.[156]

Can survival mechanisms of terrestrial extremophiles serve as models of how putative Martian extremophiles might exhibit resistance to sterilization?[157] This is a key question when designing appropriate sterilization approaches for equipment and materials returned from Mars. Workshop participants made the point that since a species's adaptation to stress evolved through natural selection, they expected

155. Race et al., *Workshop 4*, pp. 71, 110–112.
156. Race et al., *Workshop 4*, pp. 112–116.
157. Carl W. Bruch, Richard B. Setlow, and John D. Rummel, eds., *Interim Report of the Workshop Series—Workshop 2a (Sterilization) Final Report, Arlington, Virginia, November 28–30, 2000*, NASA/CP-2001-210924, Mars Sample Handling Protocol Workshop Series, June 2001, p. iii.

a life-form able to survive on Mars to have adapted to extremely hostile conditions, including a low temperature and very dry environment with an atmosphere that provided no protection against UV exposure.

Since the combination of severe conditions found on Mars does not exist anywhere on Earth, terrestrial microbes would not have adapted to it, and thus it was unlikely that Earth-microbe survival mechanisms could serve as ideal models for Martian ones when developing sterilizing approaches for returned samples and materials. Nevertheless, terrestrial microbes could be found that are very resistant to a subset of Martian conditions, and these microbes could be useful in helping to validate sterilization procedures

What sterilization procedures will best preserve the integrity of Martian soil and rock samples for future scientific analyses outside the proposed containment facility? The ideal method would destroy all replicating life-forms in Martian material but have no effect on its geological or chemical properties and leave no residue that might obfuscate further analysis. The method chosen should be supported by all the different scientific communities (geological, biomedical, and public health) and comply with all regulations.

Methods considered included moist heat, dry heat, gamma radiation, ethylene oxide, other gas sterilization methods (hydrogen peroxide plasma or vapor, chlorine dioxide gas), and combinations of the above. The aim was to find the method that came as close as possible to the ideal, although all the methods presented some problems. Heat could alter or destroy components of the material; ethylene oxide could leave a residue as well as changing the chemistry by adding ethyl groups to certain materials; and gamma radiation could make physical, chemical, and biological changes in the material.[158]

The workshop concluded that only two methods of sterilization appeared to be viable, assuming that Martian life was based on similar chemistry to that on Earth. These methods were dry heat and gamma radiation (used together or separately). These were chosen because both would penetrate the material and could provide high assurances of organism destruction while causing a minimal level of alteration to the material.[159]

---

158. Bruch et al., *Interim Report of the Workshop Series—Workshop 2a*, pp. 19–20.
159. Ibid., pp. 15–16.

When is sample material from Mars safe to be released from containment? Workshop participants identified several questions that first had to be answered, as well as technologies that could help to do so:

- *Is there anything in the samples that looks like a life-form?* Microscopy as well as high-resolution analytic probe techniques such as beam synchotron methods would be useful in answering this. Techniques were especially sought that would allow testing of contained, unsterilized samples outside of the Sample Receiving Facility.

- *Is there a chemical signature of life?* Mass spectrometry was suggested as an approach for analyzing contained samples to identify biomolecules and certain asymmetries and special bonding suggestive of life.

- *Is there evidence of replication?* Answering this question would require attempts to replicate any extant Martian microbes in cultures or in living organisms.

- *Is there any adverse effect on workers or the surrounding environment?* Medical surveillance and evaluation of living systems in proximity to the SRF were recommended.[160]

If life other than terrestrial microbes are detected in Martian samples, then what are the next steps? Workshop participants were clear that "if life is discovered in any sample material,"[161] Martian materials could not immediately be released from maximum containment. A previously constituted scientific oversight committee needed to review all actions taken, and testing had to be stopped until the adequacy of the protocol and the provisions for containment were thoroughly reviewed. Some participants also recommended a "prophylactic sterilization"[162] method that would involve applications of gamma radi-

---

160. Margaret S. Race, Donald L. DeVincenzi, John D. Rummel, and Sara E. Acevedo, eds., *Workshop 4 Final Report and Proceedings—Arlington, Virginia, June 5–7, 2001*, NASA/CP-2002-211841, Mars Sample Handling Protocol Workshop Series, June 2002, pp. 77–80.

161. Margaret S. Race, Kenneth H. Nealson, John D. Rummel, and Sara E. Acevedo, eds., *Interim Report of the Workshop Series—Workshop 3 Proceedings and Final Report—San Diego, California, March 19–21, 2001*, NASA/CP-2001-211388, Mars Sample Handling Protocol Workshop Series, December 2001, pp. 9–10.

162. Margaret S. Race, Donald L. DeVincenzi, John D. Rummel, and Sara E. Acevedo, eds., *Workshop 4 Final Report and Proceedings—Arlington, Virginia, June 5–7, 2001*, NASA/CP-2002-211841, Mars Sample Handling Protocol Workshop Series, June 2002, p. 10. Material in this paragraph is also drawn from p. 126.

ation and minimal heating. There were fears, however, that such a protocol could destroy some of the samples' scientific value. It could also erode public trust in the methodologies that had been painstakingly developed for handling potentially life-bearing samples, without adding substantively to public safety.

Although a Mars sample return is still years away, material from the planet will eventually be collected by spacecraft and returned to Earth for scientific study. Because the samples may contain specimens of microorganisms, perhaps even in a viable state, the material needs to be collected and handled in a manner that will protect it as well as our planet. Considerable behind-the-scenes planning has been under way at JPL and other Centers, at NASA Headquarters, and within the international community to make sure that reliable planetary protection controls and safeguards are built into return missions from the earliest stages of planning, to ensure compliance with all international regulations, and to protect the integrity of scientific efforts.

●●●

Continued Development of Human Mars Mission Policy

Sometime within the first several decades of the 21st century, NASA will likely send a human mission to Mars. The introduction of astronauts to the Martian environment, with all their needs and support systems, will greatly increase planetary protection concerns compared to those of robotic landers, orbiters, and flybys. To address these concerns, NASA sponsored or cosponsored a series of interorganizational meetings and studies, which included the following:

The Pingree Park Workshop

Ames Research Center sponsored a workshop in 2001, the Pingree Park workshop, held at Colorado State University, that focused on this question: *Can human exploration of the Martian surface be done effectively and without harmful contamination?* The many contamination dangers were placed within three general categories: 1) protecting Mars and Martian samples from forward contamination, 2) protecting astronaut health against risks from the Martian environment, and 3) preventing the back contamination of Earth from possible Mars contaminant sources.

The question—can nonharmful Martian exploration be performed—elicited an examination of key human and mission requirements that led to planetary protection concerns, along with strategies to mitigate or negate them. Human missions to Mars will by their nature be very complex, entailing all the planetary protection issues of robotic missions as well as others "arising through the direct involvement of humans and their accompanying microbial companions."[163] The Pingree Park workshop was the first specifically devoted to this topic since the 1990 "Planetary Protection Issues and Human Exploration of Mars" meeting, over a decade earlier.[164] That workshop preceded the Mars Global Surveyor and Mars Pathfinder missions and thus did not have the advantage of using the data they collected to better understand human exploration issues regarding Mars.

The Pingree Park workshop identified and discussed several key areas of inquiry related to human Mars missions, including these:

- Spatial dispersion of dust and contaminants on Mars by wind and other means.
- Potential impacts of a human-occupied Martian base, such as generation and management of respiratory and food-supply wastes.
- The best use of robotics for helping to conduct operations on Mars in a manner consistent with planetary protection concerns.
- Spacesuit designs consistent with planetary protection needs, especially for human activities away from pressurized habitats and rovers.
- Technology required for life detection and potential pathogen detection.

---

163. M. E. Criswell, M. S. Race, J. D. Rummel, and A. Baker, eds., "Planetary Protection Issues in the Human Exploration of Mars," NASA/CP-2005-213461, final report of a workshop held in June 2001 at Pingree Park Mountain Campus, Colorado State University, Fort Collins, CO, 9 May 2005.

164. D. L. DeVincenzi, H. P. Klein, and J. R. Bagby, eds., Planetary Protection Issues and Future Mars Missions, proceedings of the workshop "Planetary Protection Issues and Human Exploration of Mars," NASA/CP-1991-10086, final report of a workshop held at ARC, Moffett Field, CA, 7–9 March 1990.

- Site classification and biological plausibility mapping of the Martian surface and subsurface based on remote sensing data.

The Pingree Park workshop was an investigative meeting that examined vital Mars exploration issues but certainly did not have the time or personnel to completely work through these complex areas of inquiry. More meetings as well as extensive research were required. One of the Pingree Park workshop's accomplishments was to identify a range of questions that needed to be examined by future conferences. These questions included the following:

- If Martian life is discovered, what would appropriate responses be? Should it be left in place and no invasive study conducted, should it only be analyzed in situ, or should it be carried back to Earth for detailed study?
- How do human factors such as life support, work environment, and psychological stresses interact with planetary protection issues during an extended mission of many months to a few years?
- What are the best ways to prepare the public for the possible detection of nonterrestrial life? How can details regarding mission discoveries, possible impacts to Earth, and efforts to control associated risks be effectively communicated?[165]
- What might the public response be to the detection of life on Mars? Will the public accept the very small but finite risks associated with discovering and analyzing extraterrestrial life?

Mars Human Precursor Studies

In June 2004, NASA's Mars Exploration Program Analysis Group (MEPAG), whose function was to provide science input to Mars projects, chartered the Mars Human Precursor Science Steering Group

---

165. Another study that addressed societal interest in the search for extraterrestrial life and the need for public outreach and education was Margaret S. Race and Richard O. Randolph, "The Need for Operating Guidelines and a Decision Making Framework Applicable to the Discovery of Non-Intelligent Extraterrestrial Life," *Advances in Space Research* 30(6) (2002): 1583–1591.

(MHP SSG)[166] to identify priorities for investigations, measurements, and planetary protection technology demonstrations that would be carried out *prior* to the first human mission to Mars. The aim of the precursor missions was to mitigate, to the extent possible, the risks and costs of the first human mission to Mars. Recommended MHP SSG precursor activities with planetary protection–related concerns included the following:

- Determining if each Martian site to be visited by humans was free, to within acceptable risk standards, of replicating biohazards. The investigation would include analyses of atmosphere, dust, near-surface soil, deep soil, rock, and ice.
- Determining the possible toxic effects of Martian dust on humans. In particular, determining whether the dust and other Martian particulates that would be present in or on a returning spacecraft (other than deliberately collected samples, which could be held in sealed containers) had life-forms associated with them, and if so, whether that life was hazardous.
- Determining whether terrestrial life could survive and reproduce on the Martian surface.
- Determining processes by which terrestrial microbial life, or its remains, would be dispersed and/or destroyed on Mars.[167]

The MHP SSG study expanded on the work of two earlier 21st century analyses—a MEPAG 2004 report and the NRC's 2002 Safe on Mars report.[168] These studies, although important, needed to be reconsidered in light of new data from Mars robotic missions—in particular, data regarding the possibility of near-global, subsurface water and the increased biohazard that might result if liquid subsurface water was

---

166. N. W. Hinners, R. D. Braun, K. B. Joosten, C. E. Kohlhase, and R. W. Powell, *Report of the MEPAG Mars Human Precursor Science Steering Group Technology Demonstration and Infrastructure Emplacement (TI) Sub-Group*, NASA document posted July 2005 by the MEPAG at *http://mepag.jpl.nasa.gov/reports/MEPAG_MHP_SSG_TI_Sub-Group.pdf* (accessed 25 January 2011).

167. D. W. Beaty, K. Snook, C. C. Allen, D. Eppler, W. M. Farrell, J. Heldmann, P. Metzger, L. Peach, S. A. Wagner, and C. Zeitlin, "An Analysis of the Precursor Measurements of Mars Needed to Reduce the Risk of the First Human Missions to Mars," unpublished white paper posted June 2005 by MEPAG, *http://mepag.jpl.nasa.gov/reports/index.html* (accessed 26 January 2011).

168. MEPAG, "Scientific Goals, Objectives, Investigations, and Priorities: 2003," ed. G. J. Taylor, *http://mepag.jpl.nasa.gov/reports/index.html* (posted 16 July 2004); SSB, National Research Council, "Safe On Mars: Precursor Measurements Necessary to Support Human Operations on the Martian Surface" (Washington, DC: National Academies Press, 2002).

found to be present and accessible.[169] The MPG SSG study was meant to support the planning of a continuous series of Mars robotic precursor missions prior to human exploration. One of the key goals of such voyages will be to certify human mission landing sites for their safety, low risk, and scientific potential.

The NASA/ESA Workshop

Another NASA workshop was held in 2005 in conjunction with the European Space Agency (ESA). Both NASA and ESA have initiated long-term plans for the sequencing of robotic and human missions, and the sharing of planetary protection needs for these missions was very useful. The "top-level workshop goal"[170] was to map out how planetary protection requirements should be implemented during human missions and what contamination control standards should apply to human explorers. Workshop discussions considered not only the prevention of forward contamination of the Martian environment, but also the protection of the human habitat on Mars and of Earth's biosphere upon the crew's return. Future research and development needs were also identified. One of these was to investigate the impact of planetary protection requirements on various types of in situ resource utilization (ISRU), such as employing any existing Martian water resources near the landing site.

For a summary of planetary protection methodologies used on return-to-Mars as well as other missions, see Appendix F.

●●●

The Phoenix Lander Touches and Tastes Water Ice

In July 2008, the robot arm aboard NASA's Phoenix Mars Lander excavated a soil sample from a shallow trench dug next to the spacecraft and delivered the soil to an instrument in the craft's onboard laboratory

---

169. Beaty et al., "An Analysis of the Precursor Measurements," pp. 4, 73–74.
170. Margaret S. Race, Gerhard Kminek, and John D. Rummel, "Planetary Protection and Humans on Mars: NASA/ESA Workshop Results," submitted to *Advances in Space Research* in 2007.

that heated samples and analyzed the vapors given off. When the results were examined, William Boynton of the University of Arizona and lead scientist for Phoenix's Thermal and Evolved-Gas Analyzer announced, "We have water."[171] For years, remote observations such as from Mars orbiters had been picking up evidence of water ice. Phoenix too had collected data suggesting the presence of such ice. But this was the first time that "Martian water has been touched and tasted."[172]

What was of vital planetary protection importance about this find was the possibility that the ice periodically melted, forming thin films of *liquid* water around soil particles.[173] The Phoenix landing site could thus at times harbor a favorable environment for microbial life. According to Phoenix principal investigator Peter Smith, "the soil chemistry and minerals we observed lead us to believe this site had a wetter and warmer climate in the recent past—the last few million years—and could again in the future."[174]

A key lesson from the Phoenix experience is that the future exploration of Mars must proceed very carefully, in accordance with strict planetary protection procedures, because we now know that "somewhere on or under Mars it is likely that conditions exist that are amenable to at least some Earth microbes—and perhaps to martian life."[175] Future lander missions intending to search for life at Mars's highest priority astrobiological sites will require Viking-level sterilization at a minimum, and perhaps even more stringent planetary protection protocols.

171. Guy Webster, Sara Hammond, and Dwayne Brown, "NASA Spacecraft Confirms Martian Water, Mission Extended," NASA news release 08-195, 31 July 2008, http://www.nasa.gov/mission_pages/phoenix/news/phoenix-20080731.html (accessed 26 January 2011).

172. Ibid.

173. P. H. Smith, "H$_2$O at the Phoenix Landing Site," Science 325 (3 July 2009): 58–61.

174. Guy Webster and Johnny Cruz, "NASA Phoenix Results Point to Martian Climate Cycles," NASA news release 2009-106, 2 July 2009, http://www.nasa.gov/mission_pages/phoenix/news/phoenix-20090702.html (accessed 26 January 2011).

175. John D. Rummel, "Special Regions in Mars Exploration: Problems and Potential," Acta Astronautica 64 (2009): 1293–1297.

DO WE HAVE THE RIGHT TO CONTAMINATE?

The Ethical and Legal Aspects of Planetary Protection

8

It will be "a moral disaster because man will have presumed the right to inject his own contaminated material into an extraterrestrial environment where organic evolution may well be in progress."

—*Sir Bernard Lovell*[1]

The ancient Greek philosophers interpreted the study of ethics to be a search for answers to the question, how should we live? Such a question has value today, even as it did millennia ago, because as human beings, we must sometimes make decisions based on our own consciences and moral judgments in addition to simply obeying the law. The question of how to live can be extended to the topic of this book by posing the question, how shall we, the people of Earth, act in space? And in particular, how shall we act on bodies that may contain life?[2]

● ● ●

Ethical Aspects of Planetary Protection

The problem that arises in considerations of protecting extraterrestrial biota is that there are a myriad of opinions and "correct"

1. A. C. B. Lovell, *The Exploration of Outer Space* (New York: Harper & Row, 1962), as reported in James R. Newman, "Sharing the Universe," *New York Review of Books* 1(1) (February 1963).
2. Cynthia Freeland, "Virtue Ethics in the Ancient World," from Greek Moral Theory, a course taught at the University of Houston in 1992, *http://www.uh.edu/~cfreelan/courses/Virtues.html* (accessed 26 January 2011); Rosalind Hursthouse, "Virtue Ethics," Stanford Encyclopedia of Philosophy, 18 July 2003, *http://plato.stanford.edu/entries/ethics-virtue/* (accessed 26 January 2011); Paul Clancy, André Brack, and Gerda Horneck, *Looking for Life, Searching the Solar System* (New York: Cambridge University Press, 2005): 180–181.

actions. J. Baird Callicott, a founder of the field of environmental philosophy, explained that "there is general agreement that human life is a subject of moral concern but . . . there is no general agreement about the moral concern that should be considered for non-human life."[3] Nevertheless, one of the moral benefits discussed by Callicott that resulted from space travel was that "humankind saw itself as a whole, inhabitants of a small and somewhat fragile island in space,"[4] emphasizing our dependence on each other and the need for protection of our planet's biosphere.

Callicott examined an interesting combination of Kenneth E. Goodpaster's life-principle ethic,[5] which held that the fact of an organism simply being alive was a sufficient criterion for moral considerability,[6] and Albert Schweitzer's Reverence for Life Ethic, which held that "It is good to maintain and to encourage life; it is bad to destroy life or to obstruct it."[7] Schweitzer, in particular, strongly defended his belief that we should revere all "wills-to-live"[8] as we do our own. He thought that we are truly ethical only when we help all life that we are able to and shrink from harming anything that lives.

What is relevant to planetary protection is that both Goodpaster and Schweitzer refuted *sentiency* as the discriminator for protecting life, favoring instead *conativity*—having the minimal characteristics of life—which is sufficient for the entity to be a thing of value.[9] Schweitzer and Goodpaster were hardly the first to articulate these compelling sentiments. In the first chapter of Genesis, for instance, God affirmed repeatedly, in regard to grass, herbs, trees bearing fruit, ocean life, and

3. J. Baird Callicott, "Moral Considerability and Extraterrestrial Life," in *Moral Expertise*, ed. D. MacNiven (London: Routledge, 1990); Lucy Goodwin (reviewer), "J. Baird Callicott, 'Moral Considerability and Extraterrestrial Life,'" *Reviews of Ethics and Animals Literature* I (fall 1997), http://core.ecu.edu/phil/mccartyr/Animals/Real97/goodwin.htm.
4. Goodwin, "J. Baird Callicott, 'Moral Considerability and Extraterrestrial Life'."
5. Kenneth E. Goodpaster, "On Being Morally Considerable," *Journal of Philosophy* 75(6) (June 1978): 308–325.
6. Carlos Lizarraga-Celaya, "Environmental Ethics and Radical Ecology, An Overview," Laboratorio de Física Interdisciplinaria, Universidad de Sonora, Hermosillo, Sonora, México, 6 February 2001, http://labfi.fisica.uson.mx/EnvironmentalPhilosophy.html.
7. Albert Schweitzer, "The Ethics of Reverence for Life," in *The Philosophy of Civilization*, trans. C. T. Campion (Buffalo: Prometheus, 1987), Chap. 26, http://www1.chapman.edu/schweitzer/sch.reading1.html (accessed 26 January 2011).
8. Ibid.
9. Goodwin, "J. Baird Callicott, 'Moral Considerability and Extraterrestrial Life'."

every living creature that crept across Earth, *"kee-tov,"*[10] that it was good, and even *"tov m'od,"*[11] it was very good.

The conservationist Aldo Leopold took a step beyond simply identifying all life-forms as good things to be preserved. In his essay, "The Land Ethic," he presented reasons, both ethical and aesthetic, for *why* he believed this is so: "A thing is right when it tends to preserve the integrity, stability, and beauty of the biotic community. It is wrong when it tends otherwise."[12] In this statement he envisioned the connected life-skeins and interdependence necessary to maintain a bionetwork's health. In other words, he argued that an organism's importance to the web of community that surrounded it was a reason to protect it.

How Much Is Curiosity Worth?

Human beings have always longed to understand their place in the cosmos. This inquiry has been reexamined and reformulated through the centuries as our understanding of space science progressed and our technology for observing celestial bodies and making measurements improved. But a basic question that we are asking today—are we alone in the universe?—was asked ages ago. In 300 BC, the Greek philosopher Epicurus wrote to Herodotus regarding the infinite number of worlds that existed; he saw no reason why these bodies "could not contain germs of plants and animals and all the rest of what can be seen on Earth."[13] Epicurus was expressing his human curiosity to understand whether forms of life exist beyond the confines of our planet. The Roman philosopher Lucretius conveyed similar musings when he said, "Confess you must that other worlds exist in other regions of the sky, and different tribes of men, kinds of wild beasts."[14]

---

10. J. H. Hertz, ed., *The Pentateuch and Haftorahs* (London: Soncino Press, 1972), pp. 3–4.

11. Ibid., p. 5.

12. Aldo Leopold, "The Land Ethic," in *A Sand County Almanac and Sketches Here and There* (Oxford: Oxford University Press, 1949), pp. 201–226, *http://www.luminary.us/leopold/land_ethic.html*.

13. Clancy, Brack, and Horneck, *Looking for Life, Searching the Solar System*, pp. 180–181.

14. Ibid., p. 181.

But does a burning curiosity to know what lies beyond the sky justify sending ships to, and possibly contaminating forever, those other worlds? One rather arrogant position that has emerged during discussions of planetary exploration is that "the destiny of humanity is to occupy space, a destiny written in our genes."[15] This is a position reminiscent of the political philosophy of manifest destiny, held by many U.S. statesmen and business leaders in the 19th century, that our country *deserved* to, in fact was destined to, conquer the heart of North America from the Atlantic Ocean to the Pacific Ocean,[16] no matter the price paid by indigenous people or the environment. In defending the U.S. claim to new territories, the Democratic leader and influential editor John L. O'Sullivan insisted in 1845 on

> . . . the right of our manifest destiny to over spread and to possess the whole of the continent which Providence has given us for the development of the great experiment of liberty and federaltive [sic] development of self government entrusted to us. It is right such as that of the tree to the space of air and the earth suitable for the full expansion of its principle and destiny of growth.[17]

O'Sullivan's articulation of our country's supposedly God-given right to expand our sphere of influence became a rallying cry in political discourse, newspapers, and advertisements, although the philosophy behind manifest destiny surfaced even earlier in U.S. history. Andrew Jackson, for instance, appeared to act on a similar belief in 1818 when he led military forces into the Floridas to punish the Seminole Indians for taking up arms with the Spanish. Jackson's soldiers also destroyed

---

15. Ibid., p. 185.
16. Hermon Dunlap Smith Center for the History of Cartography, "Historic Maps in K–12 Classrooms," Glossary, Newberry Library Web site, 2003, *http://www.newberry.org/k12maps/ glossary/* (accessed 4 October 2006).
17. John Louis O'Sullivan, editorial supporting the annexation of Texas, in *United States Magazine and Democratic Review*, July–August 1845 ed., as reported in Michael T. Lubragge, "Manifest Destiny: The Philosophy That Created A Nation," in *From Revolution to Reconstruction*, 2003, from "A Hypertext on American History from the Colonial Period until Modern Times," Department of Humanities Computing, University of Groningen, The Netherlands, *http://www.let.rug.nl/usa/E/manifest/manif1.htm* (accessed 4 October 2006).

Spanish forces and captured several cities and forts.[18] The essence of the manifest destiny philosophy is still alive and well, I believe, and reflected within our country's program of exploring space.

While the concept of manifest destiny was once, and still may be, popular among many U.S. citizens, we were never unanimous in our support of it, either in the past or in the present. In 1837, for instance, William E. Channing wrote to Henry Clay that "We are a restless people, prone to encroachment, impatient of the ordinary laws of progress . . . forgetting that, throughout nature, noble growths are slow. . . . It is full time that we should lay on ourselves serious, resolute restraint."[19] And today, there are those who believe that "the Earth can be seen as a spaceship driven by humanity acting as a crew, and it is the destiny of a crew to stay onboard the ship . . . ."[20]

The Implications of a Second Genesis of Life in Our Solar System

Life may have begun in only one place in the solar system and possibly spread to other locations through the assistance of, for instance, asteroids and comets. Or life may have originated in two or more locations. The ethical implications are huge of discovering a life-form with an independent genesis from organisms on Earth. The existence of a second genesis would be strongly indicative that such an event probably occurred elsewhere as well and that life originating independently of Earth is widespread throughout our galaxy. Ethically, the need to protect an organism, however lowly, would be stronger if it represented "a unique life-form with an evolutionary history and origin distinct from all other manifestations of life."[21]

---

18. Michael T. Lubragge, "Manifest Destiny: The Philosophy That Created A Nation," in *From Revolution to Reconstruction*, 2003, from "A Hypertext on American History from the Colonial Period until Modern Times," Department of Humanities Computing, University of Groningen, The Netherlands, *http://www.let.rug.nl/usa/E/manifest/manif1.htm* (accessed 4 October 2006).
19. John M. Blum, William S. McFeely, Edmund S. Morgan, Arthur M. Schlesinger, Jr., and Kenneth M. Stampp, *The National Experience: A History Of The United States*, 6th ed. (New York: Harcourt Brace Jovanovich, 1985), p. 276.
20. Clancy et al., *Looking for Life*, pp. 185–186.
21. Richard O. Randolph, Margaret S. Race, and Christopher P. McKay, "Reconsidering the Theological and Ethical Implications of Extraterrestrial Life," *Center for Theology and Natural Sciences (CTNS) Bulletin* 17(3) (Berkeley, CA, summer 1997): 1–8.

Should We Explore Other Celestial Bodies as Preservers, Stewards, or Exploiters of Their Environments?

What do the various ethical positions say about appropriate guidelines for sending our ships and technologies to extraterrestrial worlds? Shall we conduct our activities on these bodies as strict environmentalists would, i.e., *preserve* the bodies in unchanged states; shall we act as *stewards* of these planets, seeking a path that maximizes the benefits to all parties concerned; or shall we *exploit* these bodies, treating them as resources that can greatly aid our species?

The Preservation Ethic

The preservation ethic suggests that human action in nature should be minimized, and this translates to the imperative to leave an extraterrestrial body unaltered—"to neither enhance its environment for the indigenous biology, if any, nor to introduce life from Earth."[22] A key question that comes up when preservation is discussed is, what exactly has intrinsic worth on the body? Is it only biological life, or should all the natural attributes of the body be preserved, including its rocks and its dirt?

I. Almar, in a paper presented at the 34th COSPAR Scientific Assembly, expressed the concern that damage caused by any human intervention on a lifeless world would be irreversible. One possible reason for protecting the lifeless space environment was its scientific aspect—areas and objects could exist of the highest scientific priority on different celestial bodies. As an example of this, much can be learned about volcanism and the impacts of tidal forces by studying the Jovian moon Io, which is most probably lifeless.[23] Almar identified

22. Randolph et al., "Reconsidering the Theological and Ethical Implications of Extraterrestrial Life."
23. Michael Meltzer, *Mission to Jupiter: A History of the Galileo Project* (Washington, DC: NASA SP-2007-4231, 2007).

the need "not to prevent any commercial utilization of Solar System resources, but to make space exploration and exploitation of resources a controlled and well-planned endeavor"[24] and recommended initiating a large-scale discussion on the possible ethical values of the lifeless environment.

The Dilemma of Planetary Stewardship: Protecting Our Species vs. Protecting Other Planets

The concept of stewardship can be understood by considering the mission of the nonprofit Association of Forest Service Employees for Environmental Ethics (AFSEEE): "to forge a socially responsible value system for the Forest Service based on a land ethic which ensures ecologically *and economically* [author's emphasis] sustainable resource management."[25] The key is on resource management rather preserving the forest in a pristine, unaltered state. Applied to extraterrestrial bodies, stewardship "would imply that the broad scientific and economic advantages from having a second planetary-scale biosphere [in addition to Earth's] would justify planetary alteration."[26]

An interesting combination of the preservationist and stewardship approaches arises from the belief that only biotic life has intrinsic value, not a body's geology. Thus, the stewardship perspective, which desires that we humans use nature wisely for our own benefit, would consider terraforming a celestial body to be ethical, even *obligatory*, if it promoted the growth of indigenous life (even as yet undiscovered life) on that body.[27]

---

24. I. Almar, "Protection of the Lifeless Environment in the Solar System" (presented at the 34th COSPAR Scientific Assembly, Second World Space Congress, 10–19 October 2002, Houston, TX, 2002).

25. Lawrence M. Hinman, "Environmental Ethics," University of San Diego Ethics Update—Environmental Ethics Resources, 17 August 2006, *http://ethics.sandiego.edu/Applied/Environment/index.asp* (accessed 13 October 2006).

26. Randolph et al., "Reconsidering the Theological and Ethical Implications of Extraterrestrial Life."

27. Ibid.

Other compelling reasons for exploring and colonizing include the following:[28]

1. Providing a long-term, unifying project on which humans, cooperating around the globe, could focus. This could prove to be an important step toward world peace.
2. Constructing active biospheres on other bodies that could serve as refuges for terrestrial life, in the event of war or some other global catastrophe.

Some voices in the space science community called for the colonization of other worlds in much stronger terms than the second point expressed above. Michael J. Rycroft of the International Space University argued that "the overarching goal of space exploration for the twenty-first century should be to send humans to Mars, with the primary objective of having them remain there,"[29] so that our human species might have a second home in the event that a disaster on Earth rendered it uninhabitable. Rycroft believed that many factors could cause such a catastrophe, including overpopulation; global terrorism; nuclear or biological war or accident; the occurrence of a super-virus; a natural disaster (e.g., from an asteroid collision, flood, volcano, and so on); the depletion of vital resources such as oil or natural gas reserves; climate change, global warming, and sea level rise; and stratospheric ozone depletion. Rycroft thought that the colonization of a habitable world was thus an imperative human endeavor of this century and an insurance policy, and he emphasized his point by quoting M. Rees's opinion that "the odds are no better than 50–50 that our present civilisation on Earth will survive to the end of the present century."[30]

But can colonization of a world containing indigenous life be performed while at the same time following the strictures of planetary protection? It could be argued that if planetary protection measures seriously delayed colonizing another world, they would be *unethical* to perform, since they would endanger the safety and future of our own species. Compulsory colonization as soon as it is feasible, on the other hand, will likely contaminate a body with Earth organisms and may

---

28. Clancy et al., *Looking for Life*, pp. 187–189; Christopher P. McKay, Owen B. Toon, and James F. Kasting, "Making Mars Habitable," *Nature* 352 (8 August 1991): 489–496.
29. Michael J. Rycroft, "Space Exploration Goals for the 21st Century," *Space Policy* 22 (2006): 158–161.
30. M. Rees, *Our Final Century* (London: William Heinemann, 2003), p. 228, as reported in Rycroft, "Space Exploration Goals for the 21st Century," 159.

well extinguish any indigenous life-forms. The human race has arguably done a terrible job of protecting its own planet's environment, so can we even imagine that we will appropriately protect other bodies we visit?[31]

Deciding which course should be followed—colonization as soon as it is feasible or waiting until a thorough search for life has been performed—depends on the intrinsic *value* we give to extraterrestrial life-forms, even nonsentient forms, and to the biotic communities of those organisms. Arguing for the planetary protection approach, however, is not always easy. It is difficult to make a convincing case for protecting the possible existence of some nonsentient microbes when their ecological niche may be required by the human race for its own survival.

Exploitation of Other Planets as Natural Resources

UNESCO discussions have viewed some issues in the ethics of outer space as similar to those in environmental ethics. The question of respect for the terrestrial environment also applies to respect for other celestial bodies. One of the ethical issues raised by the "conquest" of another planet is its possible appropriation for our own purposes. In establishing human-inhabited lunar stations or Mars outposts, does a nation have the right to mine and farm this extraterrestrial land?[32] Or should undeveloped lands of great natural beauty or, as mentioned above, of great scientific interest be carefully preserved? Our country's history has demonstrated the devastating speed at which natural resources can be exploited and destroyed, but throughout U.S. history we have also taken pride in, and placed high value on, the extraordinary beauty of our country.[33]

Lynn White's writings connect our use of a new territory's resources for our own benefit to the nature of our religious faith, which sought to destroy the pagan animism that believed "every tree, every

---

31. Clancy et al., *Looking for Life*, p. 188.
32. UNESCO, "UNESCO Activities in the Field of Ethics of Science and Technology and the Environment," *Connect—UNESCO International Science, Technology and Environmental Education Newsletter* 29 (3–4) (2004): 5.
33. Adam Rome, "Conservation, Preservation, and Environmental Activism: A Survey of the Historical Literature," History E-Library, National Park Service, *http://www.cr.nps.gov/history/hisnps/NPSThinking/nps-oah.htm* (last updated 16 January 2003, accessed 13 October 2006).

spring, every stream, every hill had its own *genius loci,* its guardian spirit."[34] According to White, our western religious tradition allowed us to develop an indifference to natural objects that then made it possible for us to exploit them.

Such an attitude seems to have been extended to parts of the outer space realm. Conquest and alteration of a celestial object certainly is acceptable to many people if the body does not harbor life. This attitude was apparent as far back as the Apollo mission. The waste that remained on the lunar surface to reduce liftoff mass was considered an acceptable price to pay for the science return. And the footprints, wheel tracks, and augur holes that were left were celebrated by some as indicative of "mankind's triumph over adversity and a symbol of its conquest of the Moon."[35]

Should We Do Away with Human Missions to Sensitive Targets?

One of the most reliable ways to reduce the risk of forward contamination during visits to extraterrestrial bodies is to make those visits only with robotic spacecraft. Sending a person to Mars would be, for some observers, more exciting. But in the view of much of the space science community, robotic missions are the way to accomplish the maximum amount of scientific inquiry since valuable fuel and shipboard power do not have to be expended in transporting and operating the equipment to keep a human crew alive and healthy. And very important to planetary protection goals, robotic craft can be thoroughly sterilized, while humans cannot. Such a difference can be critical in protecting sensitive targets, such as the special regions of Mars, from forward contamination.

Perhaps a change in the public's perspective as to just what today's robotic missions really are would be helpful in deciding what types of

---

34. Lynn White, Jr., "The Historical Roots of Our Ecologic Crisis," *Science* 155 (10 March 1967): 1203–1207.
35. Mark Williamson, "Protection of the Space Environment—Time for a Policy" (presented at the 55th International Astronautical Congress, Vancouver, BC, Canada, 4–8 October 2004).

missions are most appropriate to implement. In the opinion of Torrence Johnson, who has played a major role in many of NASA's robotic missions, including serving as the project scientist on the Galileo mission and the planned Europa Orbiter mission,[36] the term "robotic exploration" misses the point. NASA is actually conducting *human* exploration on these projects. The mission crews that sit in the control room at JPL "as well as everyone out there who can log on to the Internet" can observe in near-real time what's going on. The spacecraft instruments, in other words, are becoming more like collective sense organs for humankind. Thus, according to Johnson, when NASA conducts its so-called robotic missions, people all around the world are really "all standing on the bridge of Starship *Enterprise.*"[37] The question must thus be asked, when, if ever, is it necessary for the good of humankind to send people rather than increasingly sophisticated robots to explore other worlds?

Legal Aspects of Planetary Protection

Who Owns a Celestial Body?

Ancient Roman law spoke of the concept of *res nullius*—those physical things that have never had an owner. *Res nullius* refers to objects that have not been reduced to property because they are not, or cannot, be appropriated by individuals. Light, for example, is *res nullius.* On the other hand, the property status of something in a wild, unappropriated state was *res communis* to the Romans—literally a thing common to all. Unlike a *res nullius* object, which cannot be owned, *res communes* objects can be owned by the state.[38]

---

36. JPL, "Biographies," *http://www.jpl.nasa.gov/media/cassini-102504/bios.html* (accessed 21 November 2006).

37. Both quotes are from Torrence Johnson, foreword to *Mission to Jupiter,* by Meltzer, p. xviii.

38. *Geer v. Connecticut,* 161 U.S. 519 (1896), as reported in Eric Engle, "Economic Theory of Law and the Public Domain—When Is Piracy Economically Desirable?" lexnet site, *http://lexnet.bravepages.com/media1.html* (accessed 6 November 2006).

A key question in outer space ethics is, to which category should celestial objects belong? The *res communis* doctrine has been applied on Earth to territories that belonged to the community at large and therefore were considered accessible and exploitable by everyone. *Res communis* has been applied as well to outer space. Since the signing of the *Outer Space Treaty* of 1967, the predominant international view regarding the use of space has been aligned with *res communis*, but in the sense that space belongs to *all humankind* rather than to one country.[39]

The 1984 UN *Agreement Governing the Activities of States on the Moon and Other Celestial Bodies* (the "*Moon Treaty*") appears to consider celestial bodies to be *res communis* as well, stating in Article 4 that the *use* of the Moon shall be the "province of all mankind" and holding to a basic objective "to promote on the basis of equality the further development of co-operation among States in the exploration and use of the moon and other celestial bodies. . . . Bearing in mind the benefits which may be derived *from the exploitation of the natural resources* [author's italics] of the moon and other celestial bodies. . . ." In Article 11, the treaty states its position even more clearly regarding the human right to exploit the Moon: "The moon and its natural resources are the common heritage of mankind." But the treaty also embraces the principle of stewardship in its use of the Moon, stating in Article 7 that countries need to "take measures to prevent the disruption of the existing balance of its [the Moon's] environment, whether by introducing adverse changes in that environment, by its harmful contamination through the introduction of extra-terrestrial matter or otherwise."[40]

International Efforts To Develop a Legal Structure

In 1958, the United Nations General Assembly established an ad hoc Committee on the Peaceful Uses of Outer Space (UNCOPUOS)

39. Michael J. Listner, "It's Time To Rethink International Space Law," *Space Review* Web site, 31 May 2005, *http://www.thespacereview.com/article/381/1* (accessed 4 October 2006).
40. All quotes in this paragraph are from the United Nations *Agreement Governing the Activities of States on the Moon and Other Celestial Bodies*, adopted by the UN General Assembly in December 1979 in resolution 34/68, entered into force in July 1984.

with Resolution 1348 (VIII) in reaction to the launch of Sputnik and to the need for preventing contamination of celestial environments. The United Nations asked its ad hoc committee, as one of its first tasks, to report on the legal issues arising from human exploration of space.[41] In response, UNCOPUOS identified forward and back contamination dangers in a July 1959 report and recommended developing international agreements to "minimize the adverse effects of possible biological, radiological, and chemical contamination."[42] The UN General Assembly made UNCOPUOS a permanent committee in December 1959 with Resolution 1472 (XIV) and added new member states to it.[43]

USSR input and the UN *Declaration of Legal Principles*. The USSR's Chairman Khrushchev wrote an important letter to President Kennedy in March 1962 that built on UNCOPUOS's recommendation, advising that any outer space activities that could hinder other countries' exploration should first be discussed and agreed upon "on a proper international basis."[44] This letter as well as other USSR proposals linked the issue of outer space contamination to interference with the rights of states to conduct their activities and were reflected in the international planetary protection guidelines that the United Nations developed. In December 1963, the United Nations issued a *Declaration of Legal Principles Governing the Activities of States in the Exploration and Use of Outer Space*,[45] which included various guidelines related to planetary protection matters:

41. UN General Assembly Resolution 1348 (VIII), *Question of the Peaceful Use of Outer Space*, 792nd plenary meeting, 13 December 1958, *http://www.unoosa.org/pdf/gares/ ARES_13_1348E.pdf* (accessed 9 November 2006); L. I. Tennen, "Evolution of the Planetary Protection Policy: Conflict of Science and Jurisprudence?" *Advances in Space Research* 34 (2004): 2354–2362.

42. United Nations, "Report, Ad Hoc Committee on the Peaceful Uses of Outer Space," UN doc. A/4141 (14 July 1959), as reported in Tennen, "Evolution of the Planetary Protection Policy," 2356.

43. UN General Assembly Resolution 1472 (XIV), *International Co-operation in the Peaceful Uses of Outer Space*, 856th plenary meeting of the UN General Assembly, 12 December 1959, *http://www.unoosa.org/oosa/SpaceLaw/gares/html/gares_14_1472.html* (accessed 9 November 2006).

44. Chairman Khrushchev to President Kennedy, 21 March 1962, transmitting a letter of 20 March 1962, UN doc. A/AC,105/2 (21 March 1962), as reported in Tennen, "Evolution of the Planetary Protection Policy," 2356.

45. UN General Assembly Resolution 1962 (XVIII), *Declaration of Legal Principles Governing the Activities of States in the Exploration and Use of Outer Space*, 1280th plenary meeting, 13 December 1963, *http://www.unoosa.org/pdf/gares/ARES_18_1962E.pdf* (accessed 9 November 2006).

- Outer space exploration and use shall be carried out for the benefit and in the interests of all people.
- Outer space and celestial bodies are free for exploration by all states and are not subject to national appropriation or claims of sovereignty.
- All states will follow international legal principles, including those of the United Nations, in their use and exploration of space.
- States shall conduct all their outer space activities with due regard for the interests of other states. This declaration, which was reminiscent of the USSR's position on space activities, required that a state planning an activity that might conflict with the interests of other nations first consult them.
- States are liable for damage to the earth or air of any other nations due to objects launched into space.

The *Outer Space Treaty*. The UN *Declaration of Legal Principles* was a first step toward an international planetary protection legal structure but, as the UN Canadian delegation commented, the *Declaration* did not specifically ask states to consult with other states if an activity would harm Earth's natural environment. Also, it "did not constitute positive international law,"[46] and thus a more formal treaty document was required.

President Lyndon Johnson issued a statement on 7 May 1966 in which he proposed the development of such a treaty. The next month, the United States presented a draft treaty to UNCOPUOS, Article 10 of which provided that nations take steps "to avoid the harmful contamination of celestial bodies and adverse changes in the environment of the Earth resulting from the return of extraterrestrial matter."[47] It is noteworthy that the U.S. draft treaty extended planetary protection concerns to both forward and back contamination. The scope of the U.S. draft, however, included only celestial bodies, while the Soviet one, which was eventually accepted, covered the entire outer space environment.[48]

---

46. Tennen, "Evolution of the Planetary Protection Policy," 2357.
47. UN doc. A/AC 105/32 (17 June 1966), as reported in Tennen, "Evolution of the Planetary Protection Policy," 2358.
48. United Nations, *Treaty on Principles Governing the Activities of States in the Exploration and Use of Outer Space, Including the Moon and Other Celestial Bodies* ("*Outer Space Treaty*"), UN doc. 6347 (January 1967), http://www.oosa.unvienna.org/oosa/SpaceLaw/outerspt.html (accessed 20 January 2011).

The U.S. draft left out a powerful statement that the UN *Declaration of Legal Principles* had included: the need for international consultations for activities that may lead to interference with other nations' activities. While this omission may have been indicative of the United States' desire for autonomy in conducting its own exploration of outer space, the United States did agree to such international consultations when it signed the *Outer Space Treaty*, which was issued in January 1967.[49]

This international agreement had many similarities to another agreement—the *Antarctic Treaty*. Both sought to prevent a new type of colonial competition and self-seeking exploitation. In addition, the treaties for planetary protection of the solar system and environmental stewardship of the Antarctic were both implemented through "a political and legal forum in which science plays an advisory role."[50]

Article I of the *Outer Space Treaty* provided that extraterrestrial exploration and use, including of the Moon and other celestial bodies, must be performed for the benefit and in the interests of all countries and "shall be the province of all mankind."[51] G. M. Goh and B. Kazeminejad argued that the protection and preservation of celestial bodies such as Mars "is properly an extension of the province of Mankind principle" and that their contamination "would jeopardize the rights of all countries to use and explore outer space,"[52] although this was not explicitly stated in the treaty. Furthermore, while Article IX of the treaty directed nations to "pursue studies of outer space, including the Moon and other celestial bodies, and conduct exploration of them so as to avoid their harmful contamination and also adverse changes in the environment of the Earth

---

49. United Nations, *Treaty on Principles Governing the Activities of States in the Exploration and Use of Outer Space, Including the Moon and Other Celestial Bodies* ("Outer Space Treaty"), UN doc. 6347 (January 1967).

50. Committee on Principles of Environmental Stewardship for the Exploration and Study of Subglacial Environments, National Research Council, "Antarctic Governance and Implications for Exploration of Subglacial Aquatic Environments," in *Exploration of Antarctic Subglacial Aquatic Environments: Environmental and Scientific Stewardship* (Washington, DC: National Academies Press, 2007), Chap. 5.

51. UN *Outer Space Treaty*, Article I.

52. Both quotes are from Gerardine Meishan Goh and Bobby Kazeminejad, "Mars Through the Looking Glass: An Interdisciplinary Analysis of Forward and Backward Contamination" IAC-03-Q.3.b.05, 54th International Astronautical Congress of the International Astronautical Federation, the International Academy of Astronautics, and the International Institute of Space Law (Bremen, Germany, 29 September–3 October 2003): 4.

resulting from the introduction of extraterrestrial matter,"[53] the treaty did not explicitly define what constituted harmful contamination or an adverse change to the environment, and thus left these concepts somewhat vague and open to legal interpretation.

Was harmful contamination that which hindered the rights of all countries to use (i.e., exploit) celestial bodies, or did the term refer to actions potentially harmful to any actual biota that might reside there? SETI Institute scientist Margaret Race commented that there were strong indications the early *Outer Space Treaty* "had greater interest in protecting opportunities for science and research than it did about protecting celestial bodies per se."[54] And as discussed in Chapter 1, the planetary protection field did not in its policies have an ethical component, or one that sought to protect planets for their own sake. Planetary protection policy aimed to preserve celestial bodies for the sake of biological exploration.[55] This situation may now be changing, however. An influential 2006 document published by the National Research Council's Space Studies Board, *Preventing the Forward Contamination of Mars,* recognized the need for the space science community to address ethical issues regarding the introduction of terrestrial biota into sensitive extraterrestrial environments, to appropriately modify planetary protection policy if necessary, and to involve the public in a dialogue on this issue. Furthermore, the document strongly urged that addressing ethical concerns be done on an international level and "at the earliest opportunity."[56]

The *Moon Treaty.* A strong planetary protection statement that did not get included in the *Outer Space Treaty* was a proposed Japanese amendment that would have required nations to "exercise maximum care for the preservation and conservation of the natural resources and environment of celestial bodies."[57] The Japanese delegation to UNCOPUOS thought that other nations were afraid such

---

53. UN *Outer Space Treaty*, Article IX.
54. Margaret Race, comment to author regarding the manuscript of this book, received 18 September 2007.
55. Perry Stabekis, interview by author, Washington, DC, 9 September 2004, and comments on the manuscript, 21 June 2005.
56. Committee on Preventing the Forward Contamination of Mars, SSB, National Research Council, *Preventing the Forward Contamination of Mars* (Washington, DC: National Academies Press, 2006), pp. 6–8, 112–114.
57. Tennen, "Evolution of the Planetary Protection Policy," 2358.

wording would restrict future activities on celestial bodies.[58] While the proposed amendment was not put into the *Outer Space Treaty*, stricter planetary protection language was included in a treaty regarding the Moon. In June of 1971, the USSR proposed such a treaty to the General Assembly; it recommended "avoiding the disruption of the existing balance of the lunar environment."[59]

UNCOPUOS's Legal Subcommittee reached consensus on the final wording of the above statement and on the article in which it appeared, but the UN General Assembly did not approve the *Moon Agreement* and open it for signature until December 1979. While the planetary protection language in the *Moon Treaty* was strong, the UN General Assembly felt the need to clarify just what UNCOPUOS's intent was. It was eventually determined that the intent was definitely not to prohibit the exploitation of natural resources on celestial bodies, but rather to "protect the existing balance of the natural celestial environments."[60] The *Moon Agreement* finally entered into force in July 1984, with the clause in Article 7 that nations "shall take measures to prevent the disruption of the existing balance of its environment, whether by introducing adverse changes in that environment, by its harmful contamination through the introduction of extra-environmental matter or otherwise."[61]

This language may have been perceived by many countries as too restrictive to their future operations, because the *Moon Agreement* was far from universally accepted. As of 1 January 2006, only 12 nations had ratified it and 4 had signed it.[62] While it did provide a legal structure applicable at least to these states and possibly to other

---

58. United Nations, Legal Subcommittee of UNCOPUOS, Summary Record of the 71st Meeting, supra note 43, at pp. 13–14, as reported in Tennen, "Evolution of the Planetary Protection Policy," 2358.

59. UN doc. A/8391 and Corr. 1, annex (1971) and UN doc. A/AC.1/L.568 (5 November 1971), as reported in Tennen, "Evolution of the Planetary Protection Policy," 2359, 2362.

60. Tennen, "Evolution of the Planetary Protection Policy," 2359.

61. United Nations, *Agreement Governing the Activities of States on the Moon and Other Celestial Bodies*, Article 7, adopted by the General Assembly in December 1979 in resolution 34/68, entered into force in July 1984.

62. United Nations, "Space Law: Frequently Asked Questions," United Nations Office for Outer Space Affairs, *http://www.unoosa.org/oosa/FAQ/splawfaq.html* (accessed 28 July 2010).

nations collaborating with them on space missions, neither the United States nor Russia has ratified or signed this agreement.[63]

The *Outer Space Treaty* of 1967 had a stronger legal standing than the *Moon Treaty*. The *Outer Space Treaty* was ratified or acceded to by the majority of nations, and these included the major spacefaring nations. Widespread international support presently exists for its basic principles; as of 1 January 2006, 98 nations had ratified it and 27 had signed it.[64]

From the few treaty law documents that address planetary protection, it is clear that a very limited international legal framework exists that specifically deals with the protection of the extraterrestrial environment. As a result, outer space and celestial bodies are open to use and abuse by all nations. Legal standards will become increasingly important as outer space and celestial bodies become ever more accessible to exploration and exploitation.[65]

The Development of a U.S. Legal Structure

The National Aeronautics and Space Act of 1958 granted NASA authority over a range of activities that have bearing on the exploration and successful use of outer space. The situation becomes more complex, however, as applied to sample return missions that bring back extraterrestrial material. Earth-based activities associated with sample return—such as land, water, and air transportation; quarantine; environmental protection; hazardous material handling; occupational safety; human health; emergency preparedness; and building construction—trigger not only NASA scrutiny, but also the interest of a range of different U.S. agencies and organizations. The Environmental Protection Agency (EPA), for instance, has the responsibility for reviewing proposed activities that may

---

63. Tennen, "Evolution of the Planetary Protection Policy," 2359, 2362.
64. United Nations, "United Nations Treaties and Principles on Space Law," *http:// www.unoosa.org/oosa/SpaceLaw/treaties.html* (accessed 13 November 2006); Goh and Kazeminejad, "Mars Through the Looking Glass: An Interdisciplinary Analysis of Forward and Backward Contamination," p. 4.
65. Goh and Kazeminejad, "Mars Through the Looking Glass: An Interdisciplinary Analysis of Forward and Backward Contamination," p. 5.

impact Earth's biosphere, while the Occupational Safety & Health Administration (OSHA) examines operations with questionable human safety considerations. The Department of Health and Human Services (HHS) and its Public Health Service (PHS) have a legal charter to protect our nation's health; the Department of Agriculture (USDA)[66] has legal jurisdiction over the importation of soils and organisms that could be potential pathogens to livestock and economically important crops, while the transport of samples and other routine activities associated with a mission may come under the jurisdiction of the Federal Highway Administration (FHWA) and the Federal Aviation Administration (FAA). Local jurisdictions may also have an interest in sample return activities.

The end result of this situation is that multiple agencies "from the federal level to local zoning and permit offices"[67] become involved in sample return oversight, complicating operations with conflicting regulations and overlapping jurisdictions. A way of dealing with this potentially confusing situation in the future is to do as the Apollo program did and establish an organization akin to its Interagency Committee on Back Contamination (ICBC), which had the charter for conducting joint-agency regulatory reviews and analyzing applicable statutes.

The basic tenet of planetary protection as applied to sample return, in place since the early days of space travel, has been to avoid the harmful contamination of Earth. However, there is another complicating factor to this endeavor—the need to take action without having definitive answers on which solar system bodies, if any, harbor or have ever harbored life.

---

66. The USDA Animal Plant Health Inspection Service (APHIS) is the agency that inspects all incoming materials at airports that could potentially impact agriculture. APHIS has the authority to quarantine as well as to regulate importations of "soils," which are defined as the loose covering of earth. Questions remain how APHIS's jurisdiction will apply to materials from another planet and organisms from extraterrestrial locations. Information on the Department of Agriculture and APHIS was supplied by Margaret Race in her comments on the manuscript of this book, 7 September 2007.

67. M. S. Race, "Planetary Protection, Legal Ambiguity and the Decision Making Process for Mars Sample Return," *Advances in Space Research* 18(1/2) (1996): 345–350.

Specific U.S. Legal Requirements

U.S. legal strictures addressing issues related to extraterrestrial sample return include the following:

Code of Federal Regulations (CFR) Title 14, Part 1211— Extraterrestrial Exposure. This important U.S. code established NASA's responsibility and authority to "guard the Earth against any harmful contamination or adverse changes in its environment resulting from personnel, spacecraft, or other property returning to Earth after landing on or coming within the atmospheric envelope of a celestial body."[68] In particular, it gave the NASA Administrator the right to determine the beginning and end of quarantine periods for any U.S. space mission and to isolate a "person, property, animal, or other form of life or matter whatever is extraterrestrially exposed."[69] The validity of this U.S. code was questionable,[70] however, because in contrast to typical U.S. regulatory agencies, NASA was never granted legislative authority to promulgate or enforce quarantine regulations, even though it did develop regulations for Apollo sample return operations. Nevertheless, NASA published its lunar quarantine regulations in the Federal Register on the day of the Apollo 11 launch, 16 July 1969, as CFR Title 14 Part 1211[71] and avoided the public discussion activities that could have caused an administrative delay.

Such a unilateral decision on quarantine regulations was potential grounds for challenging them. In addition, NASA's quarantine operations involved the detention not only of extraterrestrial material, but also of human beings, which could have raised issues about NASA's authority to deprive U.S. citizens of liberty.

These regulations had been adopted largely as guidance for the Apollo program. After Apollo ended, they were not of use for many

---

68. U.S. Code of Federal Regulations (CFR) Title 14, Chap. V, Part 1211—Extraterrestrial Exposure, 1-1-88 ed., adopted in 1969 and removed in 1991, folder 16158, "Extraterrestrial Quarantine," NASA Historical Reference Collection. Also available at *http://theshadowlands. net/etlaw.txt* (accessed 8 November 2006).

69. Ibid.

70. G. S. Robinson, "Exobiological Contamination: The Evolving Law," *Annals of Air and Space Law* XVII-I (1992): 325–367.

71. Rep. Joseph McDade, Congressional Inquiries Division, NASA Office of Legislative Affairs, to Teresa Baker, 2 May 1997, folder 16158, "ET Quarantine-CFR Title 14, Part 1211, aka 'ET Law,' 1969," NASA Historical Reference Collection.

years. NASA removed CFR Title 14 part 1211 effective 26 April 1991, with the explanation that it had served its purpose and was no longer in keeping with current missions and practices.[72] This was certainly true, for no sample return missions requiring quarantine were on NASA's near horizon. But the rescinding of the regulations left no legal quarantine requirements or authority under existing federal statutes for handling extraterrestrial sample returns.[73] NASA did, however, reserve Section 1211 of the CFR for applicable extraterrestrial exposure regulations to be reinstated as appropriate if sample return or human exploration missions to other planets were initiated.

National Environmental Policy Act (NEPA). NEPA, the basic national policy instrument for protecting the human environment,[74] requires that potential environmental impacts of any kind from a NASA project be analyzed during the project's planning stages. In particular, NEPA requires that NASA provide public disclosure in an environmental assessment (EA) or environmental impact statement (EIS) of the full range of potential impacts, project alternatives, worst-case scenarios, and uncertainties for all phases of a sample return mission.[75]

These documents need to address planetary protection concerns if there is any question of a sample return impacting Earth's environment. For example, the team for the sample return Stardust mission to the Wild 2 (pronounced "vilt two") comet had to prepare an EA that analyzed and described the planned mission. It concluded in a Finding of No Significant Impact (FONSI) that "On the basis of the Stardust EA, NASA has determined that the environmental impacts associated with the mission would not individually or cumulatively have a significant impact on the quality of the human environment."[76]

---

72. Ibid.

73. Race, "Planetary Protection," 347.

74. NASA Environmental Management Division, "NASA's National Environmental Policy Act Responsibilities," in *Implementing The National Environmental Policy Act And Executive Order 12114*, Chap. 2, http://nodis3.gsfc.nasa.gov/npg_img/N_PR_8580_0001_/N_PR_8580_0001_.pdf, NPR 8580.1, 26 November 2001 (accessed 13 November 2006).

75. Donald L. DeVincenzi, Margaret S. Race, and Harold P. Klein, "Planetary Protection, Sample Return Missions and Mars Exploration: History, Status, and Future Needs," *Journal of Geophysical Research* 103(E12) (25 November 1998): 577–585.

76. Earle K. Huckins III, "National Environmental Policy Act; Stardust Mission," NASA notice 98-062, *Federal Register* 63(88) (7 May 1998): 25236–25237.

Other applicable regulations and agreements. During the Apollo era, the ICBC compiled an important regulatory analysis, *Excerpts of Federal Regulations Pertinent to Contamination Control for Lunar Sample Return Missions.*[77] The rules in this document were also relevant to *future* sample return efforts and included the following, each of which provides an example of an organization's reason for involvement in planetary protection activities:

- *Public Health Service Act (Public Law 410-78th Congress)* gives the Surgeon General authority to prevent the introduction and spread of communicable disease in the United States.
- *Department of Agriculture import regulation (CFR Title 9, Chapter I, Subchapter E, Part 122-4 March 1913)* restricts organisms from being imported into the United States without issuance of a permit and also allows quarantine of materials that are not under permit or are on prohibited lists.
- *Fish and Wildlife Act of 1956 (16 USC 742f(4) and (5))* gives the Secretary of the Interior the right "to take such steps as may be necessary for the conservation and protection of the fish and wildlife resource."[78]

These and other rules presented in the ICBC regulatory analysis were likely taken from a similar summary of relevant codes prepared the previous year by G. Briggs Phillips of the Public Health Service's Communicable Disease Center.[79] As mentioned above, an organization such as the ICBC that analyzes the missions and regulations of various U.S. organizations might prove invaluable for future NASA sample return missions in identifying the legal structure that needs to be satisfied.

Interagency agreements. The various agencies involved in Apollo sample return activities found it helpful not only to identify and examine all relevant statutes and regulations, but also to implement interagency agreements that assigned responsibilities to the appropriate

---

77. ICBC, *Excerpts of Federal Regulations Pertinent to Contamination Control for Lunar Sample Return Missions* (NASA, August 1967), folder 076-15, "Lunar Receiving Lab, August 1967," JSC Archives.
78. ICBC, *Excerpts*, pp. 6–7.
79. G. Briggs Phillips, "Summary of Laws and Regulations of Possible Applications to Returned Lunar Material," U.S. Public Health Service, Communicable Disease Center, 12 September 1966, folder 076-13, JSC Archives.

parties. For instance, a 1967 NASA agreement with the National Academy of Sciences and the Departments of Agriculture; Health, Education, and Welfare; and the Interior confirmed the non-NASA member agencies that had a seat on the ICBC as well as their authorization to analyze and advise NASA on such matters as 1) proposed quarantine protocols, 2) Lunar Receiving Laboratory specifications and construction practices, 3) recovery procedures for astronauts, 4) lunar sample protocols, 5) equipment specifications, and 6) the date and manner in which astronauts and samples could be released from quarantine.[80] Such interagency agreements could be valuable as well for future missions returning samples from outer space.

Under John Rummel's leadership as Planetary Protection Officer, an interagency Planetary Protection Advisory Committee (now called the Planetary Protection Subcommittee, or PPS) was established in February 2001 to advise the NASA Administrator on matters such as interpretation of policy, implementation plans, mission strategies, contamination risks of future missions, and research progress in such areas as sterilization technologies and life-detection approaches. PPS will likely serve a role for future Mars sample returns that is similar to what ICBC did for the Apollo missions.[81]

80. NASA, *Interagency Agreement Between the National Aeronautics and Space Administration, the Department of Agriculture, the Department of Health, Education, and Welfare, the Department of the Interior, and the National Academy of Sciences on the Protection of the Earth's Biosphere from Lunar Sources of Contamination,* 24 August 1967, folder 076-15, "Lunar Receiving Lab, August 1967," JSC Archives.
81. NASA, *Charter of the Planetary Protection Advisory Committee of the NASA Advisory Council,* 29 April 2005, *http://science.hq.nasa.gov/strategy/ppac/PPACCharter2005.pdf;* Planetary Protection Advisory Committee, *Meeting Report* (Washington, DC: NASA, 8–9 February 2005), *http://science.hq.nasa.gov/strategy/ppac/minutes/PPACmin0502.pdf;* Margaret Race, comment to author regarding the manuscript of this book, received 18 September 2007.

CONCLUSION

9

The planetary protection field comprises a complex, long-term effort by the United States and other countries to prevent forward and back contamination of celestial objects while still achieving numerous science goals. Planetary protection strategies are meant to preserve meaningful scientific investigation by keeping celestial objects as pristine as possible. The field is driven by the need to protect our quest for understanding the true nature of our universe and, in particular, the place of life in that universe.

Many approaches have been necessary in order to implement effective planetary protection into space exploration. One of the interesting aspects of the field has been the mix of political and technical efforts required to include planetary protection measures in an aggressive space exploration agenda and to do so with limited economic resources. NASA and its Apollo program, for instance, had to use both political skills and scientific and engineering know-how to develop an expensive quarantine facility with cutting-edge equipment for analyzing return samples, spacecraft, and astronauts. The battles for this facility took place in the halls of Congress and through more informal agreements with other agencies. But the effective design and operation of the facility came about through the dedicated efforts of a large community of scientists and engineers.

Implementing effective planetary protection measures into a mission necessitates the ongoing reconciliation of conflicting visions and values. As an example, the development of the Apollo postsplashdown astronaut recovery plan revealed the underlying hierarchy of ethical values that controlled operational decisions and illustrated how widely differing views needed to be considered before an operations plan could be finalized.

Planetary protection actions have had beneficial influences on other types of operations as well. After Viking, for instance, NASA managers determined that designing components to meet the mission's strict bioload limits and to withstand the rigorous terminal sterilization environment resulted in a more reliable spacecraft with a better probability of completing all mission requirements. The Viking mission provided an example of how a strict contamination reduction requirement could improve manufacturing approaches for the smallest to the largest of a spacecraft's parts.

The development of planetary protection has demonstrated the power that a small number of influential, charismatic people with persuasive ideas can have over the actions and expenditures of a much larger group. If not for the influence of two men, Joshua Lederberg and Carl Sagan, the field would probably look very different today, and the effort spent to prevent forward and backward contamination of our solar system's bodies might have been considerably smaller. These men's influence in both the political and scientific arenas was vital in establishing planetary protection as an indispensable part of our space exploration program, even for targets where the chances of finding extraterrestrial life were small.

Lederberg and Sagan were not, of course, the only people with dramatic influences on the field. The list of these individuals is long and includes such luminaries as University of California at Berkeley chemist and Nobel Laureate Melvin Calvin, British biologist J. B. S. Haldane, former NASA Director of Space Flight Programs Abe Silverstein, former Planetary Protection Officer John Rummel, Lawrence Hall, Charles Phillips, Donald DeVincenzi, Perry Stabekis, Margaret Race, and many others. Nevertheless, Lederberg's and Sagan's vision helped set the course of the planetary protection field for the next half century of effort. These people had the ear of Congress and of many managers and researchers in the space science community, and their recommendations were enormously influential.

Since the beginning of the Space Age, planetary protection considerations have been woven into missions to celestial bodies. Back contamination prevention played a major part in Apollo mission planning. Sterilization procedures for the Viking spacecraft were so rigorous that the mission's landers may have had fewer terrestrial microbes aboard than any other craft yet launched. Missions to Jupiter, Saturn, Mars, and the solar system's other bodies must minimize the chances of forward or back contamination when such occurrences might be harmful. And return sample missions to Mars, while still years in the future, have already received intense scrutiny for the purpose of preventing both forward contamination of Mars and back contamination of our planet. No mission involved in the search for extraterrestrial life can escape the demands and requirements of planetary protection.

We humans have a burning desire to increase our understanding of everything around us, but we are accountable to future generations of scientists to explore our solar system without destroying the capability of others to conduct their own investigations. The planetary protection field is thus driven by a deep regard for the importance of both present and future scientific inquiry. Careless planetary exploration in the present could forever obfuscate the answer to a vital question: Are we Earthlings alone in the universe?

APPENDICES

●●●

Appendix A:
Detailed Planetary Protection Requirements

(Source: Appendix A in *Planetary Protection Provisions for Robotic Extraterrestrial Missions*, NASA Procedural Requirement [NPR] 8020.12C, 27 April 2005)

Contents

A.1   Category-Specific List of Target Body/Mission Types (advisory only)

a. Category I: Flyby, Orbiter, Lander: Earth's Moon; Mercury; Undifferentiated, Metamorphosed Asteroids; others TBD pending National Research Council or other recommendations.

b. Category II: Flyby, Orbiter, Lander: Venus, Jupiter (exclusive of its icy moons), Saturn, Titan, Uranus, Neptune, Triton, Pluto/Charon, Kuiper belt objects, Comets, Carbonaceous Chondrite Asteroids, others TBD pending National Research Council or other recommendations.

c. Category III: Flyby, Orbiter: Mars, Europa, Ganymede, Callisto, others TBD pending National Research Council or other recommendations.

d. Category IV: Lander, Probe: Mars, Europa, Ganymede, Callisto, others TBD pending National Research Council or other recommendations.

e. Category V: Any Earth return mission. Unrestricted Earth return: Earth's Moon; Undifferentiated, Metamorphosed Asteroids; Short-Period Comets; Solar Wind; others TBD. Restricted Earth return: Mars, Europa, others TBD.

A.2 Category III/IV/V Requirements for Mars

A.2.1 Category III (Mars Orbiters)

a. Orbiter spacecraft that achieve microbial burden levels (surface, mated, and encapsulated) defined in the specification "Maximum Total Microbial Spore Burden for Category III Missions to Mars" shall not be required to meet impact or orbital lifetime requirements. The microbial burden level requirement for Mars is noted in the specification sheet "Maximum Total Microbial Spore Burden for Category III Missions to Mars." The achievement of these levels will likely require some form of active microbial reduction. Approved bioassays (see NPR 5340.1) are required to establish the microbial burden levels. Assembled spacecraft and all modules that have been bioassayed must be protected against recontamination.

b. Orbiter spacecraft that do not meet the requirements, "Maximum Total Microbial Spore Burden for Category III Missions to Mars," are required to meet a probability of impact requirement of 10^{-2} for a specified orbital lifetime limit, as noted in the specification "Orbital Lifetime Probability, Mars." Mission compliance with these requirements will consist of probability-of-impact analysis and orbital lifetime analysis. Trajectory biasing may be employed to lower the probability of impact for mission hardware, but it is not required.

c. For orbiters that meet orbital lifetime requirements, biological cleanliness is assumed by the use of International Organization for Standardization (ISO) Class 8 (or Class 100,000 under Fed. Std. 209E) cleanrooms and the associated procedures. No additional bioload quantification is generally necessary.

A.2.2 Category IV (Mars Landers)

For Mars landers, Category IV is subdivided into IVa, IVb, and IVc:

a. Category IVa missions comprise lander systems not carrying instruments for the investigation of extant Martian life. These lander systems are restricted to a total surface microbial burden no greater than Viking lander preterminal sterilization levels (see specification sheet "Maximum Surface Microbial Spore Burden for Category IVa Missions to Mars").

b. Category IVb missions comprise lander systems carrying instruments designed to investigate extant Martian life. For such missions, the following requirements apply:

    1. *Either* the entire landed system must be sterilized to the microbial burden levels defined in the specification sheet "Maximum Surface Microbial Spore Burden for Category IVb and IVc Missions to Mars" or to levels driven by the nature and sensitivity of the particular life-detection experiments, whichever are more stringent.

    2. *Or* the subsystems that are involved in the acquisition, delivery, and analysis of samples used for life detection must be sterilized to burden levels defined in the specification sheet "Maximum Surface Microbial Spore Burden for Category IVb and IVc Missions to Mars" and a method of preventing recontamination of the sterilized subsystems and the contamination of the material to be analyzed must be put in place.

c. Category IVc missions comprise lander systems that investigate Martian special regions (see definition below). For such missions, whether or not they include life detection experiments, the following requirements apply:

    1. **Case 1.** If the landing site is within the special region, the entire landed system shall be sterilized at least to the burden levels defined in the specification sheet "Maximum Surface Microbial Spore Burden for Category IVb and IVc Missions to Mars."

2. **Case 2.** If the special region is accessed through horizontal or vertical mobility, either the entire landed system shall be sterilized to the microbial burden levels defined in the specification sheet "Maximum Surface Microbial Spore Burden for Category IVb and IVc Missions to Mars," or the subsystems that directly contact the special region shall be sterilized to these levels and a method of preventing their recontamination prior to accessing the special region shall be provided.

If an off-nominal condition (such as a hard landing) would cause a high probability of inadvertent biological contamination of the special region by the spacecraft, the entire landed system must be sterilized to the Viking post-terminal sterilization microbial burden levels.

A.2.2.1 Definition of "Special Region"

A special region is defined as a region within which terrestrial organisms are likely to propagate *or* a region that is interpreted to have a high potential for the existence of extant Martian life-forms. Given the current understanding, this applies to regions where liquid water is present or may occur. Specific examples include but are not limited to the following:

a. Subsurface access in an area and to a depth where the presence of liquid water is probable.
b. Penetrations into the polar caps or other regions with significant water ice.
c. Areas of hydrothermal activity.

For all subcategories (IVa, IVb, and IVc), the following apply:

1. Achieving the prescribed levels of cleanliness will require contamination control (minimum ISO Class 8, or Class 100,000 under Fed. Std. 209E, assembly and attendant procedures), microbiological assays, and the maintenance of hardware cleanliness. Contamination-control effectiveness must be monitored and demonstrated by periodic assays. These assays must also be employed to determine the hardware's microbial burden.

2. When needed to meet the burden requirement specifications, the project must provide the facility and the means to accomplish any required microbial reduction. The facility will be subject to certification and the means of microbial reduction subject to approval and monitoring by the PPO.

3. Dry heat is the approved microbial reduction method. Alternative methods may later be certified for this purpose, but they will require a demonstration of effectiveness by the project and the approval of the PPO. Following the final predecontamination (or presterilization) microbiological assays and the microbial reduction procedure (as required), the project must demonstrate that the spacecraft or subsystem(s) are adequately protected against recontamination. This may require the use of a bioshield or shroud. Whatever the means of protection, the project must provide evidence that decontamination requirements are not compromised following the terminal treatment.

4. An organics archive is required of the bulk (>1 kg) organic constituents of all launched hardware that is intended to directly contact the target planet or that might accidentally do so. Each flight program office will provide for the collection and storage, for at least 20 years from the launch of the spacecraft, of a 50-g sample of each organic compound whose total amount in a planetary landing system exceeds 25 kg.

A.2.3 Category V (Sample Return Missions from Mars)
The Earth return mission is classified as "restricted Earth return" and is subject to the following requirements:

a. Unless specifically exempted, the outbound leg of the mission shall meet Category IVb requirements. This provision is intended to avoid false-positive indications in a life-detection and hazard-determination protocol or in the search for life in the sample after it is returned. A false-positive could prevent the distribution of the sample from containment and could lead to unnecessarily increased rigor in the requirements for all subsequent Mars missions.

b. Unless the sample to be returned is subjected to an accepted and approved sterilization process, the sample container must be sealed after the sample's acquisition. A redundant, fail-safe containment procedure with a method for the verification of its operation before Earth return shall be required. For unsterilized samples, the integrity of the flight containment system shall be maintained until the sample is transferred to containment in an appropriate receiving facility.

c. The mission and the spacecraft's design must provide a method to break the chain of contact with Mars. No uncontained hardware that contacted Mars, directly or indirectly, shall be returned to Earth. Isolation of such hardware from the Mars environment shall be provided during the sample container's loading into the containment system, its launch from Mars, and any in-flight transfer operations required by the mission.

d. Reviews and approval of the continuation of the flight mission shall be required at three stages: 1) prior to launch from Earth, 2) prior to leaving Mars for return to Earth, and 3) prior to commitment to Earth entry.

e. For unsterilized samples returned to Earth, a program of life-detection and biohazard testing or a proven sterilization process shall be undertaken as an absolute precondition for the controlled distribution of any portion of the sample.

A.3 Category III/IV/V Requirements for Europa

A.3.1 Category III/IV (Europa Orbiters and Landers)

Requirements for Europa flyby, orbiter, or lander missions, including microbial reduction, shall be applied in order to reduce the probability of inadvertent contamination of a Europan ocean to less than 1×10^{-4} per mission. These requirements will be refined in future years, but the calculation of this probability should include a conservative estimate of poorly known parameters and address the following factors, at a minimum:

a. Microbial burden at launch.
b. Cruise survival for contaminating organisms.
c. Organism survival in the radiation environment adjacent to Europa.
d. Probability of landing on Europa.
e. The mechanisms of transport to the Europan subsurface.
f. Organism survival and proliferation before, during, and after subsurface transfer.

Preliminary calculations of the probability of contamination suggest that microbial reduction will likely be necessary for Europa orbiters as well as for landers. This will require the use of cleanroom technology, the cleaning of all parts before assembly, and the monitoring of spacecraft assembly facilities to understand the bioload and its microbial diversity, including specific problematic species. Specific methods should be developed to eradicate problematic species. Methods of microbial reduction should reflect the types of environments found on Europa, focusing on Earth extremophiles most likely to survive on Europa, such as cold- and radiation-tolerant organisms.

A.3.2 Category V (Sample Return Missions from Europa)

The Earth return mission is classified as "restricted Earth return" and is subject to the following requirements:

a. Unless specifically exempted, the outbound leg of the mission shall meet Category IVb requirements. This provision is intended to avoid false-positive indications in a life-detection and hazard-determination protocol or in the search for life in the sample after it has been returned. A false-positive could prevent the distribution of the sample from containment and could lead to unnecessarily increased rigor in the requirements for all subsequent Europa missions.

b. Unless the sample to be returned is subjected to an accepted and approved sterilization process, the sample container must be sealed after the sample's acquisition. A redundant, fail-safe containment procedure with a method for the verification of its operation before Earth return shall be required. For unsterilized samples, the integrity of the flight containment system shall be maintained until the sample is transferred to containment in an appropriate receiving facility.

c. The mission and the spacecraft's design must provide a method to break the chain of contact with Europa. No uncontained hardware that contacted Europa, directly or indirectly, shall be returned to Earth. The isolation of such hardware from the Europan environment shall be provided during the sample container's loading into the containment system, its launch from Europa, and any in-flight transfer operations required by the missions.

d. Reviews and approval of the continuation of the flight mission shall be required at three stages: 1) prior to launch from Earth, 2) prior to leaving Europa for return to Earth, and 3) prior to commitment to Earth entry.

e. For unsterilized samples returned to Earth, a program of life-detection and biohazard testing or a proven sterilization process shall be undertaken as an absolute precondition for the controlled distribution of any portion of the sample.

A.4 Requirements for Small Solar System Bodies

A.4.1 Outbound Categorization

The small bodies of the solar system not elsewhere discussed in this document represent a very large class of objects. Forward contamination requirements for these missions are not warranted except on a case-by-case basis, so most such missions should adhere to Category I or II requirements (see table 7.1).

A.4.2 Sample Return Missions from Small Solar System Bodies

a. The determination as to whether a mission is classified as "restricted Earth return" or not (Category V) shall be undertaken with respect to the best multidisciplinary scientific advice, using the framework presented in the 1998 report of the U.S. National Research Council's Space Studies Board entitled *Evaluating the Biological Potential in Samples Returned from Planetary Satellites and Small Solar System Bodies: Framework for Decision Making* (SSB 1998). Specifically, such a determination shall address the following six questions for each body intended to be sampled:

    1. Does the preponderance of scientific evidence indicate that there was never liquid water in or on the target body?

    2. Does the preponderance of scientific evidence indicate that metabolically useful energy sources were never present in or on the target body?

    3. Does the preponderance of scientific evidence indicate that there was never sufficient organic matter (or CO_2

or carbonates and an appropriate source of reducing equivalents) in or on the target body to support life?

4. Does the preponderance of scientific evidence indicate that subsequent to the disappearance of liquid water, the target body was subjected to extreme temperatures (i.e., >160°C)?

5. Does the preponderance of scientific evidence indicate that there is or was sufficient radiation for biological sterilization of terrestrial life-forms?

6. Does the preponderance of scientific evidence indicate that there has been a natural influx to Earth (e.g., via meteorites) of material equivalent to a sample returned from the target body?

For containment procedures to be necessary ("restricted Earth return"), an answer of "no" or "uncertain" must be returned for all six questions.

b. For missions determined to be Category V "restricted Earth return," the following requirements shall be met:

1. Unless specifically exempted, the outbound phase of the mission shall meet contamination-control requirements to avoid false-positive indications in a life-detection and hazard-determination protocol or in the search for life in the sample after it has been returned.

2. Unless the sample to be returned is subjected to an accepted and approved sterilization process, the sample container must be sealed after the sample's acquisition. A redundant, fail-safe containment procedure with a method for the verification of its operation before Earth return shall be required. For unsterilized samples, the integrity of the flight containment system shall be maintained until the sample is transferred to containment in an appropriate receiving facility.

3. The mission and the spacecraft's design must provide a method to break the chain of contact with the target body. No uncontained hardware that contacted the target body, directly or indirectly, shall be returned to Earth. The isolation of such hardware from the body's environment shall be provided during the sample container's loading into the containment system, its launch

from the body, and any in-flight transfer operations required by the mission.

4. Reviews and approval of the continuation of the flight mission shall be required at three stages: a) prior to launch from Earth, b) prior to leaving the body or its environment for return to Earth, and c) prior to commitment to Earth entry.

5. For unsterilized samples returned to Earth, a program of life-detection and biohazard testing or a proven sterilization process shall be undertaken as an absolute precondition for the controlled distribution of any portion of the sample.

A.5 Additional Implementation Guidelines for Category V Missions

If during the course of a Category V mission there is a change in the circumstances that led to its classification or there is a mission failure—e.g., new data or scientific opinions arise that would lead to the reclassification of a mission originally classified as "unrestricted Earth return" to "restricted Earth return" and the safe return of the sample cannot be assured, or the sample containment system of a "restricted Earth return" mission is thought to be compromised and the sample's sterilization is impossible—then the sample to be returned shall be abandoned. If the sample has already been collected, the spacecraft carrying it must not be allowed to return.

Appendix B:
The Impact of "Faster, Better, Cheaper" on Planetary Protection Priorities

A lot of the new ways of doing things in industry are not the best way to go for the space business.

—*NASA Advisory Council*[1]

During the NASA Advisory Council's June 2000 meeting, Tom Young, a space-industry executive formerly of Lockheed Martin, presented results from the Mars Program Independent Assessment Team (MPIAT) that he chaired.[2] MPIAT investigated the failures of Mars Climate Orbiter (MCO), Mars Polar Lander (MPL), and Deep Space 2 (DS2), which put planetary protection objectives at risk and destroyed the eagerly awaited scientific returns from the missions. MPIAT examined the relationships between NASA Headquarters, JPL, the California Institute of Technology (Caltech), and Lockheed Martin and how breakdowns in those relationships were partially responsible for the failures. Young noted that while NASA could not go back to its old way of doing business, which involved billion-dollar missions taking 15 or more years, current industrial policies were not always compatible with the needs of the space program. These needs included protecting planetary surfaces from contamination that could result from a spacecraft crash.

---

1. Lori B. Garver (exec. secretary) and Bradford W. Parkinson (chair), NASA Advisory Council, Langley Research Center, 6–7 June 2000, "Meeting Minutes," NHRC file no. 16734, "NASA Advisory Council (NAC) 1996–."
2. Thomas Young (chair), MPIAT, *Mars Program Independent Assessment Team Summary Report*, 14 March 2000, *http://sunnyday.mit.edu/accidents/mpiat_summary.pdf* (accessed 28 November 2006).

Young talked of the requirement for a margin of safety that would allow for human operational errors without causing a mission failure. Where this margin was not implemented, either because of industry or NASA management decisions or because of insufficient mission budgets, missions sometimes failed. In particular, MPIAT found that the lack of an established, clear definition of the faster, better, cheaper (FBC) approach and its associated policies for guiding operations had resulted in both NASA and Lockheed Martin project managers having different interpretations of what were allowable risks to take.[3] While MPIAT deemed the FBC approach to building spacecraft a good one if properly applied (as evidenced by the achievements of the Pathfinder and Mars Global Surveyor missions), it found a "willingness to cut corners and a failure to follow a disciplined approach successfully," which was unacceptable in a "one strike and you're out" business such as space exploration.[4]

The situation that put some of the FBC missions and their accompanying planetary protection objectives at risk was that NASA's agenda for developing and launching these projects rigidly constrained parameters that normally had some variability in them and which could have been used to reduce the risk of failure. For instance, many of the FBC missions, which were committed to quick development and relatively small budgets, were characterized by fixed launch dates; this constrained schedules that perhaps should have been more flexible. Established launch vehicle weights, science payloads selected in large part on their competitive pricing, rigid performance standards, and fixed mission costs without adequate contingency funding also increased the risks of both mission failure and inadvertent contamination of the target planet.

Space exploration projects typically have two variables—their margins and their risks. If margins are adequate, risks can be reduced. But the above constraints severely minimized project margins. The negative results of this, as determined by MPIAT, were analysis and testing deficiencies as well as inadequate preparations

---

3. "Testimony of Thomas Young, Chairman of the Mars Program Independent Assessment Team Before the House Science Committee," *SpaceRef.com*, 12 April 2000, *http://www.spaceref.com/news/viewpr.html?pid=1444* (accessed 28 November 2006).
4. Both quotes are from Garver and Parkinson, "Meeting Minutes."

for mission operations; all of these factors led to excessive risk and contributed to the failures and to the possible forward contamination of Mars, the target body. Being managed as a single Mars '98 Project, the MCO and MPL missions were particularly constrained by significant underfunding from their inception and in meeting their inflexible performance requirements.[5]

---

5. Young, *Mars Program Independent Assessment Team Summary Report*, p. 6.

●●●

Appendix C:
Biohazard Identification:
The Synergism Between Bioterror Prevention
and Planetary Protection Research

The anthrax mailings of 2001 raised public and governmental concerns about bioterrorism threats, and this resulted in significantly increased detection and countermeasure research funding, as well as the implementation of the Public Health Security and Bioterrorism Preparedness and Response Act of 2002 and other measures.[6] The Department of Homeland Security (DHS) has made preparation against bioterrorism a priority and has developed the BioWatch program to provide early detection of mass pathogen releases. DHS has also funded research programs at national laboratories and universities for developing improved early-warning as well as decontamination technologies.[7] The technology for a nationwide network, involving mass spectrometry and modeling approaches, is being developed by Oak Ridge National Laboratory (ORNL) and the American Tower Corporation for real-time detection, identification, and assessment of bioterror as well as other threats.

The planetary protection and bioterror detection fields have markedly different objectives. Planetary protection programs will take extremely stringent *preventive* measures to ensure that any biota in future return samples does not enter our biosphere. Space scientists consider the chances that such biota even exist to be very low, but not zero. Therefore, a conservative approach is warranted in the handling of return samples.

Bioterror detection, on the other hand, seeks provide awareness of the release of hazardous agents. This may be achieved through the identification of symptoms in the victims of an attack. In contrast, planetary protection programs do not expect anyone to exhibit

---

6. Jim Kulesz, "CBRN Detection and Defense—Sensor Net," Oak Ridge National Laboratory Fact Sheet ORNL 2002-01986/jcn, *http://www.ornl.gov/~webworks/security/SensorNet.pdf* (accessed 29 November 2006).
7. Dana A. Shea and Sarah A. Lister, *The BioWatch Program: Detection of Bioterrorism,* Congressional Research Service Report No. RL 32152, 19 November 2003.

symptoms of an extraterrestrial infection, since mission personnel go to great lengths to prevent exposure.

There is a potential connection between the two fields, however, and it is this: the advanced technologies that have been developed to detect and contain potential bioterror agents can be used in the planetary protection field to search for extraterrestrial biota in return samples and, if life is detected, to reliably contain it.

Planetary protection life-detection technology may also be of use in bioterror detection. Both fields are concerned with distinguishing small quantities of microorganisms that might have large effects on human health. Data from planetary protection research on the identification of, for instance, extremophile microorganisms that are highly resistant to heat, radiation, and temperature extremes may also be useful in bioterrorism studies.

Some of the newer investigative methods of planetary protection, such as epifluorescent microscopic techniques, can detect the presence of a single microbe. The polymerase chain reaction (PCR) methodology, which enzymatically amplifies biomarkers, can also do this in addition to identifying the cell's genetic characteristics (these methodologies were discussed in Chapters 6 and 7). PCR has been examined extensively in Homeland Security studies such as ones at the Centers for Disease Control and Prevention (CDC) and at Lawrence Livermore and Los Alamos National Laboratories, as well as other institutions. But extraterrestrial back contamination prevention studies have had a different emphasis; they have focused on the approaches for identifying as yet *unknown* microbes able to survive in severely harsh environments.[8] Such planetary protection efforts may thus have applications to bioterror detection of unknown agents.

8. E. K. Wheeler, W. J. Benett, K. D. Ness, P. L. Stratton, A. A. Chen, A. T. Christian, J. B. Richards, and T. H. Weisgraber, "Disposable Polymerase Chain Reaction Device," Lawrence Livermore National Laboratory UCRL-WEB-206692, 10 December 2004, http://www.llnl. gov/nanoscience/eng_articles/polymerase.html (accessed 29 November 2006); Rupert Ruping Xu, "Next Generation Thermal Cycler for Use in Polymerase Chain Reaction (PCR) Systems," Lawrence Livermore National Laboratory UCRL-MI-124572, http://www.llnl.gov/ ipac/technology/profile/announcement/ThermalCycler.pdf (accessed 29 November 2006); "Polymerase Chain Reaction," Answers.com, http://www.answers.com/topic/polymerase-chain-reaction (accessed 29 November 2006).

Appendix D:
Committees, Organizations, and Facilities Important to the Development of Planetary Protection

Organization	Abbreviation	Parent Organization	Date Established	Importance to Planetary Protection
Ad Hoc Committee on the Lunar Sample Handling Facility	Hess Committee	NAS SSB	1965	Non-NASA review of LSRL needs. Formed by NAS SSB chairperson Harry H. Hess.
Ad Hoc Committee on the Lunar Sample Receiving Laboratory	Chao Committee	NASA	1964	Chaired by Edward Ching Te Chao, a U.S. Geological Survey geochemist. Investigated LSRL needs and established general facility concepts.
Advanced Research Projects Agency	ARPA (now DARPA)	DOD	1958	Concerned with military aspects of space, DARPA was established as a response to the Soviet launching of Sputnik.
American Institute of Biological Sciences	AIBS		1947	Cosponsor of key 1958 symposium.
Army Biological Warfare Laboratories		DOD	1943	Methods for managing biologically hazardous materials were used to help develop quarantine and handling procedures for lunar samples.
Baylor University College of Medicine			1900	Published an in-depth set of procedures for operating the Lunar Receiving Laboratory and quarantining lunar samples and astronauts.
Bionetics Corporation			1969	Founded specifically to perform planetary quarantine support for NASA's Viking mission.
Bioscience Advisory Committee	Kety Committee	NASA	1959	Studied NASA's capabilities and future roles in bioscience.
Committee on Contamination by Extraterrestrial Exploration	CETEX	ICSU	1958	Early work on planetary protection policy. Superseded by COSPAR.

Organization	Abbreviation	Parent Organization	Date Established	Importance to Planetary Protection
Committee on Planetary and Lunar Exploration	COMPLEX	NAS SSB	early 1970s	Advisory committee to NASA.
Committee on Space Research	COSPAR	ICSU	October 1958	Major contributor over the years to planetary protection policy.
Communicable Disease Center	CDC	PHS	1946	Developed protective procedures for handling and quarantining lunar samples. Later renamed the Centers for Disease Control and Prevention.
Department of Agriculture	USDA		1862	Responsibility concerning the health of U.S. crops and animals of economic importance.
Department of Defense	DOD		1789	Military and strategic perspective.
Goddard Space Flight Center	GSFC	NASA	1959	Contributor to many NASA missions.
Interagency Committee on Back Contamination	ICBC		January 1966	Advised and guided NASA on planetary protection matters, with a charter to protect public health, agriculture, and other resources.
International Council of Scientific Unions	ICSU		1931	Contributor to international planetary protection policy. In 1958, ICSU's ad hoc Committee on Contamination by Extraterrestrial Exploration (CETEX) developed a code of behavior for space exploration.
Jet Propulsion Laboratory	JPL	NASA	early 1930s	NASA Center managing robotic space missions.
Kennedy Space Center	KSC	NASA	1962	Provided launch services for many NASA missions. Also conducted terminal sterilization of the Viking craft. Began as the Launch Operations Center and was renamed after President Kennedy's death.
Langley Research Center	LaRC	NASA	1917	Managed NASA's Viking project and contributed to many other NASA missions.
Lewis Research Center	LeRC	NASA	1941	Designed the Titan-Centaur launch vehicles for the Viking mission. Renamed Glenn Research Center in 1999.

Organization	Abbreviation	Parent Organization	Date Established	Importance to Planetary Protection
Lunar Sample Receiving Laboratory	LSRL	NASA	July 1969	Quarantined and examined Apollo lunar samples as well as astronauts. Later renamed the Lunar Receiving Laboratory (LRL). The last of LRL's certifications were not officially signed until less than 24 hours before the Apollo 11 Command Module splashed down in the Pacific Ocean.
Manned Spacecraft Center	MSC	NASA	1961	Managed the Apollo program. Renamed the Lyndon B. Johnson Space Center (JSC) in 1973.
Mars Exploration Program Analysis Group	MEPAG	NASA JPL	ca. 2000	Considered a 100-year timeframe for Martian special regions in which they could be visited by any given mission. The analysis took into account past and present climate and included the study of current special regions as well as the possibility that a locale could become a special region within a century after a spacecraft's arrival due to a natural event such as a volcanic eruption.
Mars Human Precursor Science Steering Group	MHP SSG	MEPAG	2004	Identified priorities for investigations, measurements, and planetary protection technology demonstrations to be carried out prior to the first human mission to Mars.
Marshall Space Flight Center	MSFC	NASA	1960	Contributor to many NASA missions.
Martin Marietta Corporation (later Lockheed Martin)	MMC		1961	Built the Viking spacecraft. Later became Lockheed Martin Space Systems and assembled many robotic spacecraft.
Mobile Quarantine Facility	MQF	NASA	1966	Housed personnel, equipment, and material returning from the Moon during their trip to the LSRL. The MQF was constructed by MelPar, a subsidiary of American Standard, and resembled an Airstream housetrailer.
National Academy of Sciences	NAS		March 1863	Key adviser to NASA.

Organization	Abbreviation	Parent Organization	Date Established	Importance to Planetary Protection
National Aeronautics and Space Administration	NASA		October 1958	
National Research Council Committee on Preventing the Forward Contamination of Mars	NRC PREVCOM	NAS SSB	ca. 2005	Conducted the NASA-requested study *Preventing the Forward Contamination of Mars*.
National Science Foundation	NSF		1950	Adviser to NASA.
Pickering Committee		NASA MSC	March 1966	A site survey board, led by NASA's Col. John Pickering, which screened 27 existing facilities around the U.S. for possible use as a lunar sample receiving facility.
Planetary Protection Advisory Committee (now the Planetary Protection Subcommittee)	PPAC (now PPS)	NASA	2001	Advised NASA Administrator on planetary protection policy and technical issues.
Planetary Quarantine Office	PQO	NASA	1963	Later renamed the Planetary Protection Office, this is the NASA entity overseeing protection of solar system bodies, including planets, moons, comets, and asteroids, from contamination by Earth life, as well as protecting Earth from possible life-forms that may be returned from other solar system bodies.
Public Health Service	PHS	Now part of HHS	July 1798	Had responsibility for the health of the United States and for any potential threat to that health from extraterrestrial life, particularly from back contamination.
Space Science Board	SSB	NAS	1958	Major adviser to NASA on all interplanetary contamination issues. Later renamed the Space Studies Board.

Organization	Abbreviation	Parent Organization	Date Established	Importance to Planetary Protection
Space Science Steering Committee		NASA	February 1960	Active during the Apollo program. Developed specifications for a lunar sample receiving facility.
Task Group on Issues in Sample Return	TGISR	NAS SSB	ca. 1996	Examined issues related to bringing back material from solar system bodies.
Task Group on Sample Return from Small Solar System Bodies		NAS SSB	ca. 1998	Assessed the potential for a living entity to be contained in or on samples returned from planetary satellites and other small solar system bodies such as asteroids and comets.
UN Committee on the Peaceful Uses of Outer Space	UNCOPUOS	United Nations	December 1958	Concerned with contamination dangers to celestial bodies. Dealt with the legal aspects of planetary protection.
West Coast Committee on Extraterrestrial Life	WESTEX	NAS	February 1959	Formed by Joshua Lederberg to address the protection and preservation of planetary surfaces during space exploration.

Appendix E:
Timeline of Important
Planetary Protection–Related Events

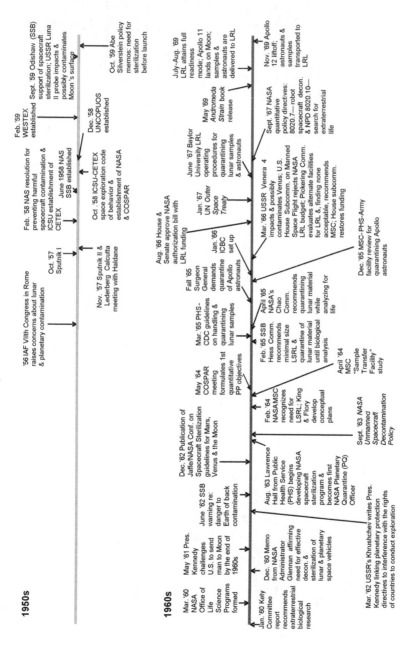

1950s

'56 IAF VIIth Congress in Rome raises concerns about lunar & planetary contamination

Oct. '57 Sputnik I

Nov. '57 Sputnik II & Lederberg Calcutta meeting with Haldane

Feb. '58 NAS resolution for preventing harmful spacecraft contamination & ICSU establishment of CETEX   June 1958 NAS SSB established

Oct. '58 ICSU-CETEX space exploration code of behavior & establishment of NASA & COSPAR

Dec. '58 UNCOPUOS established

Feb. '59 WESTEX established

Sept. '59 Odishaw (SSB) support of spacecraft sterilization; USSR Luna II probe impacts & possibly contaminates Moon 's surface

Oct. '59 Abe Silverstein policy memos: need for sterilization before launch

1960s

Jan. '60 Kety Committee report recommends extraterrestrial biological research

Mar. '60 NASA Office of Life Science Programs formed

Dec. '60 Memo from NASA Administrator Glennan affirming need for effective decon. & sterilization of lunar & planetary space vehicles

May '61 Pres. Kennedy challenges U.S. to send man to Moon by the end of 1960s

Mar. '62 USSR's Khrushchev writes Pres. Kennedy linking planetary protection directives to interference with the rights of countries to conduct exploration

June '62 SSB warning re: danger to Earth of back contamination

Dec. '62 Publication of Jaffe/NASA Conf. on Spacecraft Sterilization guidelines for Mars, Venus & the Moon

Sept. '63 NASA *Unmanned Spacecraft Decontamination Policy*

Aug. '63 Lawrence Hall for Public Health Service (PHS) begins developing NASA spacecraft sterilization program & becomes first NASA Planetary Quarantine (PQ) Officer

Feb. '64 NASA/MSC recognizes need for LSRL; King & Flory develop conceptual plans

April '64 MSC "Sample Transfer Facility" study

May '64 COSPAR meeting formulates 1st quantitative PP objectives

Mar. '65 PHS-CDC guidelines on handling & quarantining lunar samples

Feb. '65 SSB Hess Comm. recommends minimal size LSRL & quarantine of lunar material until biological analysis

April '65 NASA's Chao Comm. recommends quarantining lunar material while analyzing for life

Fall '65 Surgeon General demands quarantine of Apollo astronauts

Dec. '65 MSC-PHS-Army facility review for quarantining Apollo astronauts

Jan. '66 ICBC set up

Mar. '66 USSR Venera 4 impacts & possibly contaminates Venus; U.S. House Subcomm. on Manned Space Flight rejects NASA LRL budget; Pickering Comm. evaluates alternate facilities for LRL &, finding none acceptable, recommends MSC; House subcomm. restores funding

Aug. '66 House & Senate approve NASA authorization bill with LRL funding

Jan. '67 UN Outer Space Treaty

June '67 Baylor University LRL operating procedures for quarantining lunar samples & astronauts

Sept. '67 NASA quantitative policy directives: 8020.7—robot spacecraft decon. & NPD 8020.10—search for extraterrestrial life

May '69 *Andromeda Strain* book release

July-Aug. '69 LRL attains full readiness mode; Apollo 11 lands on Moon; samples & astronauts are delivered to LRL

Nov. '69 Apollo 12 liftoff; astronauts & samples transported to LRL

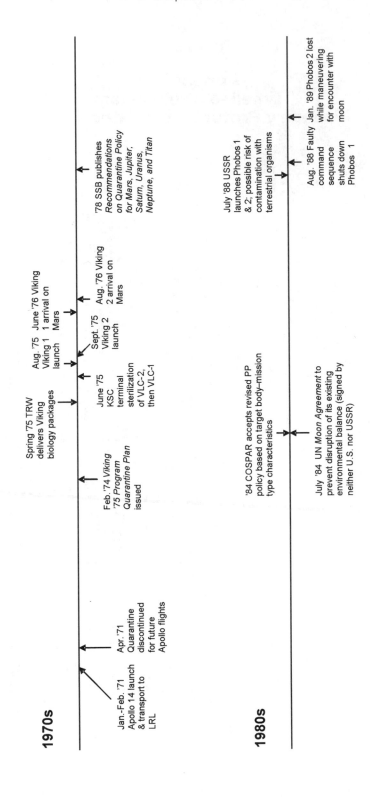

1970s

Jan.–Feb. '71 Apollo 14 launch & transport to LRL

Apr. '71 Quarantine discontinued for future Apollo flights

Feb. '74 *Viking '75 Program Quarantine Plan* issued

Spring '75 TRW delivers Viking biology packages

June '75 KSC terminal sterilization of VLC-2, then VLC-1

Aug. '75 Viking 1 launch

Sept. '75 Viking 2 launch

June '76 Viking 1 arrival on Mars

Aug. '76 Viking 2 arrival on Mars

'78 SSB publishes *Recommendations on Quarantine Policy for Mars, Jupiter, Saturn, Uranus, Neptune, and Titan*

1980s

July '84 UN *Moon Agreement* to prevent disruption of its existing environmental balance (signed by neither U.S. nor USSR)

'84 COSPAR accepts revised PP policy based on target body–mission type characteristics

July '88 USSR launches Phobos 1 & 2; possible risk of contamination with terrestrial organisms

Aug. '88 Faulty command sequence shuts down Phobos 1

Jan. '89 Phobos 2 lost while maneuvering for encounter with moon

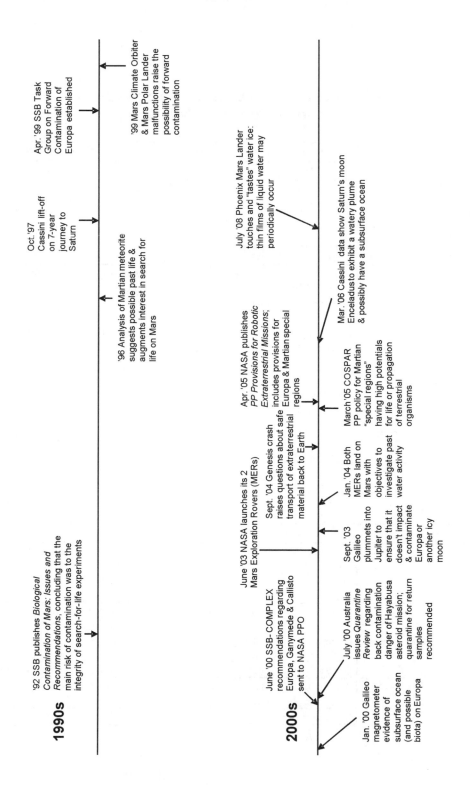

'92 SSB publishes *Biological Contamination of Mars: Issues and Recommendations*, concluding that the main risk of contamination was to the integrity of search-for-life experiments

1990s

'96 Analysis of Martian meteorite suggests possible past life & augments interest in search for life on Mars

Oct. '97 Cassini lift-off on 7-year journey to Saturn

Apr. '99 SSB Task Group on Forward Contamination of Europa established

'99 Mars Climate Orbiter & Mars Polar Lander malfunctions raise the possibility of forward contamination

Jan. '00 Galileo magnetometer evidence of subsurface ocean (and possible biota) on Europa

July '00 Australia issues *Quarantine Review* regarding back contamination danger of Hayabusa asteroid mission; quarantine for return samples recommended

June '00 SSB-COMPLEX recommendations regarding Europa, Ganymede & Callisto sent to NASA PPO

2000s

June '03 NASA launches its 2 Mars Exploration Rovers (MERs)

Sept. '03 Galileo plummets into Jupiter to ensure that it doesn't impact & contaminate Europa or another icy moon

Jan. '04 Both MERs land on Mars with objectives to investigate past water activity

Sept. '04 Genesis crash raises questions about safe transport of extraterrestrial material back to Earth

March '05 COSPAR PP policy for Martian "special regions" having high potentials for life or propagation of terrestrial organisms

Apr. '05 NASA publishes *PP Provisions for Robotic Extraterrestrial Missions*; includes provisions for Europa & Martian special regions

Mar. '06 Cassini data show Saturn's moon Enceladus to exhibit a watery plume & possibly have a subsurface ocean

July '08 Phoenix Mars Lander touches and "tastes" water ice: thin films of liquid water may periodically occur

Appendix F:
Planetary Protection Approaches
Used on Various Missions

Chapter	Mission	Planetary Protection Approach
General Approaches Used on Many Missions		
3		Payload capsule deflection maneuvers to prevent unsterilized parts of spacecraft, such as propulsion sections, from impacting target body.
4		Spacecraft assembly in cleanrooms.
Apollo		
Planetary Protection While Apollo Was on Moon Flight		
4	Apollo 11	Astronaut reentry to spacecraft after Moon walk: brushed off as much loose lunar material from their spacesuits as they could and left their outer shoe coverings and backpacks behind on the Moon.
4	Apollo 11	Before docking Lunar Module (LM) with Command Module (CM) orbiting the Moon, astronauts who had visited the lunar surface cleaned LM's insides with vacuum brush to minimize quantity of material carried into CM. Vacuum did not work that well; lunar material clung tenaciously to surfaces.
4	Apollo 11	Astronauts extensively cleaned LM while it was docked with CM in order to minimize contamination of CM.
4	Apollo 11	LM cabin gas continuously circulated through lithium hydroxide (LiOH) filter beds intended to remove virtually all particulate matter.
4	Apollo 11	Atmospheric pressure in CM kept higher than in LM so that gas would flow out of rather than into CM and exit spacecraft through LM's cabin relief valve.
4	Apollo 11	During the trip back to Earth, astronauts repeatedly vacuumed and wiped CM interior.
4	Apollo 11	CM atmosphere constantly circulated through LiOH filters to trap particles as small as bacteria.
After Splashdown		
4	Apollo 11	Biological isolation garments (BIGs) were passed into CM and astronauts donned them before exiting and boarding raft. BIGs trapped any organisms astronauts exhaled. Astronauts would wear BIGs until they entered shipboard quarantine unit.
4	Apollo 11	Recovery crew scrubbed area around CM's escape hatch as well as around postlanding vents with Betadine (an iodine solution) to kill any microorganisms that might have escaped the inside of the module and might be adhering to the vent area. The Betadine would not, of course, have killed any organisms that had escaped into the atmosphere or the ocean.
4	Apollo 11	Astronauts and recovery crew decontaminated each other's protective suits with a sodium hypochlorite (bleach) solution.
4	Apollo 11	CM was sealed until arrival at Lunar Receiving Laboratory (LRL).

Chapter	Mission	Planetary Protection Approach
4	Apollo 11	Astronauts were hoisted aboard a helicopter and taken to recovery vessel. Helicopter crew wore oxygen masks at all times to guard against inhaling any germs from the astronauts. The helicopter itself was later decontaminated with formalin.
4	Apollo 11	Astronauts walked across the recovery ship's deck between helicopter and Mobile Quarantine Facility (MQF). NASA personnel sprayed their path with glutaraldehyde, a sterilant typically used in hospitals to disinfect equipment.
Mobile Quarantine Facility		
4	Apollo 11	Astronauts entered MQF, which was capable of biologically isolating them until arrival at LRL. Quarantine conditions were maintained on MQFs through negative internal air pressure, filtration of effluent air, internal capture and storage of wastes in holding tanks, and a transfer lock system that could bring materials into or out of the facility without letting potentially infected air out.
4	Apollo 11	The recovery ship contained a second MQF that could be used to quarantine any of the ship's crew who might have been directly exposed to either astronauts or spacecraft. If a major biocontainment breach had occurred, the recovery ship would itself have become the isolation unit and would have had to remain at sea for the entire astronaut and lunar sample quarantine period.
4	Apollo 11	After the astronauts were helicoptered to recovery ship, CM was hoisted aboard and connected by plastic tunnel to MQF.
4	Apollo 11	Film shot on Moon and lunar sample containers were carried through the tunnel into the MQF since they, too, were potential sources of back contamination. These items were then passed out of MQF through a decontamination lock and flown to LRL.
Lunar Receiving Laboratory		
4	Apollo 11	Quarantine of lunar material, astronauts, spacecraft, and equipment took place at LRL until analyses behind biological barriers indicated no danger to Earth.
4	Apollo 11	LRL was kept at negative air pressure.
4	Apollo 11	All effluents and emissions from LRL were sterilized before release.
	Apollo 11	Staff transferred materials out of isolated areas using two-door sterilization cabinets, in which steam and/or ethylene oxide gas was employed.
4	Apollo 11	Glove boxes were employed in the quarantine facility so that lunar samples could be analyzed without contaminating the user. Technicians did not directly touch the samples but worked from the outside boxes and reached into interior by putting their hands in glove ports. The glove boxes were enclosures designed so that a range of tools inside the chambers could be manipulated by staff through a pair of impermeable gloves attached to ports in the chamber wall. The glove boxes were critical for preventing contamination of LRL staff and the outside environment with Moon material.
4	Apollo 11	A leak in a glove exposed two technicians to the Moon rocks, and these personnel had to go into quarantine.

Chapter	Mission	Planetary Protection Approach
4	Apollo 11	The MQF was flown to Ellington Air Force Base near Houston, then transferred onto a special flatbed truck designed for it and driven to the Crew Reception Area (CRA) of the LRL. The truck backed up to the CRA's loading dock, and technicians taped a shroud tightly in place that formed a sort of airlock preventing outside air from being contaminated when the MQF was opened and the astronauts entered the CRA.
4	Apollo 11	Medical staff watched astronauts, support personnel, and technicians put in quarantine after breach of biobarrier for any sign of infection and illness, conducting frequent examinations and blood tests. Staff conducted clinical observations and chemical, biological, and immunological analyses. Mission plans called for astronauts to remain in quarantine for at least 21 days after exposure to lunar material. Period of quarantine might have been extended if anyone had exhibited signs of infection in the vacuum laboratory.
4	Apollo 11	CM was quarantined and decontaminated at LRL in its spacecraft storage room. All stowed equipment, including clothing, was also quarantined. CM's water and waste management system was decontaminated by a piped-in formalin solution containing 40 percent formaldehyde, which remained in the system for 24 hours. A recovery engineer opened all compartments and wiped them down with disinfectant. LRL staff heated interior of CM to 110°F (43°C), evacuated its pressure to 8.5 psi, and filled it with formaldehyde gas for 24 hours. Because recovery crew that performed the decontamination might have gotten contaminated themselves, they had to be quarantined afterward.
Viking		
General Planetary Protection Approaches for Viking Mission		
5	Viking	Selected or developed materials for spacecraft components that were as heat-resistant as possible in order to resist damage from thermal sterilization.
5	Viking	Engineered manufacturing processes that minimized biological loads accruing on the Viking Lander Capsule (VLC).
5	Viking	Applied sterilizing heat to a range of components and assemblies.
5	Viking	Conducted terminal sterilization of the entire VLC, then hermetically sealed it in order to maintain sterility.
Viking Orbiter (VO)		
5	Viking	Assembly procedures generating a minimum of contamination.
5	Viking	Periodic bioassays of VO surfaces.
5	Viking	Designed VO trajectory to avoid impact with, and contamination of, Mars.
Viking Lander Capsule		
5	Viking	Testing and development program to determine which off-the-shelf materials and components withstood sterilization heating environments. Testing was conducted on electronic parts, solder, coatings, fasteners, gyroscopes, recorders, computers, batteries, bioshield, and other parts.
Clean Manufacturing Processes		
5	Viking	Maximized cleanliness of raw materials and basic parts.
5	Viking	Reduced dust and contamination in work space and on personnel.
5	Viking	Regular cleaning of subassemblies, including wiping down, vacuuming, and blowing off all accessible surfaces.

Chapter	Mission	Planetary Protection Approach
5	Viking	Applying standard spacecraft cleaning procedures to all surfaces prior to mating.
5	Viking	Implementing component packaging requirements that preserved their level of cleanliness.
Subcontractor Planetary Protection Responsibilities		
5	Viking	Applied flight acceptance heat cycle that verified parts' abilities to withstand heat sterilization environments and reduced bioburdens, especially of encapsulated and mated-surface microbes. These heat cycles helped minimize the necessary exposure time and thermal stresses of the terminal sterilization cycle on the complete spacecraft.
5	Viking	Biology package probability of contamination had to be kept below 1×10^{-6}, which required additional cleaning and sterilization. Plasma cleaning of surface contaminants was employed as well as sterilization in 120°C (248°F) dry-nitrogen atmosphere environment for 54 hours. Biology package was sealed in biological barrier during its sterilization and transport procedures in order to lower the risk of recontamination.
Prime Contractor (Martin Marietta) Planetary Protection Responsibilities		
5	Viking	Built and heat-sterilized certain electronics assemblies, then tested them to ensure they were functional.
5	Viking	Heat-sterilized loose, nonmetallic hardware parts.
5	Viking	Periodic surface cleaning of spacecraft was carried out as it was being assembled.
5	Viking	All cleaning, assembly, and test operations of VLC systems as well as of the capsule itself were conducted in cleanrooms with personnel working under strict handling constraints and procedures. This minimized the bioburden accumulation on the VLC.
5	Viking	Conducted heat-compatibility test on spacecraft antenna.
5	Viking	Conducted heat-compatibility flight acceptance test (FAT) after assembly of entire VLC, before sending it to KSC.
5	Viking	Monitored activities of Viking subcontractors who were manufacturing key parts of spacecraft.
Kennedy Space Center (KSC) Terminal Sterilization of VLC		
5	Viking	Verification tests performed in which KSC staff disassembled, inspected, and functionally checked each VLC.
5	Viking	Performed detailed bioassays to estimate microbe load on each VLC. Bioassays were key in determining appropriate durations of terminal sterilization cycles.
5	Viking	Applied terminal sterilization with VLC bioshield sealed and pressurized. Procedure carried out in thermal test chamber using inert gas atmosphere of 97 percent nitrogen, 2.5 percent oxygen, and 0.5 percent other gases. To bring interior VLC temperatures up more quickly and reduce temperature gradients (and thermal stresses they could cause), heated nitrogen gas mixture was injected directly into VLC, including into its heating and coolant lines. Residence times ran from 43 to 50 hours, with each VLC being exposed to 111.7°C for 30 hours after the coldest contaminated point reached that temperature.
5	Viking	Preventing recontamination of the VLC: electrical and mechanical connections from lander (within VLC) interfaced with bioshield, which in turn interfaced with VO. This prevented contamination from VO reaching inside VLC.

Chapter	Mission	Planetary Protection Approach
5	Viking	Positive air pressure within the bioshield during and following terminal sterilization also helped prevent recontamination.
5	Viking	Pressurized gas and propellant loading: Pressurized gas fill lines for VLC and line filters went through terminal sterilization. Pressurant gases themselves were sterilized by being passed through the filters. Hydrazine propellant was self-sterilizing, in that it was toxic to terrestrial microbes and thus did not need additional sterilization.
5	Viking	Alcohol-formaldehyde sporicide was pumped into coolant loop for radioisotope thermal generators (RTGs), a source of shipboard power.
5	Viking	Hydrazine propellant was subjected to a repeated freezing and thawing process to remove aniline contamination. When aniline was burned, hydrogen cyanide was produced, a substance often involved in amino acid reactions that could interfere with search for life.
5	Viking	Lander engine nozzles redesigned to minimize chemical contamination and severe disturbance of Martian surface.
5	Viking	All equipment and materials used for bioassays (test tubes, racks, nutrient materials, alcohols, and squeeze bottles) were themselves thoroughly sterilized to prevent recontaminating parts of the spacecraft being analyzed.
Planetary Protection Procedures During Launch and Cruise Phases		
5	Viking	Excess pressure buildup within VLC during launch ascent, when outside pressure rapidly dropped, was vented through biofilter that did not allow microbes to reenter. This avoided danger of bioshield rupture.
5	Viking	After spacecraft separated from Centaur upper stage rocket, which was not sterilized, Centaur was deflected off its flight path in order to prevent an eventual impact with, and possible contamination of, Mars.
5	Viking	Bioshield was removed from VLC only after spacecraft exited atmosphere, eliminating danger of recontamination by atmospheric molecules.
5	Viking	Particles, chemicals, and possibly biotic material from VO, which was unsterilized, as well as gases emitted by its coatings, greases, and paints and propellants from its rocket engines were all capable of contaminating VLC, although NASA calculated this was very unlikely. Nevertheless, before Viking reached Mars, mission control turned it so that VLC faced toward the Sun, allowing ultraviolet solar radiation to provide a sterilant function on the capsule's exposed surfaces.
Planetary Protection During Arrival at and Descent to Mars		
5	Viking	After separation of lander, trajectory of unsterilized VO had to be carefully controlled so it didn't impact Mars.
5	Viking	After touchdown, lander's biology package, which was sensitive enough to detect a single microbe, confirmed no contamination within the biology package itself.
Missions to Small Bodies of the Solar System		
6	Galileo	To prevent an accidental impact with Europa or another Jovian icy moon, Galileo orbiter used last of its fuel to impact and burn up in Jupiter's atmosphere.

Chapter	Mission	Planetary Protection Approach
6	Cassini	Cassini was deemed to have low probability of contaminating Titan, and so PPO did not impose planetary protection requirements other than assembly of spacecraft in a cleanroom, limited mission documentation, and assurances that spacecraft was not likely to accidentally impact Titan. However, with the presence of water detected on Enceladus, most planetary protection requirements may be added.
6	Hayabusa	Japanese return mission to Itokawa asteroid had landing site in Australia. Although back contamination risk to Earth was extremely low to negligible, Australian Department of Agriculture, Fisheries and Forestry recognized that environmental contamination could occur and required that return samples be immediately placed into secure, impervious containers and exported.
6	Return samples from P- and D-type asteroids	Since little is known about them, SSB recommends analysis in dedicated asteroid sample curation laboratory with cleanroom, subfreezing storage and processing, dedicated processing cabinets for samples from each parent body, with positive-pressure nitrogen atmospheres, and continuous monitoring for inorganic, organic, and biological contamination.
6	Stardust	An unrestricted Earth return mission because this mission, which sampled particles in the Wild 2 comet's tail as well as interstellar dust sweeping through the solar system, was considered by scientific opinion to be carrying no extraterrestrial life-forms.
6	Genesis	An unrestricted Earth return mission because this mission, that sampled solar wind particles, was considered by scientific opinion to be carrying no extraterrestrial life-forms.
Return to Mars		
General Bioload Reduction Procedures Typical of Post-Viking Mars Missions		
7	Post-Viking Mars missions	Spacecraft contamination involves both molecular and particulate types. Molecular contamination includes greases and oils, while particulate includes typically micrometer-sized conglomerations of matter such as dust. Subcontractors and vendors followed precision-cleaning protocols and assembly of components in relatively clean areas in order to reduce bioburdens on their parts. Typical precision-cleaning procedures targeted removal of both particles and molecular films and consisted of wiping gross contamination off part followed by application of organic solvents and aqueous rinses with or without ultrasonic treatment. The part was then sometimes subjected to vapor degreasing, followed by final rinsing typically conducted with isopropyl or other alcohol solutions.
7	Post-Viking Mars missions	After parts arrived at Lockheed Martin, they were usually subjected to further bioload reduction using cloth wipes saturated with 100 percent isopropyl alcohol (IPA) solution. Cleaning action of wipes was due mainly to IPA's solvent properties, but a frictional action was applied as well in order to lift contamination off part surfaces. Cloth wipes were repeatedly folded to keep only the clean sections of cloth touching parts.
7	Post-Viking Mars missions	Numerous bioassays were performed using two moistened, absorbent cotton swabs mounted on wood applicator sticks and moistened polyester wipe cloths. Swabs and wipes were first sterilized in autoclave.

Chapter	Mission	Planetary Protection Approach
Specific Missions		
7	Mars Observer	Category III (no direct contact of space vehicle with target object).
7	Mars Observer	Launch phase: trajectory of launch vehicle's upper stage was carefully biased so as to prevent its eventual impact with Mars.
7	Mars Observer	Upper stage separated from the spacecraft using maneuver designed to minimize probability that spacecraft would get contaminated, either through collision with the upper stage or through materials given off by it. This maneuver also further biased upper stage away from a Mars impact trajectory.
7	Mars Observer	Cruise phase: aimpoint of spacecraft trajectory biased so as to be suitable for orbit insertion but not lead to an impact with Mars if Earth lost control of the craft.
7	Mars Observer	Approximately 90 days before orbit insertion at Mars, increased tracking efforts implemented to reduce navigation delivery errors and ensure as accurate targeting of spacecraft as possible.
7	Mars Observer	Orbit insertion objective was for spacecraft to enter three-day capture orbit. Care had to be taken during spacecraft engine firing and first periapsis, when there was a small probability of impact with Mars. All intermediate orbits between insertion maneuver and eventual mapping orbit had to be rigorously analyzed to make sure that their lifetimes were of sufficient length for the maneuver to be completed. Factors that decayed these orbits and shortened their lifetimes included atmospheric drag and nonsphericity of Martian gravitational field.
7	Mars Observer	Mapping orbit needed to be designed to have an adequate lifetime for the phase to be completed.
7	Mars Observer	A quarantine orbit needed to be designed such that the craft could be raised into it and have a probability of impact with Mars that would not exceed 1×10^{-2} by the year 2009 and 5×10^{-2} by the year 2039.
7	Mars Surveyor Program	As with other programs, trajectory course maneuvers (TCMs) were employed to minimize chances of an accidental impact with Mars.
7	Mars Global Surveyor (MGS)	Not sterilized. External surfaces needed only to be visibly clean at payload encapsulation.
7	MGS	Orbit selected to be low enough for instruments to make close-range observations of Martian surface, but not so low that friction from planet's atmosphere would slow spacecraft down enough for an impact.
7	MGS	When spacecraft completed its mapping work, propulsion system was designed to raise the average altitude to a quarantine orbit that would greatly reduce frictional drag. Quarantine orbits were intended to provide a contamination-free period of decades in which to conduct missions examining the biotic nature of the planet.

Chapter	Mission	Planetary Protection Approach
7	Mars Pathfinder	Category IVa: landers without life-detection instruments. Required detailed documentation, bioassays, probability-of-contamination analysis, an inventory of bulk constituent organics, trajectory biasing, use of cleanrooms during spacecraft assembly and testing, and bioload reduction to levels on Viking before terminal sterilization: 300 spores per square meter and total spore burden no higher than 3.0×10^5.
7	Mars Pathfinder	Aimpoint of the upper stage and spacecraft was biased away from Mars to ensure that any unsterilized part of Pathfinder would not impact Martian surface.
7	Mars Pathfinder	Trajectory after TCMs 1 and 2 was such that if control of spacecraft was subsequently lost, the craft would enter Martian atmosphere at a shallow-enough entry angle for atmospheric braking to slow down the craft to below 1,000 feet per second at impact, even if craft's parachute did not deploy. This was to minimize contamination of Mars from parts of spacecraft in the event of a crash landing.
7	Nozomi	Japanese spacecraft to Mars had thruster problems and damage from solar flares that raised danger of a crash and contamination of planet. Mission scientists sent commands to spacecraft to fire small alternate thrusters and avoid orbit insertion, minimizing possibility of a Mars impact.
7	Mars Climate Orbiter (MCO)	Small errors in trajectory estimates became more important by Mars orbit insertion. Spacecraft flew too low and burned up in Mars atmosphere or left atmosphere and reentered interplanetary space but did not hit or contaminate planet.
7	Mars Polar Lander (MPL)	MPL did not have specific life-detection experiments on board and was thus cleaned to level of the Viking landers before they received terminal sterilization (same as Pathfinder).
7	MPL	MPL did not land softly as planned. Engines shut down prematurely. It crashed at about 50 miles per hour and probably broke up, exposing surfaces to Mars that may have contained Earth microbes.
7	Mars Exploration Rover (MER)	New planetary protection focus: protecting integrity of equipment for making critical biologically related measurements as well as preventing contamination of planetary surface.
7	MER	Category IVa mission—landers without life-detection instruments. Science instruments could, however, detect organic compounds that might suggest presence of life.
7	MER	Planetary protection measures were applied not only to spacecraft but also to facilities in which it was handled. Before MERs were shipped to KSC, its highbay facility and ground support equipment were thoroughly cleaned and sampled to reduce biological contamination.
7	MER	After MERs arrived at KSC, they were disassembled and cleaned to remove any contaminants received during their cross-country trip.
7	MER	Dry heat microbial reduction was performed on large components that could tolerate 125°C (257°F) for 5 hours and were difficult to adequately wipe clean. These components included thermal blankets, airbags, honeycombed parts (such as the core structures of the rovers), and parachutes packed in cans.

Chapter	Mission	Planetary Protection Approach
7	MER	Spacecraft propellant lines were precision-cleaned.
7	MER	High-efficiency particulate air (HEPA) filters were employed on the vents of each rover's core box, which contained computers as well as other critical electronics. The aim was to keep an estimated 99.97 percent of particles 0.3 microns or larger inside and not contaminate the Martian surface.
7	MER	Spores detected in acoustic foam backing of fairing enclosing MER-1 spacecraft led to additional cleanliness and bioburden requirements in launch vehicle and spacecraft manufacturing specifications.
7	MER	As with other missions, trajectories of various components were carefully designed to prevent unwanted impacts. When third stage of MER's unsterilized Delta launch vehicle separated from spacecraft, both objects were on nearly identical paths that would miss Mars. Approximately 10 days later, spacecraft performed its first TCM in order to readjust its path to arrive at Mars.
7	MER	Aeroshells for landers were subjected to dry heat microbial reduction (DHMR). Necessary for parts with bioloads difficult to reduce adequately in other ways due to factors such as sizes, shapes, and materials.
7	Phoenix Mars Lander	Parachute subjected to DHMR involving bakeout above 110°C. Procedure was necessary because of difficulty cleaning large components.
7	Phoenix Mars Lander	To prevent recontamination, parachute was double bagged before transport and packed with sample strips, which were then archived upon arrival at Lockheed Martin, available if needed to confirm parachute cleanliness. If there was a life indication on Mars, it would be compared with the sample strips to hopefully prove that life-forms had not been carried to Mars on parachutes.

ACRONYMS AND ABBREVIATIONS

AEC	Atomic Energy Commission
AFCOB	National Research Council-Armed Forces Committee on Bioastronautics
AFSEEE	Association of Forest Service Employees for Environmental Ethics
AgZn	Silver-zinc
AIBS	American Institute of Biological Sciences
ALSRC	Apollo Lunar Sample Return Container
APHIS	USDA Animal Plant Health Inspection Service
ARC	Ames Research Center
ARPA	Advanced Research Projects Agency (now DARPA)
AU	Astronomical unit
BG	*Bacillus globigii* (also known as *Bacillus subtilis* variety niger)
BIG	Biological isolation garment
BPA	Bioshield Power Assembly
Caltech	California Institute of Technology
CDC	PHS's Communicable Disease Center (later renamed the Centers for Disease Control and Prevention)
CETEX	Committee on Contamination by Extraterrestrial Exploration
CFR	Code of Federal Regulations
CM	Command Module
CM2	Carbonaceous chondrite type 2 (for example, the Murchison meteorite)
CO_2	Carbon dioxide
COMPLEX	Committee on Planetary and Lunar Exploration
COSPAR	Committee on Space Research
CRA	Crew Reception Area
CSIRO	Commonwealth Industrial and Scientific Research Organization
DARPA	Defense Advanced Research Projects Agency
DHMR	Dry heat microbial reduction
DHS	Department of Homeland Security
DNA	Deoxyribonucleic acid, a material inside the nucleus of cells that carries genetic information
DOD	U.S. Department of Defense
DOE	U.S. Department of Energy
DOI	Department of the Interior
DS2	Deep Space 2
EA	Environmental Assessment
EASTEX	East Coast Committee on Extraterrestrial Life
EDT	Eastern daylight time
EIS	Environmental Impact Statement
EOS	Electro-Optical Systems, a subsidiary of Xerox Corporation

EPA	U.S. Environmental Protection Agency
ESA	European Space Agency
ESB	Electric Storage Battery Company
EST	Eastern standard time
FAA	Federal Aviation Administration
FAT	Flight acceptance test
FBC	Faster, better, cheaper
FEP	Fluorinated ethylene propylene
FHWA	Federal Highway Administration
FONSI	Finding of No Significant Impact
GCMS	Gas chromatograph mass spectrometer
GRC	Glenn Research Center at Lewis Field
GSFC	Goddard Space Flight Center
HEPA	High-efficiency particulate air (a type of filter)
HHS	Department of Health and Human Services
HZE	High atomic number and energy
IAF	International Astronautical Federation
ICBC	Interagency Committee on Back Contamination
ICSU	International Council of Scientific Unions
IDP	Interplanetary dust particles
IGY	International Geophysical Year
IPA	Isopropyl alcohol
ISO	International Organization for Standardization
ISRU	In situ resource utilization
JAXA	Japan Aerospace Exploration Agency
JPL	Jet Propulsion Laboratory
JSC	Johnson Space Center
KSC	Kennedy Space Center
LaRC	Langley Research Center
LCROSS	Lunar CRater Observation and Sensing Satellite
LeRC	Lewis Research Center (now GRC)
LiOH	Lithium hydroxide
LM	Lunar Module
LRL	Lunar Receiving Laboratory
LSRL	Lunar Sample Receiving Laboratory (now LRL)
MCO	Mars Climate Orbiter
MEPAG	Mars Exploration Program Analysis Group
MER	Mars Exploration Rover
$(Mg,Fe)_2SiO_4$	Olivine
MGS	Mars Global Surveyor
MHP SSG	Mars Human Precursor Science Steering Group
μm	Micrometer
MMC	Martin Marietta Corporation (later Lockheed Martin)
MPIAT	Mars Program Independent Assessment Team
MPL	Mars Polar Lander
MQF	Mobile Quarantine Facility
MSC	Manned Spacecraft Center
MSFC	Marshall Space Flight Center
MSL	Mars Science Laboratory

NAI	NASA Astrobiology Institute
NAS	National Academy of Sciences
NASA	National Aeronautics and Space Administration
NEPA	National Environmental Policy Act
NHB	NASA Handbook
NHRC	NASA Historical Reference Collection
NiCd	Nickel-cadmium
NIH	National Institutes of Health
NIMS	Near-infrared mapping spectrometer
NMI	NASA Management Instruction
NPD	NASA Policy Directive
NPR	NASA Procedural Requirement
NRC	National Research Council
NRC PREVCOM	National Research Council Committee on Preventing the Forward Contamination of Mars
NSF	National Science Foundation
NSSDC	National Space Science Data Center
O_2	Molecular oxygen
OMB	Office of Management and Budget
ORNL	Oak Ridge National Laboratory
OSHA	Occupational Safety & Health Administration
OSSA	NASA Office of Space Science and Applications
PCR	Polymerase chain reaction
PHS	U.S. Public Health Service
PI	Principal investigator
PP	Planetary protection
PPAC	Planetary Protection Advisory Committee (now PPS)
PPO	Planetary Protection Officer
PPS	Planetary Protection Subcommittee
PQ	Planetary Quarantine
PQO	Planetary Quarantine Office
PREVCOM	See "NRC PREVCOM"
PST	Pacific standard time
PTC	Proof test capsule
R&D	Research and development
RNA	Ribonucleic acid, a molecule similar to DNA but containing ribose rather than deoxyribose
RTG	Radioisotope thermal generator
SAEB	Spacecraft Assembly and Encapsulation Building
SETI	Search for Extraterrestrial Intelligence
SRC	Sample return capsule
SRF	Sample receiving facility
SRI	Stanford Research Institute
SR-SAG	Special Regions Science Analysis Group
SSB	Space Studies Board (previously Space Science Board)
TCM	Trajectory course maneuver
TGISR	Task Group on Issues in Sample Return
TRW	Thompson Ramo Wooldridge Inc., from the 1958 merger of Thompson Products with the Ramo-Wooldridge Corporation

UCLA	University of California at Los Angeles
UN	United Nations
UNCOPUOS	United Nations Committee on the Peaceful Uses of Outer Space
UNESCO	United Nations Educational, Scientific and Cultural Organization
USAF	United States Air Force
USAMRIID	U.S. Army Medical Research Institute of Infectious Diseases
USDA	U.S. Department of Agriculture
USGS	U.S. Geological Survey
UV	Ultraviolet
VLC	Viking Lander Capsule
VO	Viking Orbiter
WESTEX	West Coast Committee on Extraterrestrial Life

THE NASA HISTORY SERIES

REFERENCE WORKS, NASA SP-4000:

Grimwood, James M. *Project Mercury: A Chronology.* NASA SP-4001, 1963.

Grimwood, James M., and Barton C. Hacker, with Peter J. Vorzimmer. *Project Gemini Technology and Operations: A Chronology.* NASA SP-4002, 1969.

Link, Mae Mills. *Space Medicine in Project Mercury.* NASA SP-4003, 1965.

Astronautics and Aeronautics, 1963: Chronology of Science, Technology, and Policy. NASA SP-4004, 1964.

Astronautics and Aeronautics, 1964: Chronology of Science, Technology, and Policy. NASA SP-4005, 1965.

Astronautics and Aeronautics, 1965: Chronology of Science, Technology, and Policy. NASA SP-4006, 1966.

Astronautics and Aeronautics, 1966: Chronology of Science, Technology, and Policy. NASA SP-4007, 1967.

Astronautics and Aeronautics, 1967: Chronology of Science, Technology, and Policy. NASA SP-4008, 1968.

Ertel, Ivan D., and Mary Louise Morse. *The Apollo Spacecraft: A Chronology, Volume I, Through November 7, 1962.* NASA SP-4009, 1969.

Morse, Mary Louise, and Jean Kernahan Bays. *The Apollo Spacecraft: A Chronology, Volume II, November 8, 1962–September 30, 1964.* NASA SP-4009, 1973.

Brooks, Courtney G., and Ivan D. Ertel. *The Apollo Spacecraft: A Chronology, Volume III, October 1, 1964–January 20, 1966.* NASA SP-4009, 1973.

Ertel, Ivan D., and Roland W. Newkirk, with Courtney G. Brooks. *The Apollo Spacecraft: A Chronology, Volume IV, January 21, 1966–July 13, 1974.* NASA SP-4009, 1978.

Astronautics and Aeronautics, 1968: Chronology of Science, Technology, and Policy. NASA SP-4010, 1969.

Newkirk, Roland W., and Ivan D. Ertel, with Courtney G. Brooks. *Skylab: A Chronology.* NASA SP-4011, 1977.

Van Nimmen, Jane, and Leonard C. Bruno, with Robert L. Rosholt. *NASA Historical Data Book, Volume I: NASA Resources, 1958–1968.* NASA SP-4012, 1976; rep. ed. 1988.

Ezell, Linda Neuman. *NASA Historical Data Book, Volume II: Programs and Projects, 1958–1968.* NASA SP-4012, 1988.

Ezell, Linda Neuman. *NASA Historical Data Book, Volume III: Programs and Projects, 1969–1978.* NASA SP-4012, 1988.

Gawdiak, Ihor, with Helen Fedor. *NASA Historical Data Book, Volume IV: NASA Resources, 1969–1978.* NASA SP-4012, 1994.

Rumerman, Judy A. *NASA Historical Data Book, Volume V: NASA Launch Systems, Space Transportation, Human Spaceflight, and Space Science, 1979–1988.* NASA SP-4012, 1999.

Rumerman, Judy A. *NASA Historical Data Book, Volume VI: NASA Space Applications, Aeronautics and Space Research and Technology, Tracking and Data Acquisition/Support Operations,*

Commercial Programs, and Resources, 1979–1988. NASA SP-4012, 1999.

Rumerman, Judy A. *NASA Historical Data Book, Volume VII: NASA Launch Systems, Space Transportation, Human Spaceflight, and Space Science, 1989–1998*. NASA SP-2009-4012, 2009.

No SP-4013.

Astronautics and Aeronautics, 1969: Chronology of Science, Technology, and Policy. NASA SP-4014, 1970.

Astronautics and Aeronautics, 1970: Chronology of Science, Technology, and Policy. NASA SP-4015, 1972.

Astronautics and Aeronautics, 1971: Chronology of Science, Technology, and Policy. NASA SP-4016, 1972.

Astronautics and Aeronautics, 1972: Chronology of Science, Technology, and Policy. NASA SP-4017, 1974.

Astronautics and Aeronautics, 1973: Chronology of Science, Technology, and Policy. NASA SP-4018, 1975.

Astronautics and Aeronautics, 1974: Chronology of Science, Technology, and Policy. NASA SP-4019, 1977.

Astronautics and Aeronautics, 1975: Chronology of Science, Technology, and Policy. NASA SP-4020, 1979.

Astronautics and Aeronautics, 1976: Chronology of Science, Technology, and Policy. NASA SP-4021, 1984.

Astronautics and Aeronautics, 1977: Chronology of Science, Technology, and Policy. NASA SP-4022, 1986.

Astronautics and Aeronautics, 1978: Chronology of Science, Technology, and Policy. NASA SP-4023, 1986.

Astronautics and Aeronautics, 1979–1984: Chronology of Science, Technology, and Policy. NASA SP-4024, 1988.

Astronautics and Aeronautics, 1985: Chronology of Science, Technology, and Policy. NASA SP-4025, 1990.

Noordung, Hermann. *The Problem of Space Travel: The Rocket Motor.* Edited by Ernst Stuhlinger and J. D. Hunley, with Jennifer Garland. NASA SP-4026, 1995.

Astronautics and Aeronautics, 1986–1990: A Chronology. NASA SP-4027, 1997.

Astronautics and Aeronautics, 1991–1995: A Chronology. NASA SP-2000-4028, 2000.

Orloff, Richard W. *Apollo by the Numbers: A Statistical Reference.* NASA SP-2000-4029, 2000.

Lewis, Marieke, and Ryan Swanson. *Astronautics and Aeronautics: A Chronology, 1996–2000.* NASA SP-2009-4030, 2009.

Ivey, William Noel, and Ryan Swanson. *Astronautics and Aeronautics: A Chronology, 2001–2005.* NASA SP-2010-4031, 2010.

MANAGEMENT HISTORIES, NASA SP-4100:

Rosholt, Robert L. *An Administrative History of NASA, 1958–1963.* NASA SP-4101, 1966.

Levine, Arnold S. *Managing NASA in the Apollo Era.* NASA SP-4102, 1982.

Roland, Alex. *Model Research: The National Advisory Committee for Aeronautics, 1915–1958.* NASA SP-4103, 1985.

Fries, Sylvia D. *NASA Engineers and the Age of Apollo*. NASA SP-4104, 1992.

Glennan, T. Keith. *The Birth of NASA: The Diary of T. Keith Glennan*. Edited by J. D. Hunley. NASA SP-4105, 1993.

Seamans, Robert C. *Aiming at Targets: The Autobiography of Robert C. Seamans*. NASA SP-4106, 1996.

Garber, Stephen J., ed. *Looking Backward, Looking Forward: Forty Years of Human Spaceflight Symposium*. NASA SP-2002-4107, 2002.

Mallick, Donald L., with Peter W. Merlin. *The Smell of Kerosene: A Test Pilot's Odyssey*. NASA SP-4108, 2003.

Iliff, Kenneth W., and Curtis L. Peebles. *From Runway to Orbit: Reflections of a NASA Engineer*. NASA SP-2004-4109, 2004.

Chertok, Boris. *Rockets and People, Volume I*. NASA SP-2005-4110, 2005.

Chertok, Boris. *Rockets and People: Creating a Rocket Industry, Volume II*. NASA SP-2006-4110, 2006.

Chertok, Boris. *Rockets and People: Hot Days of the Cold War, Volume III*. NASA SP-2009-4110, 2009.

Laufer, Alexander, Todd Post, and Edward Hoffman. *Shared Voyage: Learning and Unlearning from Remarkable Projects*. NASA SP-2005-4111, 2005.

Dawson, Virginia P., and Mark D. Bowles. *Realizing the Dream of Flight: Biographical Essays in Honor of the Centennial of Flight, 1903–2003*. NASA SP-2005-4112, 2005.

Mudgway, Douglas J. *William H. Pickering: America's Deep Space Pioneer*. NASA SP-2008-4113, 2008.

PROJECT HISTORIES, NASA SP-4200:

Swenson, Loyd S., Jr., James M. Grimwood, and Charles C. Alexander. *This New Ocean: A History of Project Mercury.* NASA SP-4201, 1966; rep. ed. 1999.

Green, Constance McLaughlin, and Milton Lomask. *Vanguard: A History.* NASA SP-4202, 1970; rep. ed. Smithsonian Institution Press, 1971.

Hacker, Barton C., and James M. Grimwood. *On the Shoulders of Titans: A History of Project Gemini.* NASA SP-4203, 1977; rep. ed. 2002.

Benson, Charles D., and William Barnaby Faherty. *Moonport: A History of Apollo Launch Facilities and Operations.* NASA SP-4204, 1978.

Brooks, Courtney G., James M. Grimwood, and Loyd S. Swenson, Jr. *Chariots for Apollo: A History of Manned Lunar Spacecraft.* NASA SP-4205, 1979.

Bilstein, Roger E. *Stages to Saturn: A Technological History of the Apollo/Saturn Launch Vehicles.* NASA SP-4206, 1980 and 1996.

No SP-4207.

Compton, W. David, and Charles D. Benson. *Living and Working in Space: A History of Skylab.* NASA SP-4208, 1983.

Ezell, Edward Clinton, and Linda Neuman Ezell. *The Partnership: A History of the Apollo-Soyuz Test Project.* NASA SP-4209, 1978.

Hall, R. Cargill. *Lunar Impact: A History of Project Ranger.* NASA SP-4210, 1977.

Newell, Homer E. *Beyond the Atmosphere: Early Years of Space Science.* NASA SP-4211, 1980.

Ezell, Edward Clinton, and Linda Neuman Ezell. *On Mars: Exploration of the Red Planet, 1958–1978*. NASA SP-4212, 1984.

Pitts, John A. *The Human Factor: Biomedicine in the Manned Space Program to 1980*. NASA SP-4213, 1985.

Compton, W. David. *Where No Man Has Gone Before: A History of Apollo Lunar Exploration Missions*. NASA SP-4214, 1989.

Naugle, John E. *First Among Equals: The Selection of NASA Space Science Experiments*. NASA SP-4215, 1991.

Wallace, Lane E. *Airborne Trailblazer: Two Decades with NASA Langley's 737 Flying Laboratory*. NASA SP-4216, 1994.

Butrica, Andrew J., ed. *Beyond the Ionosphere: Fifty Years of Satellite Communications*. NASA SP-4217, 1997.

Butrica, Andrew J. *To See the Unseen: A History of Planetary Radar Astronomy*. NASA SP-4218, 1996.

Mack, Pamela E., ed. *From Engineering Science to Big Science: The NACA and NASA Collier Trophy Research Project Winners*. NASA SP-4219, 1998.

Reed, R. Dale. *Wingless Flight: The Lifting Body Story*. NASA SP-4220, 1998.

Heppenheimer, T. A. *The Space Shuttle Decision: NASA's Search for a Reusable Space Vehicle*. NASA SP-4221, 1999.

Hunley, J. D., ed. *Toward Mach 2: The Douglas D-558 Program*. NASA SP-4222, 1999.

Swanson, Glen E., ed. *"Before This Decade Is Out . . ." Personal Reflections on the Apollo Program*. NASA SP-4223, 1999.

Tomayko, James E. *Computers Take Flight: A History of NASA's Pioneering Digital Fly-By-Wire Project*. NASA SP-4224, 2000.

Morgan, Clay. *Shuttle-Mir: The United States and Russia Share History's Highest Stage*. NASA SP-2001-4225, 2001.

Leary, William M. *"We Freeze to Please": A History of NASA's Icing Research Tunnel and the Quest for Safety*. NASA SP-2002-4226, 2002.

Mudgway, Douglas J. *Uplink-Downlink: A History of the Deep Space Network, 1957–1997*. NASA SP-2001-4227, 2001.

No SP-4228 or SP-4229.

Dawson, Virginia P., and Mark D. Bowles. *Taming Liquid Hydrogen: The Centaur Upper Stage Rocket, 1958–2002*. NASA SP-2004-4230, 2004.

Meltzer, Michael. *Mission to Jupiter: A History of the Galileo Project*. NASA SP-2007-4231, 2007.

Heppenheimer, T. A. *Facing the Heat Barrier: A History of Hypersonics*. NASA SP-2007-4232, 2007.

Tsiao, Sunny. *"Read You Loud and Clear!" The Story of NASA's Spaceflight Tracking and Data Network*. NASA SP-2007-4233, 2007.

CENTER HISTORIES, NASA SP-4300:

Rosenthal, Alfred. *Venture into Space: Early Years of Goddard Space Flight Center*. NASA SP-4301, 1985.

Hartman, Edwin P. *Adventures in Research: A History of Ames Research Center, 1940–1965*. NASA SP-4302, 1970.

Hallion, Richard P. *On the Frontier: Flight Research at Dryden, 1946–1981*. NASA SP-4303, 1984.

Muenger, Elizabeth A. *Searching the Horizon: A History of Ames Research Center, 1940–1976*. NASA SP-4304, 1985.

Hansen, James R. *Engineer in Charge: A History of the Langley Aeronautical Laboratory, 1917–1958*. NASA SP-4305, 1987.

Dawson, Virginia P. *Engines and Innovation: Lewis Laboratory and American Propulsion Technology*. NASA SP-4306, 1991.

Dethloff, Henry C. *"Suddenly Tomorrow Came . . .": A History of the Johnson Space Center, 1957–1990*. NASA SP-4307, 1993.

Hansen, James R. *Spaceflight Revolution: NASA Langley Research Center from Sputnik to Apollo*. NASA SP-4308, 1995.

Wallace, Lane E. *Flights of Discovery: An Illustrated History of the Dryden Flight Research Center*. NASA SP-4309, 1996.

Herring, Mack R. *Way Station to Space: A History of the John C. Stennis Space Center*. NASA SP-4310, 1997.

Wallace, Harold D., Jr. *Wallops Station and the Creation of an American Space Program*. NASA SP-4311, 1997.

Wallace, Lane E. *Dreams, Hopes, Realities. NASA's Goddard Space Flight Center: The First Forty Years*. NASA SP-4312, 1999.

Dunar, Andrew J., and Stephen P. Waring. *Power to Explore: A History of Marshall Space Flight Center, 1960–1990*. NASA SP-4313, 1999.

Bugos, Glenn E. *Atmosphere of Freedom: Sixty Years at the NASA Ames Research Center*. NASA SP-2000-4314, 2000.

No SP-4315.

Schultz, James. *Crafting Flight: Aircraft Pioneers and the Contributions of the Men and Women of NASA Langley Research Center*. NASA SP-2003-4316, 2003.

Bowles, Mark D. *Science in Flux: NASA's Nuclear Program at Plum Brook Station, 1955–2005*. NASA SP-2006-4317, 2006.

Wallace, Lane E. *Flights of Discovery: An Illustrated History of the Dryden Flight Research Center*. NASA SP-2007-4318, 2007. Revised version of NASA SP-4309.

Arrighi, Robert S. *Revolutionary Atmosphere: The Story of the Altitude Wind Tunnel and the Space Power Chambers*. NASA SP-2010-4319, 2010.

GENERAL HISTORIES, NASA SP-4400:

Corliss, William R. *NASA Sounding Rockets, 1958–1968: A Historical Summary*. NASA SP-4401, 1971.

Wells, Helen T., Susan H. Whiteley, and Carrie Karegeannes. *Origins of NASA Names*. NASA SP-4402, 1976.

Anderson, Frank W., Jr. *Orders of Magnitude: A History of NACA and NASA, 1915–1980*. NASA SP-4403, 1981.

Sloop, John L. *Liquid Hydrogen as a Propulsion Fuel, 1945–1959*. NASA SP-4404, 1978.

Roland, Alex. *A Spacefaring People: Perspectives on Early Spaceflight*. NASA SP-4405, 1985.

Bilstein, Roger E. *Orders of Magnitude: A History of the NACA and NASA, 1915–1990*. NASA SP-4406, 1989.

Logsdon, John M., ed., with Linda J. Lear, Jannelle Warren Findley, Ray A. Williamson, and Dwayne A. Day. *Exploring the Unknown: Selected Documents in the History of the U.S. Civil Space Program, Volume I: Organizing for Exploration*. NASA SP-4407, 1995.

Logsdon, John M., ed., with Dwayne A. Day and Roger D. Launius. *Exploring the Unknown: Selected Documents in the History of the U.S. Civil Space Program, Volume II: External Relationships.* NASA SP-4407, 1996.

Logsdon, John M., ed., with Roger D. Launius, David H. Onkst, and Stephen J. Garber. *Exploring the Unknown: Selected Documents in the History of the U.S. Civil Space Program, Volume III: Using Space.* NASA SP-4407, 1998.

Logsdon, John M., ed., with Ray A. Williamson, Roger D. Launius, Russell J. Acker, Stephen J. Garber, and Jonathan L. Friedman. *Exploring the Unknown: Selected Documents in the History of the U.S. Civil Space Program, Volume IV: Accessing Space.* NASA SP-4407, 1999.

Logsdon, John M., ed., with Amy Paige Snyder, Roger D. Launius, Stephen J. Garber, and Regan Anne Newport. *Exploring the Unknown: Selected Documents in the History of the U.S. Civil Space Program, Volume V: Exploring the Cosmos.* NASA SP-2001-4407, 2001.

Logsdon, John M., ed., with Stephen J. Garber, Roger D. Launius, and Ray A. Williamson. *Exploring the Unknown: Selected Documents in the History of the U.S. Civil Space Program, Volume VI: Space and Earth Science.* NASA SP-2004-4407, 2004.

Logsdon, John M., ed., with Roger D. Launius. *Exploring the Unknown: Selected Documents in the History of the U.S. Civil Space Program, Volume VII: Human Spaceflight: Projects Mercury, Gemini, and Apollo.* NASA SP-2008-4407, 2008.

Siddiqi, Asif A., *Challenge to Apollo: The Soviet Union and the Space Race, 1945–1974.* NASA SP-2000-4408, 2000.

Hansen, James R., ed. *The Wind and Beyond: Journey into the History of Aerodynamics in America, Volume 1: The Ascent of the Airplane.* NASA SP-2003-4409, 2003.

Hansen, James R., ed. *The Wind and Beyond: Journey into the History of Aerodynamics in America, Volume 2: Reinventing the Airplane.* NASA SP-2007-4409, 2007.

Hogan, Thor. *Mars Wars: The Rise and Fall of the Space Exploration Initiative.* NASA SP-2007-4410, 2007.

MONOGRAPHS IN AEROSPACE HISTORY, NASA SP-4500:

Launius, Roger D., and Aaron K. Gillette, comps. *Toward a History of the Space Shuttle: An Annotated Bibliography.* Monographs in Aerospace History, No. 1, 1992.

Launius, Roger D., and J. D. Hunley, comps. *An Annotated Bibliography of the Apollo Program.* Monographs in Aerospace History, No. 2, 1994.

Launius, Roger D. *Apollo: A Retrospective Analysis.* Monographs in Aerospace History, No. 3, 1994.

Hansen, James R. *Enchanted Rendezvous: John C. Houbolt and the Genesis of the Lunar-Orbit Rendezvous Concept.* Monographs in Aerospace History, No. 4, 1995.

Gorn, Michael H. *Hugh L. Dryden's Career in Aviation and Space.* Monographs in Aerospace History, No. 5, 1996.

Powers, Sheryll Goecke. *Women in Flight Research at NASA Dryden Flight Research Center from 1946 to 1995.* Monographs in Aerospace History, No. 6, 1997.

Portree, David S. F., and Robert C. Trevino. *Walking to Olympus: An EVA Chronology.* Monographs in Aerospace History, No. 7, 1997.

Logsdon, John M., moderator. *Legislative Origins of the National Aeronautics and Space Act of 1958: Proceedings of an Oral History Workshop.* Monographs in Aerospace History, No. 8, 1998.

Rumerman, Judy A., comp. *U.S. Human Spaceflight: A Record of Achievement, 1961–1998.* Monographs in Aerospace History, No. 9, 1998.

Portree, David S. F. *NASA's Origins and the Dawn of the Space Age.* Monographs in Aerospace History, No. 10, 1998.

Logsdon, John M. *Together in Orbit: The Origins of International Cooperation in the Space Station.* Monographs in Aerospace History, No. 11, 1998.

Phillips, W. Hewitt. *Journey in Aeronautical Research: A Career at NASA Langley Research Center.* Monographs in Aerospace History, No. 12, 1998.

Braslow, Albert L. *A History of Suction-Type Laminar-Flow Control with Emphasis on Flight Research.* Monographs in Aerospace History, No. 13, 1999.

Logsdon, John M., moderator. *Managing the Moon Program: Lessons Learned from Apollo.* Monographs in Aerospace History, No. 14, 1999.

Perminov, V. G. *The Difficult Road to Mars: A Brief History of Mars Exploration in the Soviet Union.* Monographs in Aerospace History, No. 15, 1999.

Tucker, Tom. *Touchdown: The Development of Propulsion Controlled Aircraft at NASA Dryden.* Monographs in Aerospace History, No. 16, 1999.

Maisel, Martin, Demo J. Giulanetti, and Daniel C. Dugan. *The History of the XV-15 Tilt Rotor Research Aircraft: From Concept to Flight.* Monographs in Aerospace History, No. 17, 2000. NASA SP-2000-4517.

Jenkins, Dennis R. *Hypersonics Before the Shuttle: A Concise History of the X-15 Research Airplane.* Monographs in Aerospace History, No. 18, 2000. NASA SP-2000-4518.

Chambers, Joseph R. *Partners in Freedom: Contributions of the Langley Research Center to U.S. Military Aircraft of the 1990s*. Monographs in Aerospace History, No. 19, 2000. NASA SP-2000-4519.

Waltman, Gene L. *Black Magic and Gremlins: Analog Flight Simulations at NASA's Flight Research Center*. Monographs in Aerospace History, No. 20, 2000. NASA SP-2000-4520.

Portree, David S. F. *Humans to Mars: Fifty Years of Mission Planning, 1950–2000*. Monographs in Aerospace History, No. 21, 2001. NASA SP-2001-4521.

Thompson, Milton O., with J. D. Hunley. *Flight Research: Problems Encountered and What They Should Teach Us*. Monographs in Aerospace History, No. 22, 2001. NASA SP-2001-4522.

Tucker, Tom. *The Eclipse Project*. Monographs in Aerospace History, No. 23, 2001. NASA SP-2001-4523.

Siddiqi, Asif A. *Deep Space Chronicle: A Chronology of Deep Space and Planetary Probes, 1958–2000*. Monographs in Aerospace History, No. 24, 2002. NASA SP-2002-4524.

Merlin, Peter W. *Mach 3+: NASA/USAF YF-12 Flight Research, 1969–1979*. Monographs in Aerospace History, No. 25, 2001. NASA SP-2001-4525.

Anderson, Seth B. *Memoirs of an Aeronautical Engineer: Flight Tests at Ames Research Center: 1940–1970*. Monographs in Aerospace History, No. 26, 2002. NASA SP-2002-4526.

Renstrom, Arthur G. *Wilbur and Orville Wright: A Bibliography Commemorating the One-Hundredth Anniversary of the First Powered Flight on December 17, 1903*. Monographs in Aerospace History, No. 27, 2002. NASA SP-2002-4527.

No monograph 28.

Chambers, Joseph R. *Concept to Reality: Contributions of the NASA Langley Research Center to U.S. Civil Aircraft of the 1990s.* Monographs in Aerospace History, No. 29, 2003. NASA SP-2003-4529.

Peebles, Curtis, ed. *The Spoken Word: Recollections of Dryden History, The Early Years.* Monographs in Aerospace History, No. 30, 2003. NASA SP-2003-4530.

Jenkins, Dennis R., Tony Landis, and Jay Miller. *American X-Vehicles: An Inventory—X-1 to X-50.* Monographs in Aerospace History, No. 31, 2003. NASA SP-2003-4531.

Renstrom, Arthur G. *Wilbur and Orville Wright: A Chronology Commemorating the One-Hundredth Anniversary of the First Powered Flight on December 17, 1903.* Monographs in Aerospace History, No. 32, 2003. NASA SP-2003-4532.

Bowles, Mark D., and Robert S. Arrighi. *NASA's Nuclear Frontier: The Plum Brook Research Reactor.* Monographs in Aerospace History, No. 33, 2004. NASA SP-2004-4533.

Wallace, Lane, and Christian Gelzer. *Nose Up: High Angle-of-Attack and Thrust Vectoring Research at NASA Dryden, 1979–2001.* Monographs in Aerospace History, No. 34, 2009. NASA SP-2009-4534.

Matranga, Gene J., C. Wayne Ottinger, Calvin R. Jarvis, and D. Christian Gelzer. *Unconventional, Contrary, and Ugly: The Lunar Landing Research Vehicle.* Monographs in Aerospace History, No. 35, 2006. NASA SP-2004-4535.

McCurdy, Howard E. *Low-Cost Innovation in Spaceflight: The History of the Near Earth Asteroid Rendezvous (NEAR) Mission.* Monographs in Aerospace History, No. 36, 2005. NASA SP-2005-4536.

Seamans, Robert C., Jr. *Project Apollo: The Tough Decisions.* Monographs in Aerospace History, No. 37, 2005. NASA SP-2005-4537.

Lambright, W. Henry. *NASA and the Environment: The Case of Ozone Depletion*. Monographs in Aerospace History, No. 38, 2005. NASA SP-2005-4538.

Chambers, Joseph R. *Innovation in Flight: Research of the NASA Langley Research Center on Revolutionary Advanced Concepts for Aeronautics*. Monographs in Aerospace History, No. 39, 2005. NASA SP-2005-4539.

Phillips, W. Hewitt. *Journey into Space Research: Continuation of a Career at NASA Langley Research Center*. Monographs in Aerospace History, No. 40, 2005. NASA SP-2005-4540.

Rumerman, Judy A., Chris Gamble, and Gabriel Okolski, comps. *U.S. Human Spaceflight: A Record of Achievement, 1961–2006*. Monographs in Aerospace History, No. 41, 2007. NASA SP-2007-4541.

Dick, Steven J., Stephen J. Garber, and Jane H. Odom. *Research in NASA History*. Monographs in Aerospace History, No. 43, 2009. NASA SP-2009-4543.

Merlin, Peter W. *Ikhana: Unmanned Aircraft System Western States Fire Missions*. Monographs in Aerospace History, No. 44, 2009. NASA SP-2009-4544.

Fisher, Steven C., and Shamim A. Rahman. *Remembering the Giants: Apollo Rocket Propulsion Development*. Monographs in Aerospace History, No. 45, 2009. NASA SP-2009-4545.

ELECTRONIC MEDIA, NASA SP-4600:

Remembering Apollo 11: The 30th Anniversary Data Archive CD-ROM. NASA SP-4601, 1999.

Remembering Apollo 11: The 35th Anniversary Data Archive CD-ROM. NASA SP-2004-4601, 2004. This is an update of the 1999 edition.

The Mission Transcript Collection: U.S. Human Spaceflight Missions from Mercury Redstone 3 to Apollo 17. NASA SP-2000-4602, 2001.

Shuttle-Mir: The United States and Russia Share History's Highest Stage. NASA SP-2001-4603, 2002.

U.S. Centennial of Flight Commission Presents Born of Dreams— Inspired by Freedom. NASA SP-2004-4604, 2004.

Of Ashes and Atoms: A Documentary on the NASA Plum Brook Reactor Facility. NASA SP-2005-4605, 2005.

Taming Liquid Hydrogen: The Centaur Upper Stage Rocket Interactive CD-ROM. NASA SP-2004-4606, 2004.

Fueling Space Exploration: The History of NASA's Rocket Engine Test Facility DVD. NASA SP-2005-4607, 2005.

Altitude Wind Tunnel at NASA Glenn Research Center: An Interactive History CD-ROM. NASA SP-2008-4608, 2008.

A Tunnel Through Time: The History of NASA's Altitude Wind Tunnel. NASA SP-2010-4609, 2010.

CONFERENCE PROCEEDINGS, NASA SP-4700:

Dick, Steven J., and Keith Cowing, eds. *Risk and Exploration: Earth, Sea and the Stars.* NASA SP-2005-4701, 2005.

Dick, Steven J., and Roger D. Launius. *Critical Issues in the History of Spaceflight.* NASA SP-2006-4702, 2006.

Dick, Steven J., ed. *Remembering the Space Age: Proceedings of the 50th Anniversary Conference*. NASA SP-2008-4703, 2008.

Dick, Steven J., ed. *NASA's First 50 Years: Historical Perspectives*. NASA SP-2010-4704, 2010.

SOCIETAL IMPACT, NASA SP-4800:

Dick, Steven J., and Roger D. Launius. *Societal Impact of Spaceflight*. NASA SP-2007-4801, 2007.

Dick, Steven J., and Mark L. Lupisella. *Cosmos and Culture: Cultural Evolution in a Cosmic Context*. NASA SP-2009-4802, 2009.

INDEX

G:O U.S. GOVERNMENT PRINTING OFFICE: 2011—368-770